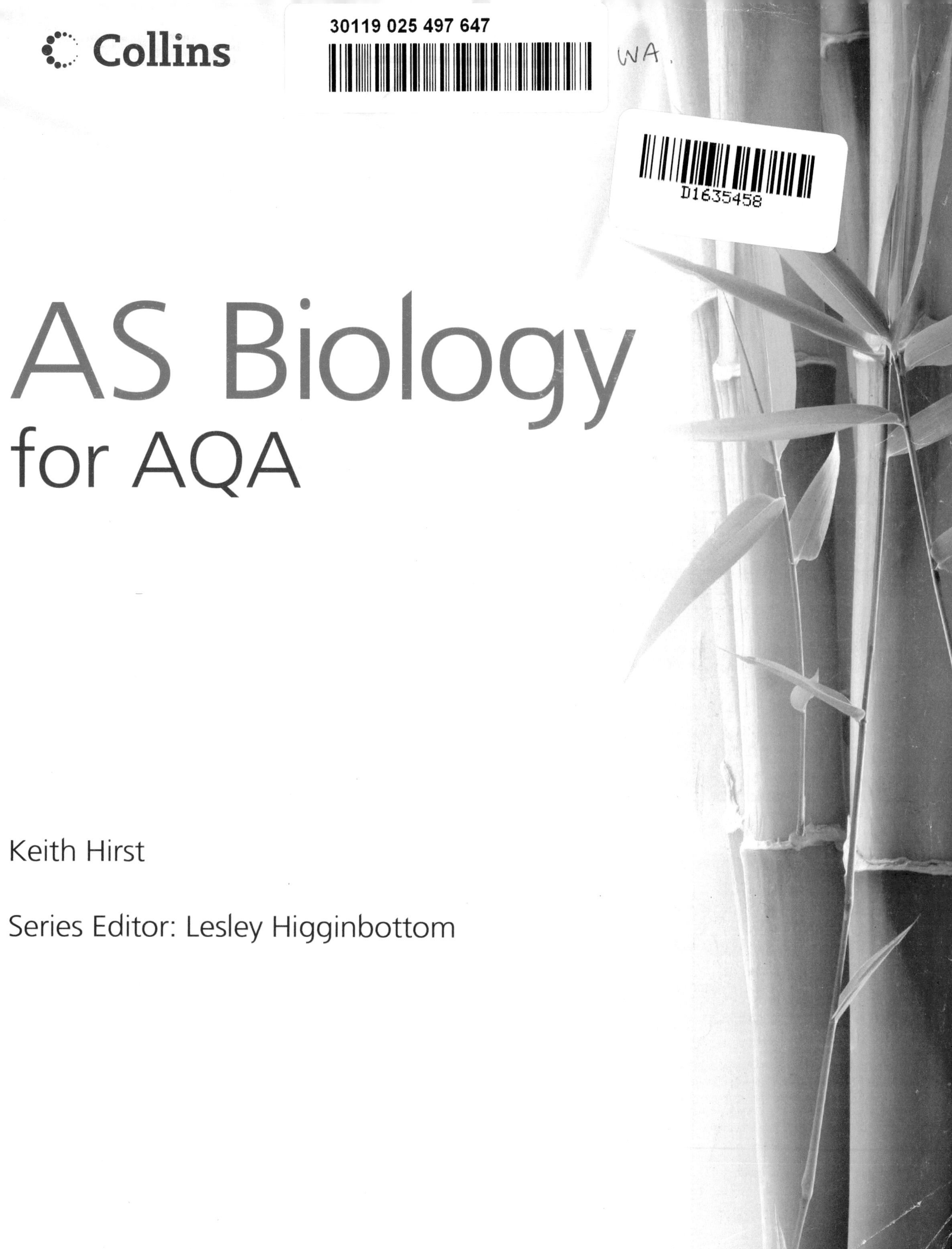

AS Biology
for AQA

Keith Hirst

Series Editor: Lesley Higginbottom

William Collins's dream of knowledge for all began with the publication of his first book in 1819. A self-educated mill worker, he not only enriched millions of lives, but also founded a flourishing publishing house. Today, staying true to this spirit, Collins books are packed with inspiration, innovation and practical expertise. They place you at the centre of a world of possibility and give you exactly what you need to explore it.

Collins. Freedom to teach.

Published by Collins
An imprint of HarperCollinsPublishers
77-85 Fulham Palace Road
Hammersmith
London
W6 8JB

Browse the complete Collins catalogue at
www.collinseducation.com

10 9 8 7 6 5 4 3 2 1

ISBN-13 978-0-00-726821-4
ISBN-10 0-00726821-1

Keith Hirst asserts his moral right to be identified as the author of this work.

British Library Cataloguing in Publication Data. A Catalogue record for this publication is available from the British Library.

Commissioned by Penny Fowler
Series Editor Lesley Higginbottom
Edited by Rosie Parrish
Index by Dr Laurence Errington
Proof read by Helen Barham
Design by Bookcraft Ltd.
Cover design by Angela English
Production by Arjen Jansen
Printed and bound in Hong Kong by Printing Express

FSC is a non-profit international organisation established to promote the responsible management of the world's forests. Products carrying the FSC label are independently certified to assure consumers that they come from forests that are managed to meet the social, economic and ecological needs of present and future generations.

Find out more about HarperCollins and the environment at
www.harpercollins.co.uk/green

Unit One Biology and Disease

Unit Two The Variety of Living Organisms

Acknowledgments

Text and diagrams reproduced by kind permission of:
Wiley-Blackwell Publishing Ltd., The Health Protection Agency, The Daily Express, The Daily Telegraph, The Daily Mirror, The Random House Group Ltd., Science Daily, The University of Sheffield, Cengage Learning Inc., Wiley New York, Guardian News and Media 1993, English Nature Somerset Team, The Royal Botanic Gardens, Kew.

Every effort has been made to contact the holders of copyright material, but if any have been inadvertently overlooked the publishers will be pleased to make the necessary arrangements at the first opportunity.

The publishers would like to thank the following for permission to reproduce photographs
(T = Top, B = Bottom, C = Centre, L = Left, R= Right):

Andrew Lambert 40, 57
Animal Photography/Sally Anne Thompson 186B, 187BL&BR
Anthony Blake Photo Library 210BL
Aquarius Picture Agency 139
Australia Photosynthesis Research Organisation 150
BBC Picture Archives 120R
Beechams Pharmaceuticals 201
Beverley Hills Chamber of Commerce 120L
Biophoto Associates, 23, 25T&C, 27, 30, 31, 32, 89L&R, 141B4, 151, 153BL, 154, 165, 177CL, 178
Boca Raton Community Hospital 149
Bruce Coleman Collection 123, 159, 160, 173, 176L, 177T, 187TL&TR, 195BC
C. Corbis and J. Lenars 195T
Flickr Creative Commons 25B, 50; Aidan Jones 84, Andrea-on-the-go 78D, Catdirt Record 122, Foraggio 83, Koepke 38, Michael McLaughlin 93, Ninjapoodles 112, Renfield 195BR
Hulton Getty Picture Collection 200L
iStock 6, 49; Jurga Rubinovaite 13BL
J. Allen Cash Ltd 65L
Jupiter Images 13R, 13TL, 16, 17, 56, 216T
NHPA 210TR,BR,BC; B. Beehler 195BL, G. Bernard 188, J. Shaw 176R, N. Wu 186T
Oxford Scientific Films/B. Wells 158BR, Dguravich 158BL, M. Colebeck 158T
Royal Botanic Gardens Kew 217
Royal Holloway College 33
Science Photo Library 14T, 15, 22, 36, 37, 39, 45, 62, 71, 79, 98, 99, 106, 107, 108, 121, 128, 130, 131, 132, 141T, 153TR, 161, 167
Scoop NZ, Open Rescue Collective 219, 220
South Eastern Technology Centre, Augusta, Georgia 77
Steve Donald 209
Still Pictures/J. Cancalosi 214
The Children's Hospital, Denver 86
The Stock Market 65R
Wikipedia 196
Woodfall Wild Image 211, 213

Cover photograph © Steve Dibblee/istockphoto.com

To the student

This book aims to make your study of advanced science successful and interesting. Science is constantly evolving and, wherever possible, modern issues and problems have been used to make your study stimulating and to encourage you to continue studying science after you complete your current course.

Using the book

Don't try to achieve too much in one reading session. Science is complex and some demanding ideas need to be supported with a lot of facts. Trying to take in too much at one time can make you lose sight of the most important ideas – all you see is a mass of information.

Each chapter starts by showing how the science you will learn is applied somewhere in the world. At other points in the chapter you will find the *How Science Works* boxes. These will help you to pose scientific questions and analyse, interpret and evaluate evidence and data. Using these boxes and the *How Science Works* assignments at the end of each chapter will help you to tackle questions in your final examination with a *How Science Works* element.

The numbered questions in the main text allow you to check that you have understood what is being explained. These are all short and straightforward in style – there are no trick questions. Don't be tempted to pass over these questions, they will give you new insights into the work. Answers are given in the back of the book.

This book covers the content needed for AQA Biology at AS-level. The Key Facts for each section summarise the information you will need in your examination. However, the examination will test your ability to apply these facts rather than simply to remember then. The *How Science Works* boxes will help you to develop this skill.

Words written in bold type appear in the glossary at the end of the book. If you don't know the meaning of one of these words, check it out immediately – don't persevere, hoping all will become clear.

Past paper questions are included at the end of each chapter. These will help you to test yourself against the sorts of questions that will come up in your examination. You can find the answers to these questions on the website www.collinseducation.com/advancedscienceaqa

The website also provides sample student answers to these questions – a stronger and a weaker answer for each question – to help you to improve your own answers. On this website you will also find mathematical and examination technique guidance to help you to prepare for your examinations and practice ISAs, which your teacher may do with you or which you may want to look at yourself.

1 Diseases caused by pathogens and lifestyle

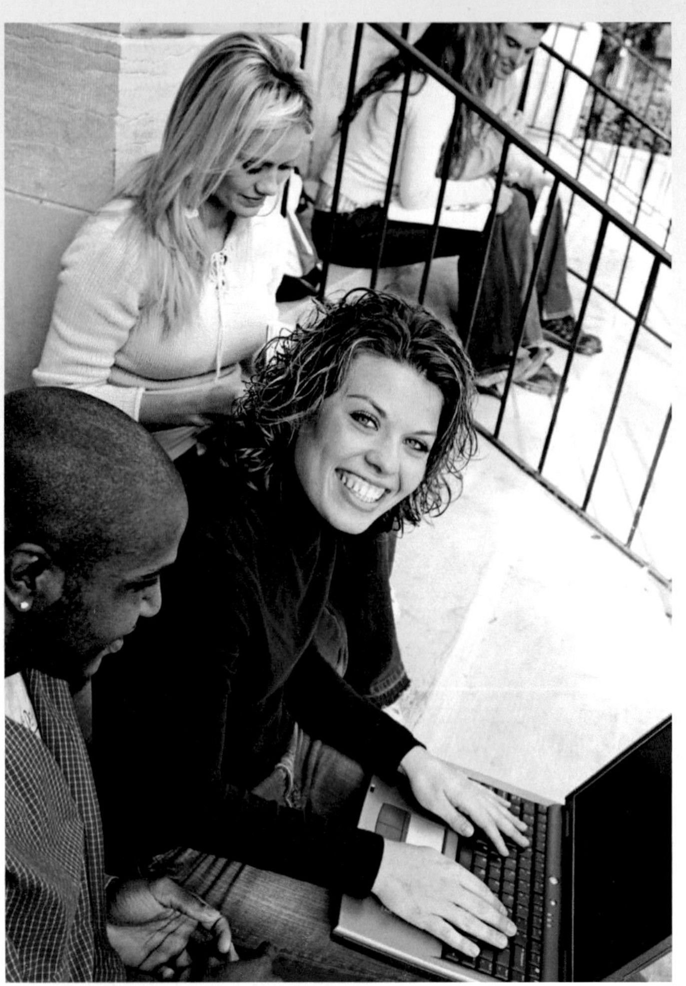
These young people can expect a longer, healthier life than their parents.

We have been fighting a long war with disease throughout human history. Widespread outbreaks of disease are called **epidemics**, and epidemics that spread internationally are called **pandemics**. Pandemics of the 'black death' in pre-industrial Europe reduced the population by 30–50% in the 14th century. Nowadays, in developed countries clean drinking water, sewage treatment and high standards of food hygiene help prevent the spread of disease. Immunisation provides protection from many infectious diseases, and health services provide expertise and drugs in the event of illness. The average life expectancy in these nations has risen.

In overcrowded conditions, infectious diseases spread rapidly and contribute to a high death rate. But the spread of infectious disease is neither a problem of the past, nor limited to developing countries today. Influenza pandemics regularly kill people all over the world. Other **communicable diseases** currently causing concern include HIV **infection**, and infection by a virulent form of the bacterium that causes tuberculosis (TB).

There are now 39.5 million people living with HIV infection, according to the annual Joint United Nations (UN) programme on HIV/AIDS report, 4.3 million of those were infected in 2006. That is 400 000 more than were infected in 2004.

In Western Europe, the gains made by programmes aimed at preventing infection have not been maintained. The report says, "The largest increases have been reported in the UK, where HIV remains one of the principal communicable disease threats". New diagnoses are increasing in areas other than London, which has the most cases. Most of those with HIV were infected when living in or visiting sub-Saharan Africa. Fear of stigma and discrimination is discouraging Africans in the UK from being tested, says the UN.

In this chapter you will learn about the nature and transmission of infectious disease.

1.1 What causes infectious disease?

Communicable diseases are caused by micro-organisms, commonly **bacteria**, **viruses** and **fungi**. Fig. 1 shows a comparison of these three types of microorganisms.

Infection occurs when microorganisms get past the external defence mechanisms and enter the body tissues. Infection may occur through cuts or abrasions in the skin, and via natural openings of the breathing, digestive and urino-genitary systems. Not all infections cause disease. The microorganisms that cause disease are known as **pathogens**.

1 What is the main difference between a bacterium and a virus?

2 What parts of the body are mainly infected by fungi?

Fig. 1 Biology of viruses, bacteria and fungi

Organism	Characteristic features	Pathogen and disease
Viruses Lipid envelope Helical nucleocapsid consisting of RNA Protein matrix 150 nm	• Small (ultra-microscopic) 20–3000 nm but typically they are 100–300 nm • Acellular • Ability to reproduce only within the host cell • Reproduction exploits metabolism and materials of host cell	Influenza virus – influenza (flu) Mumps virus – mumps Human immunodeficiency virus (HIV) – acquired immune deficiency syndrome (AIDS) Rubella virus – german measles Measles virus – measles
Bacteria Capsule Stored food reserve Cell wall Cell membrane Ribosomes Nuclear material – a loop of DNA Plasmid 1 μm	• Small (0.5–10 μm) • Prokaryotic (lack nuclei, nucleic acid not membrane-bound) • Asexual reproduction by binary fission • Sexual reproduction by transfer of genetic material from one cell to another	*Salmonella enteritidis* – Salmonella food poisoning *Mycobacterium tuberculosis* – tuberculosis (TB) *Corynebacterium diphtheriae* – diphtheria *Clostridium tetani* – tetanus *Bordetella pertussis* – pertussis (whooping cough)
Fungi Endoplasmic reticulum Cell wall Mitochondrion Hyphae Nucleolus Vacuole Nucleus Double nuclear membrane Cell membrane 1 μm	• Thread-like growth (hyphae) in filamentous forms of fungi • Cell walls made of chitin • Eukaryotic with nucleus and membrane-bound organelles • Reproduction by spores	*Candida* – thrush *Epidermophyton* – athlete's foot *Epidermophyton* – ringworm

1.2 Transmission of pathogens

Fig. 2 Transmission of pathogens

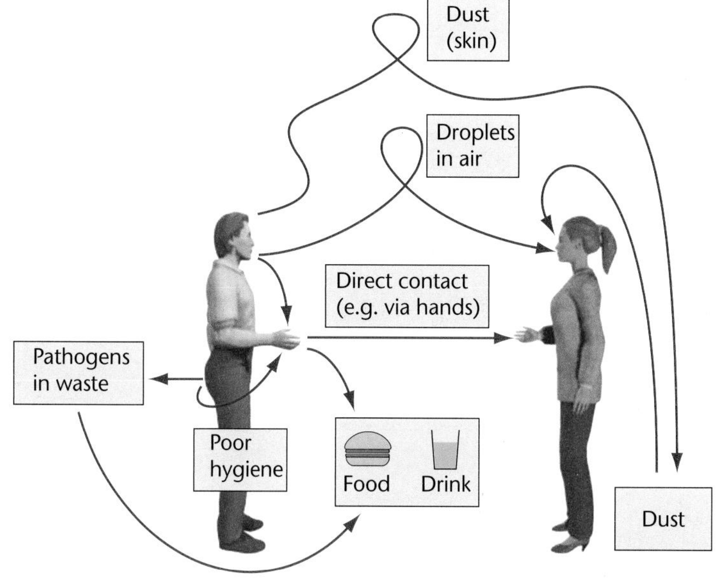

Pathogens can infect an individual in a number of different ways.

Transmission may be by:

- air
- water
- contaminated food
- direct contact (including sexual intercourse)
- animals (vectors).

3 Through which two parts of the body do most pathogens enter?

how science works

The Black Death – what caused it?

Illustration of the Black Death from the Toggenburg Bible (1411).

The Black Death was one of the most devastating pandemics in human history. It began in Asia and spread to Europe by the late 1340s. It is estimated that, worldwide, a total of 75 million people died from the Black Death, 20 million of which were in Europe. The Black Death is estimated to have killed between a third and two-thirds of Europe's population.

The Black Death was characterised by buboes (swellings in lymph nodes). Similar buboes appeared in patients with Asian Bubonic plague in the 19th century. Scientists and historians assumed at the beginning of the twentieth century that the Black Death was an outbreak of the same disease, caused by the bacterium *Yersinia pestis* and spread by fleas with the help of animals like the black rat (*Rattus rattus*). However, this view has recently been questioned.

- In 2000, Gunnar Karlsson pointed out that the Black Death killed between half and two-thirds of the population of Iceland, although there were no rats in Iceland at this time.
- Some research indicated that tooth pulp tissue from a fourteenth-century plague cemetery in Montpellier, France, tested positive for molecules associated with *Y. pestis*. But in September 2003, a team of researchers from Oxford University tested 121 teeth from 66 skeletons found in fourteenth-century plague graves. The remains showed no genetic trace of *Y. pestis*, and the researchers suspect that the Montpellier study was flawed.
- In 1984, Graham Twigg argued that the climate and ecology of Europe, and particularly England, made it nearly impossible for rats and fleas to have transmitted bubonic plague. He combined information on the biology of *R. rattus*, *R. norvegicus*, and the common fleas *Xenopsylla cheopis* and *Pulex irritans* with modern studies of plague **epidemiology**. From these studies Twigg concludes that it would have been nearly impossible for *Y. pestis* to have been the causative agent of the beginning of the plague, let alone its explosive spread across all of Europe. Twigg proposes that the Black Death may actually have been an epidemic of pulmonary anthrax caused by *Bacillus anthracis*.
- In 2001, Susan Scott and Christopher Duncan from Liverpool University proposed the **theory** that the Black Death might have been caused by an Ebola-like virus, not a bacterium. Their rationale was that this plague spread much faster and the incubation period was much longer than other confirmed *Y. pestis* plagues. A longer period of incubation will allow carriers of the infection to travel farther and infect more people than a shorter one. The plague also appeared in areas of Europe where rats were uncommon, such as Iceland.
- Historian Norman F. Cantor suggests the Black Death might have been a combination of pandemics, including a form of anthrax, a cattle murrain. He cites many forms of evidence including: reported disease symptoms not in keeping with the known effects of either bubonic or pneumonic plague, the discovery of anthrax spores in a plague pit in Scotland, and the fact that meat from infected cattle was known to have been sold in many rural English areas prior to the onset of the plague.

Summarise the evidence for and against the transmission of bubonic plague by fleas from rats.

1.3 Why we feel ill

Pathogens colonise and reproduce in tissues and body fluids, causing physical damage to cell structure, disrupting cell functions, releasing **toxins**, and stimulating the response of the body's **immune system**. For an infection to take hold and cause disease, an organism must do three things:

- attach itself to host tissues;
- penetrate the host cells;
- colonise and reproduce within the host tissue.

The effects of these are shown in Fig. 3. The actions shown in Fig. 3 produce the signs and symptoms we call disease. The signs of a disease are the visible features of the disease. For example, somebody with measles will develop a rash and a fever (a high temperature). The symptoms of a disease are not visible to another person. For example, somebody with measles may feel nauseous and very weak, or have various aches and pains.

Fig. 3 Infection by a pathogen

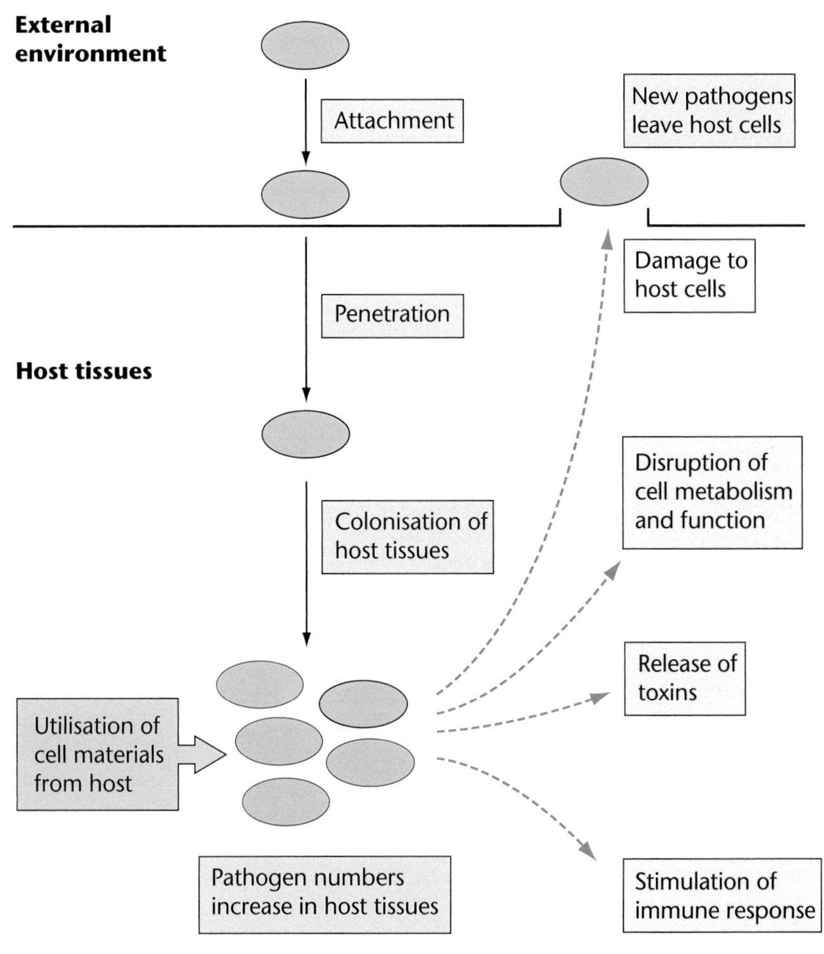

Pathogens can recognise and attach to host cells. Receptor binding protein molecules, called **ligands**, are found in the wall of microbes or viral coat. These bind with specific protein receptor molecules on the host cell membrane. Many organisms may bind to cells but do not enter. Pathogens enter host cells by **endocytosis** (an infolding of the cell membrane) or by producing enzymes that breach the host cell membrane. Once inside, colonisation depends on the ability of the pathogen to reproduce in the host tissues. This may take time; the period of time between infection and the appearance of the signs and symptoms of a disease is called the incubation period.

5 What is a pathogen?

6 Distinguish between the terms infection and disease.

7 Give three ways in which pathogens cause disease.

Toxins

Many bacterial pathogens produce toxins – poisons that can have a harmful effect on the body. There are two types of bacterial toxin: **exotoxins** and **endotoxins**. Some toxins are mild and do not produce serious illness. Others, such as the diphtheria toxin, are highly toxic and cause life-threatening disease.

Exotoxins are proteins that are secreted by, or leak from, bacteria. Different bacteria produce different exotoxins, each with a different effect on the body. The symptoms of many bacterial diseases are caused by the production of exotoxins, rather than the growth of the infecting bacteria. Diarrhoea caused by *Escherichia coli* infection is a result of exotoxins affecting the lining of the intestines. Tetanus, caused by the bacterium *Clostridium tetani*, is a disease of the nervous system, which results in the painful, involuntary contraction of muscles, particularly in the face and neck. The characteristic open-mouthed facial expression of someone with tetanus gave rise to another name for the disease – *lockjaw*. Tetanus is caused by toxins from *C. tetani* affecting the function of nerve cells that normally prevent muscle contraction, resulting in what is known as spastic paralysis. People can be vaccinated against *C. tetani*, in what is commonly referred to as a 'tetanus jab'. The bacterium is very common, and one place it is found is in the soil. This is one reason why it is best to wear protective gloves when gardening, to prevent the bacteria getting into your skin through small scratches.

Endotoxins are complex compounds usually found in bacterial cell walls. Endotoxins do not affect the host organism whilst they are attached to living bacteria. However, they are released when the bacterial cell dies and the cell wall breaks down, releasing the complex compounds. These are picked up by a type of white blood cell called a macrophage. The endotoxins cause the macrophages to produce proteins that alter the body's temperature-regulating mechanisms, setting the body's 'thermostat' to a higher level, resulting in raised body temperature, or fever. Other effects caused by the proteins include weakness and aching. These symptoms are common to all diseases resulting from endotoxins.

8 What is the difference between an endotoxin and an exotoxin?

9 What causes a rise in body temperature during some illnesses?

key facts

- Infectious diseases are caused by microorganisms called pathogens.
- Pathogens include bacteria, viruses and fungi.
- Pathogens enter the body mainly via the digestive and gas exchange systems.
- Pathogens cause disease by damaging the cells of the host and by producing toxins.

hsw how science works

Koch's postulates

In the nineteenth century, Louis Pasteur and Robert Koch worked on understanding the relationship between disease and microorganisms. Koch summed up the conditions for proving that a specific microorganism causes a specific disease in a list known as Koch's postulates. These are that:

- The organism thought to be causing the disease should always be present in animals suffering from the disease and be absent from healthy ones.
- The organism must be cultivated in pure culture outside the body of the infected animal.
- When inoculated into healthy animals, the culture should cause the characteristic symptoms of the disease.
- The organism should be re-isolated from the experimental animals and be cultured again in the laboratory. This new culture should be the same as the original one.

Koch's work was carried out on sheep using anthrax bacteria, a fatal disease for both sheep and humans. He used mice as the experimental animals, and showed that the bacteria were the causative agents for the disease. His work, and that of Pasteur, was ground breaking and was important in establishing the scientific basis for a wide range of public health developments, such as water and sewage treatment, and food hygiene.

10 Why can't Koch's postulates be applied to diseases that affect only humans?

11 How can a link between a specific pathogenic organism and a specific 'human-only' disease be made?

12 Why did the development of water treatment systems in the nineteenth century reduce the occurrence of certain infectious diseases such as cholera?

how science works

Testing a hypothesis

Fig. 4 summarises the stages in scientific research. Progress in science is made when a hypothesis is tested by an experiment. Contrary to popular belief, scientists do not just do experiments to see what happens. Fun though it might be, they don't just mix chemicals together and watch the results. An experiment must be designed to test one possible explanation of an observation. A good hypothesis is one that an experiment can either support or disprove. Strictly, experiments can never prove that a hypothesis is absolutely definitely correct. There is always the possibility that some other explanation, which nobody has thought of, could fit the evidence equally well. However, an experiment can prove a hypothesis to be definitely wrong.

Humans seem to be uncomfortable when they are unable to find an explanation for a phenomenon. Even scientists have a tendency to be biased towards finding evidence to support their hypothesis. As a student doing an investigation you may well have been disappointed to get results in an investigation that disproved your hypothesis or were not what was expected. When this happens, students often suggest that their experiment has 'gone wrong', but, in scientific research, negative results are just as important as positive ones.

Fig. 4 The stages of observation, hypothesis, experiment

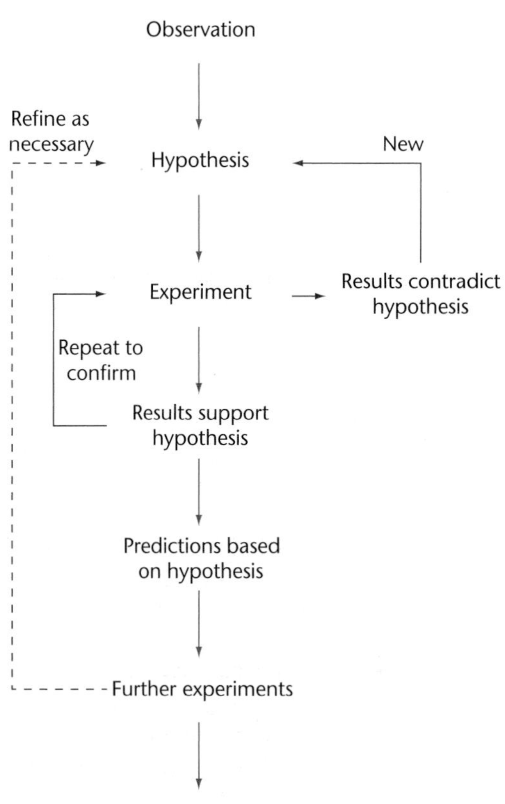

When a hypothesis becomes an accepted theory

When a hypothesis is supported by experimental results, it may be accepted as the best explanation of an observation. The explanation may suggest predictions that, in turn, can be tested by further experiments and observations. Other scientists may try to think of alternative interpretations of the results. It is normal practice for one scientist to be critical of another's published results. To ensure that published work is of sufficient quality, journals practise peer review – a submitted paper is reviewed by two or three other experts in the field to make sure the experiments have been carried out well, and that the interpretations of the results are reasonable.

It should also be possible to repeat an experiment and get the same results. Only after many confirmatory experiments is it likely that a new idea will be accepted. For example, for many years it was thought that cell membranes have a structure rather like a sandwich with protein 'bread' and phospholipid 'filling'. After many experiments this hypothesis was shown to be false and it has been replaced by the fluid-mosaic explanation that you will come across in the AS course. This idea is now so well supported that it is described as the theory of plasma membrane structure.

A theory is a well-established hypothesis that is supported by a substantial body of evidence. The theory of natural selection, for example, is based on huge numbers of observations, predictions and experiments that support the underlying hypothesis.

13 If you breathe in and out of a large plastic bag several times, so that you keep re-breathing the same air, your rate of breathing increases. Give two different hypotheses that could explain this observation.

14 Suggest hypotheses that could explain each of the following observations.

a Having a hot drink can make you sweat, even on a cold day.

b Only sunflower seeds can germinate in the area close to a sunflower plant.

c Cats have narrow vertical pupils in their eyes, rather than round ones as humans do.

key facts

- A hypothesis is a suggested explanation for a phenomenon.
- A theory is a hypothesis that is well supported by scientific data.

Epidemics

Epidemiology is the study of disease in populations. Epidemiologists study:

- the aetiology of a disease, that is, the characteristic causes of the disease
- patterns in the distribution of a disease
- how the disease spreads.

An epidemic is an outbreak of a disease affecting a large number of people in a given population, such as the flu epidemic of 1999–2000. If the outbreak is worldwide, as in the case of AIDS, then it is called a pandemic.

The 'attack rate' of a pathogen – its potential to cause an epidemic – is partly determined by how many people in a given population have already acquired immunity through having been exposed to the pathogen in the past. In general, older people have come into contact with more pathogens than young people. They will have built up a greater degree of immunity, although the immune system starts to break down in advanced age. As a result, populations with a high proportion of older people are less prone to epidemics than populations with high proportions of young people.

15 hsw Imagine an outbreak of *Salmonella* food poisoning in a small town, where 20 people developed symptoms. List two sets of data that an epidemiologist would need to collect in order to understand the cause of the small epidemic.

16 Apart from the level of acquired immunity in a given population, list three other factors that might influence the development of a *Salmonella* epidemic.

17 Explain why fewer epidemics are found in populations with a high proportion of older people than ones with a high proportion of young people.

18 Epidemics caused by bacterial pathogens are quite rare in modern Western Europe and the USA, although they were quite common over 100 years ago. Suggest two key reasons for this decrease.

19 *Salmonella* outbreaks spread quite quickly in care homes for the elderly. Suggest two reasons why this may be so.

1.4 Lifestyle and health

How we live affects our health. The food we eat, the type or amount of exercise we take, air pollution and smoking, ultraviolet light, our working environment, drinking alcohol, and stress are all factors that affect our health and well-being. To some extent we are in control of most of these, so health promotion campaigns try to encourage us to do the right things to improve our quality of life.

Are there health issues we are certain about? On what is health advice based? What can we do about our own health?

These are all possible risks to our health.

1.5 Diet-related disease

In developed countries, diet-related diseases are usually caused by eating too much. People put on weight and become obese if they consume too much energy-rich food and do too little exercise. However, there are other health risks if the nutrient balance is incorrect.

Maintaining the health of the **cardiovascular system** is very important. People can do much to improve and maintain their cardiovascular health by:

- eating a low-fat diet;
- eating a low-salt diet;
- not smoking;
- reducing stress;
- taking exercise.

Diseases of the heart and circulation are called cardiovascular disease; a blocked or burst blood vessel can have serious consequences.

Atherosclerosis

Atherosclerosis is a disease caused by the build-up of fatty deposits, known as **atheroma**, on the inner lining of arteries. Atheroma can occur in any blood vessel but the usual sites are the aorta, the **cerebral** arteries in the brain, and the coronary arteries in cardiac muscle.

Factors that contribute to atheroma formation include:

- high-fat diet, leading to obesity;
- high-salt diet, leading to high blood pressure;
- stress, leading to high blood pressure;
- smoking;
- age (and menopause in women);
- diabetes.

Atheroma restricts blood flow and increases the chance of a blood clot forming. A **thrombus** is a stationary blood clot. An **embolus** is a mobile clot. A thrombus restricts the flow of blood; an embolus is carried round the body by the blood, but may eventually lodge somewhere and restrict blood flow.

20 Explain how to reduce the risk of developing atheroma.

This light micrograph of a healthy artery (top right) has red blood cells in the centre. There is plenty of space for blood flow. In the coloured light micrograph of an unhealthy artery (bottom right), the build up of material (red and yellow) that has reduced the central space by nearly half, is called atheroma. It is the result of cholesterol being deposited inside the blood vessel. The large mass at the centre (red) is a blood clot attached to the atheroma. Together they almost block the artery. Atheroma in a coronary artery can lead to a heart attack. Cerebral atheroma can cause cerebral haemorrhage (stroke).

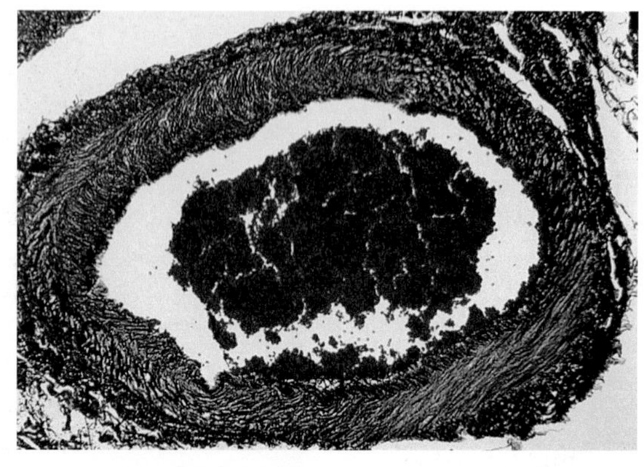

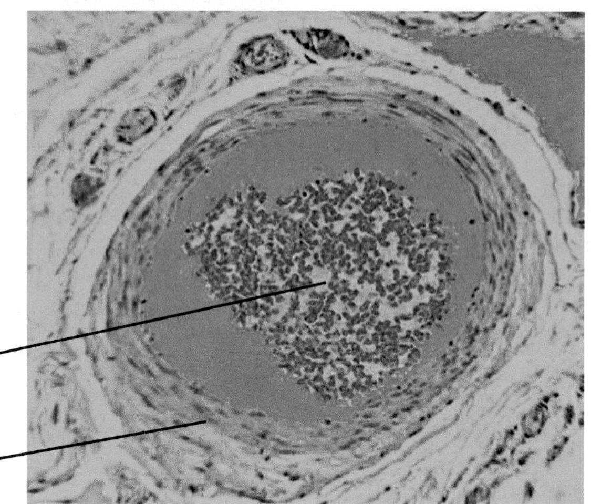

Coronary heart disease

Atheroma in coronary arteries is called coronary heart disease. Narrowing of the coronary arteries may restrict blood flow and starve an area of cardiac muscle of oxygen. This causes a condition called **angina**, which results in severe pain, usually felt in the chest in men, and in the neck and arms in women.

Blood clots may form in these narrowed blood vessels and block the artery completely, thus depriving the cardiac muscle of its blood supply. Embolisms can form somewhere else in the body and then become lodged in a coronary artery. After such a blockage, areas of heart muscle do not function properly and may die. This is called a **myocardial infarction** and can lead to a heart attack when the cardiac muscle fails to contract. This can kill, although about half of patients will survive a heart attack if they are treated quickly.

21 How might a blood clot forming in an artery in the leg eventually cause a heart attack?

22 What is this type of blood clot called?

key facts

- Diseases of the heart and circulation are called cardiovascular diseases.
- Atherosclerosis is a build up of atheroma in blood vessels.
- A blood clot may block blood vessels, causing embolism.
- Coronary heart disease is caused by a blockage in the coronary arteries.

1.6 Cancer

Cancer occurs when mutations occur in the genes that control cell division. This results in unchecked, irregular growth and a **tumour** forms. There are many factors that increase the risk of cancer.

A risk factor is anything that increases a person's chance of getting a disease. Some **risk factors** can be changed, and others cannot. Risk factors for cancer can include a person's age, sex and family medical history. Others are linked to cancer-causing factors in the environment. Still others are related to lifestyle choices such as tobacco and alcohol use, diet, and sun exposure.

Having a risk factor for cancer means that a person is more likely to develop the disease at some point in their lives. However, having one or more risk factors does not necessarily mean

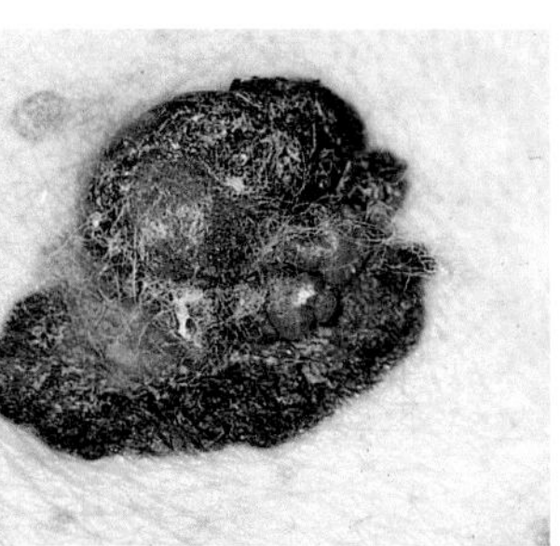

A skin cancer begins as a localised growth of skin cells.

that a person will get cancer. Some people with one or more risk factors never develop the disease, while other people who do develop cancer have no apparent risk factors. Even when a person who has a risk factor is diagnosed with cancer, there is no way to prove that the risk factor actually caused the cancer.

Different kinds of cancer have different risk factors.

- Tobacco use is related to cancers of the lung, mouth, larynx, bladder, kidney, cervix, oesophagus and pancreas. Smoking alone causes one-third of all cancer deaths.
- Skin cancer is related to unprotected exposure to strong sunlight.
- Breast cancer risk factors include: age; changes in hormone levels throughout life, such as age at first menstruation, number of pregnancies, and age at menopause; obesity; and physical activity. Women with a mother or sister who have had breast cancer are more likely to develop the disease themselves.
- All men are at risk of prostate cancer, but several factors can increase the chances of developing the disease, such as age, race, and diet. The chance of getting prostate cancer goes up with age. Men with a father or brother who have had prostate cancer are more likely to get prostate cancer themselves.

Overall, environmental factors, which include tobacco use, diet and infectious diseases, as well as chemicals and radiation, cause an estimated 75% of all cancer cases. Research shows that about one-third of all cancer deaths are related to dietary factors and lack of physical activity in adulthood.

Certain cancers, such as cervical cancer, are related to viral infections and could be prevented by behaviour changes and vaccines. More than 1 million skin cancers diagnosed in 2003 could have been prevented by protection from the sun's rays.

23 What is a cancer?

24 What is a risk factor?

25 What type of factor is responsible for the majority of cancer cases?

- Lifestyle affects human health.
- Factors affecting human health include diet, exercise and use of recreational drugs.
- Factors which increase the risk of coronary heart disease include:
  - high-fat diet, leading to obesity
  - high-salt diet, leading to high blood pressure
  - stress, leading to high blood pressure
  - smoking
  - age (and menopause in women)
  - diabetes.
- Factors which increase the risk of cancer include:
  - smoking
  - exposure of the skin to strong sunlight
  - ageing.
- Changes in lifestyle such as diet and exercise may reduce the risk of contracting these conditions.

1.7 The benefits of exercise

People who exercise regularly are better off both physically and psychologically. Regular exercise increases muscle performance; it also affects the circulation and breathing systems of the body.

Regular exercise leads to a thickening of the cardiac muscle, accompanied by an increase in the size of the heart's chambers. This means that there is an increase in the volume of blood pumped by each stroke of the heart. If more blood is pumped with each beat, fewer beats per minute are needed to supply the body with the necessary oxygen and nutrients. So, the average **heart rate** of a very active person is lower than that of an inactive person. For most people, the resting rate is 65–80 beats per minute, but the resting rate for athletes can be as low as 40–50 beats per minute.

Regular activity also lowers blood pressure. Blood pressure is the force exerted by blood on the walls of the vessels that contain it. Exercise enlarges the blood vessels and improves the blood supply, so reducing blood pressure.

Exercise improves the effectiveness of the breathing mechanism by strengthening the **diaphragm** and **intercostal muscles** that control breathing. Gaseous exchange is more effective in both supplying oxygen and removing carbon dioxide. At the cellular level, the **metabolic rate** is higher, so deposits of atheroma are less likely and less fat is stored in the body, reducing the risk of cardiovascular disease.

26 Which two systems of the body benefit from regular exercise?

key facts

Regular exercise:

- improves muscle performance
- increases the heart's efficiency
- lowers blood pressure
- strengthens the muscles that control breathing.

how science works

Benefits of exercise

Training increases heart efficiency. The amount of blood pumped by the heart per minute is called the **cardiac output** and it varies with activity. As activity increases, cardiac output increases.

cardiac output = heart rate × **stroke volume**

The difference between the resting output and the maximum the heart can achieve is called the **cardiac reserve**.

27 A 25-year-old athlete has a resting heart rate of 45 beats per minute and a cardiac output of 5.25 $dm^3\ min^{-1}$. Calculate the stroke volume.

28 An average adult has a stroke volume of 0.06 dm^3. Explain the difference between this figure and your answer to Question 27.

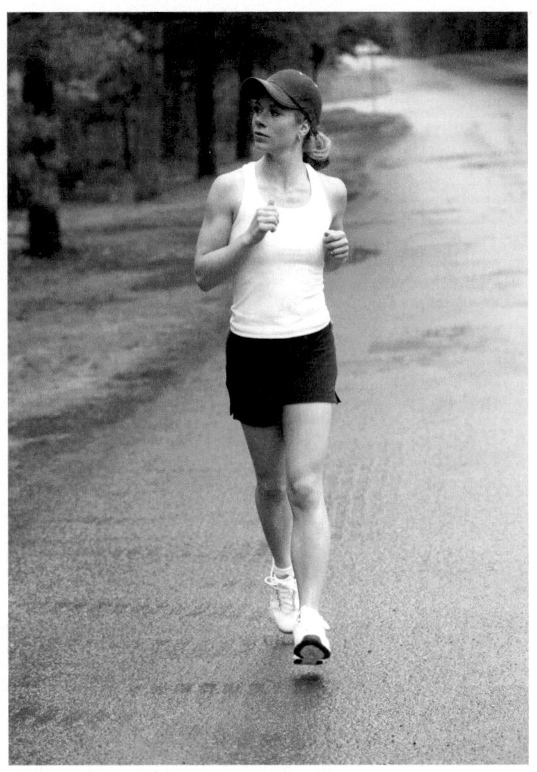

Exercise is good for you.

29 Study the data in the table.

	Cardiac output ($dm^3\ min^{-1}$)	
Individual	**at rest**	**vigorous activity**
average adult	5.0	21.0
athlete	5.25	30.0

a Compare the difference in cardiac output of the athlete and average adult at rest and in vigorous activity.

b Calculate the cardiac reserve of the average adult and the athlete.

c Explain the effect of regular exercise on the cardiac reserve.

30 Explain why transport of oxygen to active tissues is more rapid in an athlete due to:

a the heart;

b blood vessels;

c breathing system.

examination questions

1

a Describe how an atheroma is formed in a blood vessel. (2)

Cholesterol is carried around the body by lipoproteins. HDL is one type of lipoprotein.

The table below shows the effect (very low – very high) of different concentrations of HDL and cholesterol in the blood on the risk of heart disease

HDL concentration in arbitrary units	Total cholesterol concentration in arbitrary units					
	161–170	171–180	181–190	191–200	201–210	211–220
below 21	low	average	average	very high	very high	very high
21–30	low	average	high	high	high	very high
31–35	very low	low	average	average	high	high
36–40	very low	low	low	average	average	high
41–45	very low	very low	low	average	average	average
46–50	very low	very low	low	low	average	average
51–55	very low	very low	very low	low	low	average
56–60	very low	very low	very low	low	low	low
61–65	very low	very low	very low	low	low	low
66–70	very low	very low	very low	very low	low	low
71–75	very low	very low	very low	very low	very low	low

b Describe the relationships shown in the table between the concentrations of HDL and cholesterol in the blood and the risk of heart disease. (3)

Total 5

2 The table below shows the results of an investigation into the link between smoking and lung cancer

Smoking status	Number of control subjects in sample	Number of people in sample who had developed lung cancer	Relative risk of getting lung cancer
Non smoker	90	3	1.0%
Past smoker	161	87	3.9%
Current smokers – duration of smoking in years			
1–20 years	13	2	5.6%
21–39 years	137	52	17.4%
40+ years	90	144	29.8%
Current smokers – number of cigarettes smoked per day			
1–20 cigarettes	104	69	3.0%
21–39 cigarettes	88	73	7.1%
40+ cigarettes	37	43	12.5%

a Explain what the data in the table shows about the link between smoking and lung cancer (3)

b Does the data show a causal link or an association between smoking and lung cancer? Explain the reasons for your answer. (3)

Total 6

3 Life expectancy is the number of years a person can expect to live. The chart below shows some of the causes of death in populations with different life expectancies.

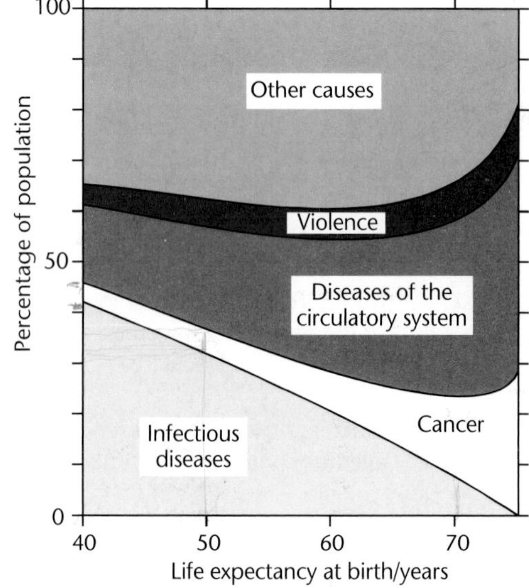

a i What is the main cause of death in a population with a life expectancy at birth of 40 years? (1)

ii Suggest two explanations for this. (2)

b i What proportion of the population with a life expectancy at birth of 50 years die from cancer? (1)

ii Give two factors that increase the risk of cancer. (1)

c i Describe the relationship between death from diseases of the circulatory system, and life expectancy at birth. (2)

ii Suggest explanations for this relationship. (3)

Total 10

4 Atheromas can form in coronary arteries and produce a condition called angina. An angina sufferer experiences pains in the chest, usually during exercise. The pain is caused by a build-up of lactate in heart tissues. Sometimes an atheroma gets damaged and a blood clot forms which blocks the coronary artery. This causes myocardial infarction which can be fatal.

Angina can be diagnosed from a type of X-ray called an angiogram. A small tube is inserted into an artery at the top of the leg and pushed up to the opening of the coronary artery. A special dye is injected into the artery and an X-ray is taken.

a Explain the meaning of

i atheroma; (1)

ii myocardial infarction. (1)

b Describe **one** way in which the lifestyle of a person could increase the risk of developing angina. (1)

c Suggest why angina sufferers experience a build-up of lactate during exercise. (3)

d Suggest why it is necessary to inject a special dye before taking an angiogram. (1)

Total 7

AQA, June 2002, Unit 8, Question 5

5

a Describe how the growth of a malignant tumour in one organ can lead to tumours developing elsewhere in the body. (1)

The table below shows the numbers of deaths of people aged under 65 years due to different cancers, in the UK in 1997.

Men		**Women**	
Cancer	**Number of deaths**	**Cancer**	**Number of deaths**
Lung	4013	Breast	4118
Colorectum	1691	Lung	2157
Lymphatic tissue	1607	Ovary	1313
Oesophagus	971	Colorectum	1133
Brain	845	Lymphatic tissue	1103
Others	6501	Others	5323
Total	15628	Total	15147

b i Calculate the percentage of deaths due to lung cancer. Show your working. (2)

ii Suggest **one** reason for the difference between men and women in the number of deaths due to lung cancer. (1)

c Explain how excessive sunbathing can lead to malignant skin tumours. (3)

Total 7

AQA, June 2003, Unit 8, Question 3

6

a With reference to typhoid fever and salmonella food poisoning, explain what is meant by infectivity. (2)

b Typhoid fever is mainly a water-borne infection. Salmonella food poisoning is caused by contaminated food. The graph below shows changes in the incidence of these diseases in the USA in the twentieth century.

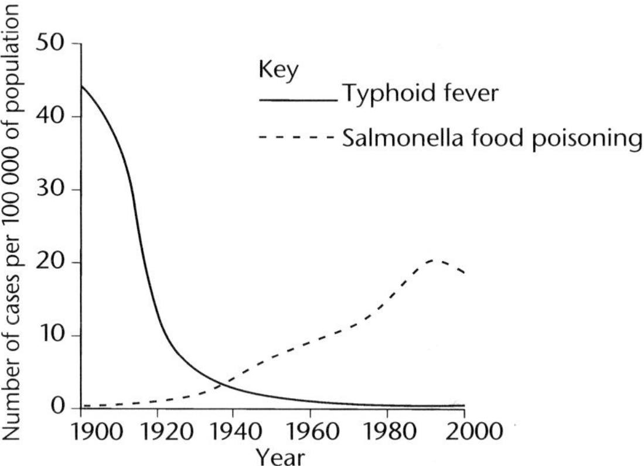

Suggest an explanation for the overall trend in the number of cases of

i typhoid fever; (1)

ii salmonella food poisoning. (1)

Total 4

AQA, June 2005, Unit 7, Question 3

7 The death rate from malignant skin tumours was investigated in the USA. The graph below shows the results for fair-skinned men in different age groups.

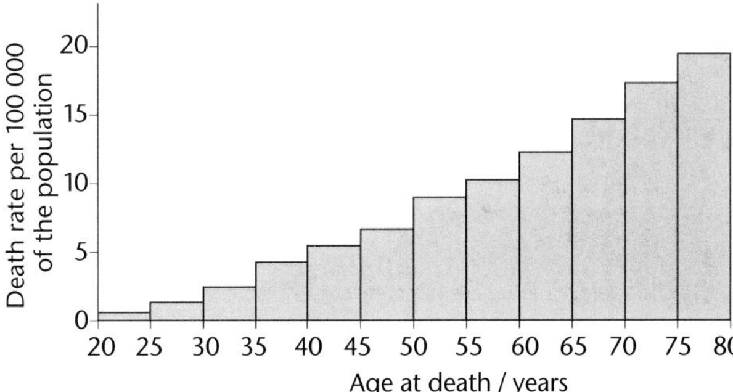

a Describe what is meant by a *malignant tumour.* (3)

b Give **one** reason for the change in death rate from malignant skin tumours with increasing age. (1)

c The data for fair-skinned and dark-skinned people were collected separately. Explain why skin colour was a factor likely to affect the death rate. (2)

Total 6

AQA, June 2007, Unit 8, Question 2

how science works **assignment**

Correlation and association

Many biological investigations depend on a combination of observation and data analysis rather than on actual experiments. This is because it is often not practical to carry out proper controlled experiments with living organisms in the field, sometimes because of the complexity of interrelationships between organisms and the environment and sometimes for ethical reasons. You can't experiment on the effects of smoking by taking two groups of people and making one group smoke while keeping all other factors the same.

Investigators therefore have to look for associations that occur in the normal course of events. However, care needs to be taken when drawing conclusions. The number of fish may decline in a lake affected by acid rain or some other pollutant, but this association does not necessarily prove that the pollution has caused the decline, or even that the two are connected. Further investigations could look for data on natural populations of particular fish species in water of different acidity. It would also be possible to carry out laboratory experiments to determine fish survival rates in water of different acidity. Results might well show that the lower the pH, the lower the survival rate. In this case there would be a **correlation** between pH and fish survival. This would still not prove that the decline in fish numbers in the lake was actually caused by the acidity.

If you counted, say, the number of nightclubs and pubs and the number of churches in several towns and cities and then plotted a graph of one against the other you would almost certainly find a correlation. But this would obviously not prove that churches cause nightclubs and pubs to be built or the other way round. The **correlation** is likely to be the result of completely separate factors, such as the size of the town or city.

Nevertheless, it is only by searching for correlations and investigating them further that biologists can increase their understanding. A correlation can be either positive or negative. When one factor increases as another increases it is a positive correlation; when one increases as another decreases it is a negative correlation.

Epidemiology

Links between human afflictions and factors such as environmental pollutants, diet, smoking and other aspects of lifestyle are equally hard to establish by experiment. Most associations have been established by studies of the incidence of disease or disorder on large groups of people. Looking for patterns in the occurrence of disease in human populations is called epidemiology. Many of the suggested links have been matters of controversy and some have caused considerable confusion in the minds of the public. Consider issues such as the danger of using mobile phones; the possible link between taking contraceptive pills and various cancers; or fat consumption and heart disease. There are still some people who refuse to accept the association between smoking and cancer, despite the overwhelming statistical evidence.

The stages in establishing the cause of a non-infectious disease are:

- searching for a correlation between a disease and a specific factor;
- developing hypotheses as to how the factor might have its effect;
- testing these hypotheses to determine whether the factor can induce the disease.

To establish a correlation means collecting data from large numbers of people. Because of the huge variability between people and their lifestyles, it requires comparisons to be made as far as possible between matched groups. For example, suppose you were looking for a correlation between beer consumption and heart disease. It would not be sufficient just to compare the rates of heart disease between 1000 beer drinkers and 1000 non-drinkers. The ideal comparison would be between groups of people where the only difference in lifestyle was whether or not they drank beer. In practice this would be virtually impossible to achieve. However, at least a much more valid comparison could be made between groups matched for age, sex, amount of exercise taken and major features of diet. The difficulty is to eliminate the possibility that any correlation found is not due to some other linked factor, such as that people who are tempted to indulge in beer-drinking are also consumers of excessive quantities of fish and chips, or simply that they have some genetic predisposition to heart disease. The latter is a particularly difficult argument to refute, and has regularly been used as an excuse for shedding doubt on the smoking/lung cancer correlation.

Once a correlation has been found, the next stage is to try to determine how exactly the factor actually causes its effect. This is often much more difficult. Many diseases, such as cardiovascular disorders and cancers, develop as a result of several interacting factors. The correlation between smoking and incidence of lung cancer has been established for many years. The search to isolate a specific **carcinogen** in cigarette smoke has still not been successful. However, the tar inhaled into the lungs contains a massive cocktail of organic compounds, many of which may have carcinogenic properties. Some of the effects may be additive – a combination of substances and other factors may be the main cause. Also, individuals differ in their susceptibility, probably due to genetic factors. Research has involved detailed chemical analyses of tar, experiments on animals and with tissue cultures, and comparisons between many genetically distinct groupings. Until a precise mechanism is discovered, the arguments will no doubt continue, and

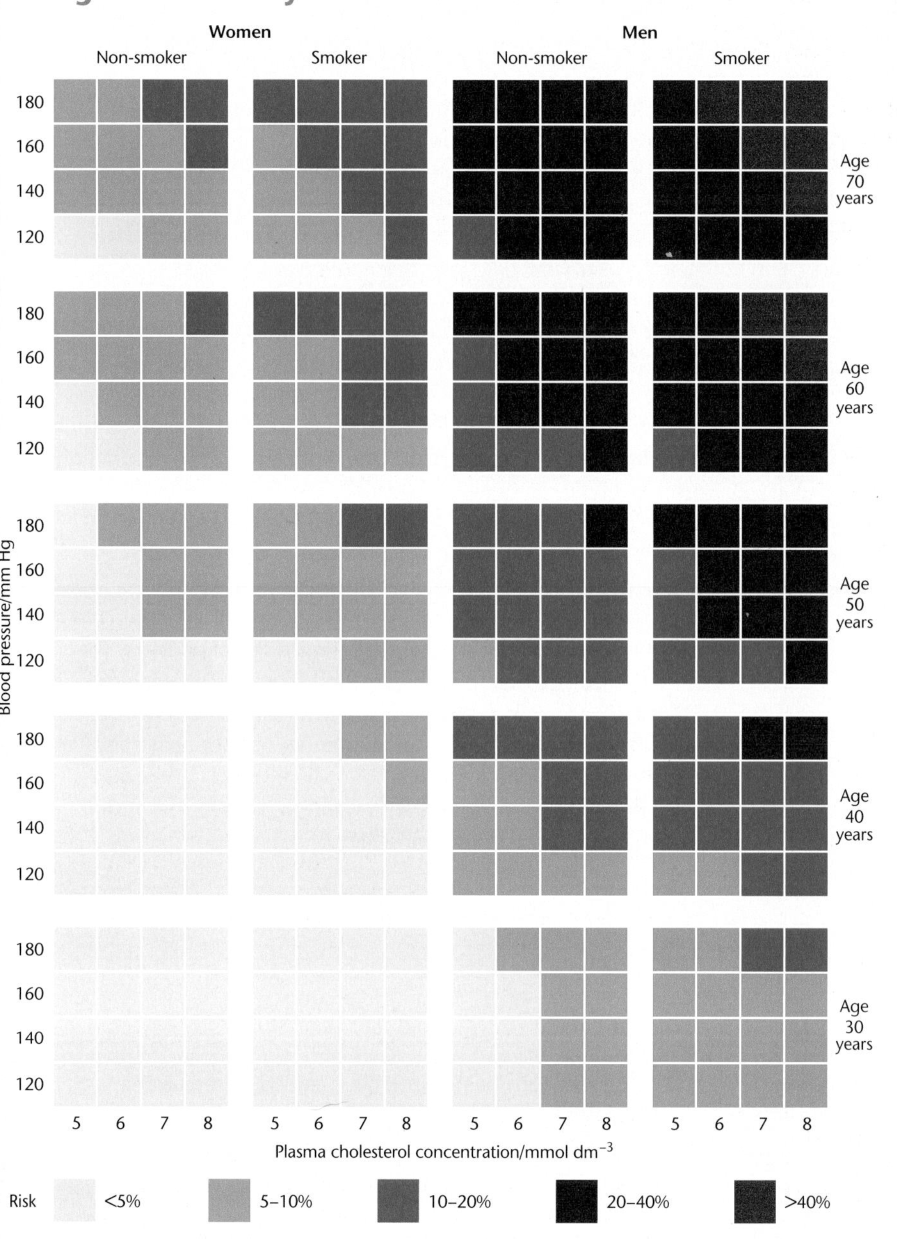

Fig. 5 Coronary heart disease and risk factors

only after that will it be possible to produce effective countermeasures apart, of course, from not smoking!

The chart in Fig. 5 shows the risk of developing coronary heart disease during the next 10 years in relation to a number of risk factors.

A1 Which risk factors have been taken into account in this chart?

A2 According to the chart, what are the characteristics of the people with the highest risk of developing coronary heart disease?

A3 A group of 500 men aged between 40 and 50 years is studied. They all have high blood pressure of over 180 mm of mercury, and plasma cholesterol concentration between 7 and 8 mmol dm^{-3}. They all smoke. How many of these men would be expected to get coronary heart disease over the next 10 years? How many would get it if they were all non-smokers?

A4 Explain the limitations of using the chart as a way of predicting whether an individual will develop coronary heart disease.

A5 The data used to construct the chart - have a number of limitations as a means of predicting risk. Suggest three weaknesses in the data.

key facts

- An **association** in data is where there appears to be a link between one factor and another, such as between the number of fish in a lake and its acidity.
- Where a change in one factor is associated with a corresponding pattern of change in another factor, the two factors are **correlated**. For example, one factor may increase as the other increases (a **positive correlation**), or one may decrease while the other increases (a **negative correlation**).
- A correlation is not proof that a change in one factor is the direct cause of the change in the other. This can only be supported by experimental evidence.
- In complex situations, such as in ecosystems and human populations, it is difficult to establish cause and effects because many factors may interact.
- In human populations **epidemiologists** look for associations between factors and try to establish the likely causes of diseases by studying data from matched groups.

2 Cells

Screening for **cervical cancer** started in 1964 and had only a modest effect on death rates during the 14 years that followed. However, a national call system was introduced in 1988, which doubled the proportion of eligible women tested to 85 per cent and trebled the rate of fall in death rates. By 2005 only 45 per cent as many women were dying of cervical cancer compared with 1950. A recent report in the *British Medical Journal* said that in 2005 alone, screening probably prevented 800 deaths from cervical cancer in women aged 25–54 years.

Most women have a smear test every 3–5 years. A small brush is used to remove a few cells from the surface of the cervix. These are transferred to a microscope slide, stained and examined with a light microscope. A trained technician can spot cells showing early cancerous changes by their enlarged appearance and bigger nuclei. Prompt treatment of a woman whose smear contains cancerous or pre-cancerous cells can prevent a life-threatening cancer.

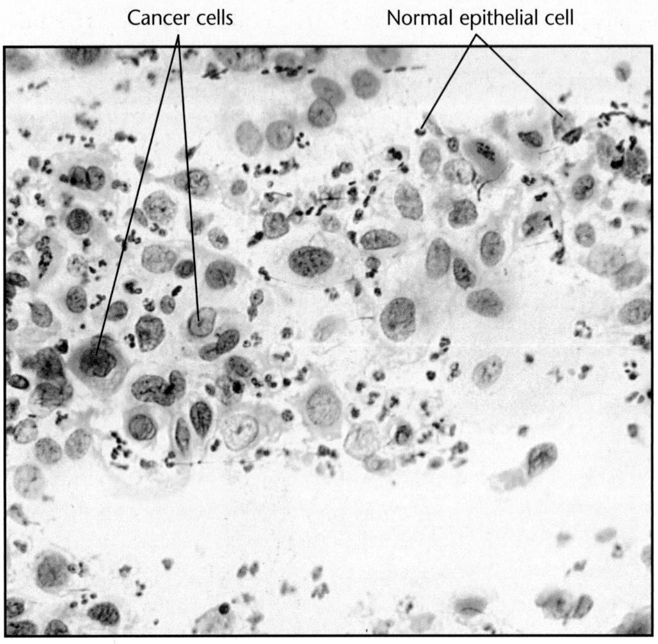

The photograph shows cells obtained by a cervical smear test. The orange-stained cancerous cells are larger than normal epithelial cells and their nuclei are more noticeable. Cancer occurs when the normal process of cell division and differentiation goes wrong and cells divide out of control.

The human papilloma virus (HPV) is responsible for 70 per cent of cervical cancer. The virus is transmitted by sexual intercourse. Condoms do not give protection against infection by HPV. A **vaccine** has been developed which gives protection against most forms of HPV and will therefore protect against cervical cancer. To be most effective, the vaccine needs to be given to girls before they become sexually active and it has therefore been proposed that the vaccine is given to girls at about 12 years of age. However, this has presented something of an ethical dilemma: some people are concerned that vaccination against what is essentially a sexually transmitted infection may encourage girls to be complacent about the use of barrier contraceptive methods that protect against other sexually transmitted infections, including Chlamydia and HIV/AIDS.

HPV is causing other ethical dilemmas. Scientists are researching how HPV is passed from men to their female sexual partners. The aim is to learn whether men should also be vaccinated against HPV. The men in the study won't be told whether they are infected or not. 'There is no treatment for HPV, so we are not doing any harm by not disclosing infections,' one scientist said. 'There also is no strategy for prevention of transmission to partners, because condoms aren't protective against HPV.'

1 Evaluate the pros and cons of:

a vaccinating 12-year-old girls against HPV

b the methods being used by scientists to research transmission of HPV.

2.1 Looking at cells

At birth, a human baby has about 20 billion (20 000 000 000) cells. All have developed from a single cell, the fertilised egg. Normal growth over the nine months of pregnancy depends on two processes: cell division to increase the numbers of cells and then **differentiation**, which enables cells to specialise to carry out a particular function. When control of either of these crucial processes breaks down, the result can be cancer. Cancer cells lose the special features of the tissue they arise from and they divide much faster than the cells around them.

The light microscope

You have probably used a light microscope to look at cells from the lining of the cheek. Cheek cells are epithelial cells very like the normal cervical cells in the photograph above. Fig. 1 shows the appearance of cells seen through a light microscope.

Staining techniques usually show the cell **nucleus** as a blob, and the **cytoplasm** as a paler substance that looks empty. In fact the cytoplasm is packed with organelles, such as **mitochondria** and **ribosomes**.

Units of measurement

The two most common units used to describe the size of organelles are:

- the **micrometre (μm)**, one thousandth of a millimetre (10^{-6} metre),
- the **nanometre (nm)**, one thousandth of a micrometre (10^{-9} metre).

Resolving power and cell size

You can see only the larger **organelles** in a cell with a light microscope because even the larger organelles are near the limit of a light microscope's **resolving power**.

Resolving power should not be confused with magnification. Resolving power is the ability to distinguish between two objects. Think of a car at night. When it is far away, the two headlights appear as a single light. As the car gets nearer you can see two headlights; at some point your eyes are able to resolve two lights, not one. With binoculars you could resolve the light into two headlights when the car is much further away.

The limitations of the light microscope are not due to the construction of the equipment, but to the nature of light itself. Even the most powerful lens cannot resolve two dots that are separated by less than 250 nm. This is because no lens system can ever resolve two dots that are closer together than half the wavelength of the light used to view them. The wavelength of visible light is 500–650 nm.

Since the lenses in a light microscope will not allow you to distinguish between two objects that are smaller than 250 nm (0.25 μm), the maximum magnification of a light microscope is × 1500. This is adequate to see animal cells (diameter 30–50 μm) and individual bacteria (length 5–8 μm). It is also sufficient to detect the changes in a cell that might be the beginning of cancer and to view larger organelles such as the nucleus (diameter 10 μm). But it is not powerful enough to resolve the structure of small organelles such as ribosomes (diameter 20 nm) or cell membranes (thickness 7–10 nm).

The basic structure of a standard light microscope is shown in Fig. 2. This is the sort of microscope that is commonly used for teaching purposes and that you have probably used in

Fig. 2 The light microscope

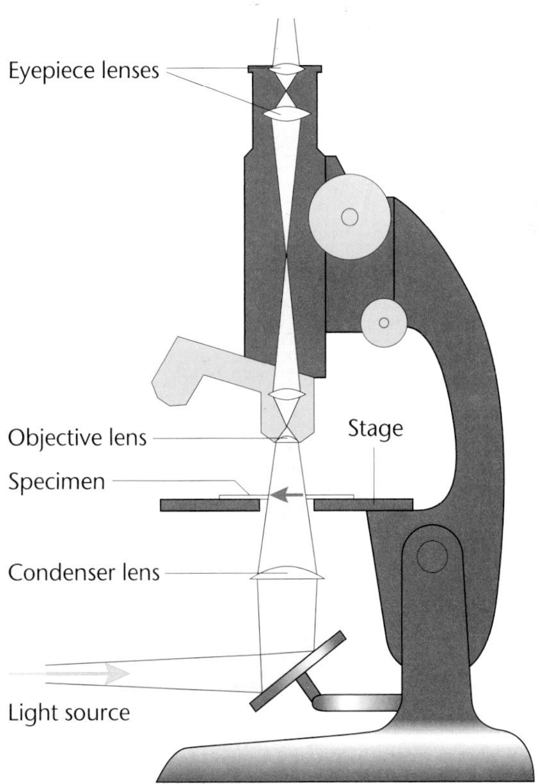

Fig. 1 What can you see with a light microscope?

A light micrograph of epithelial cells from the human small intestine with a diagram showing the main features visible.

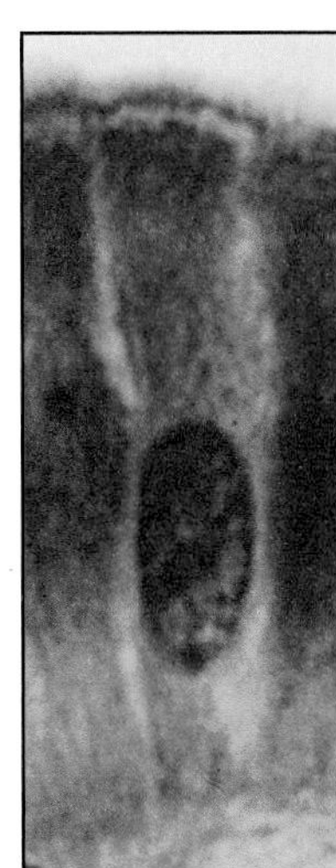

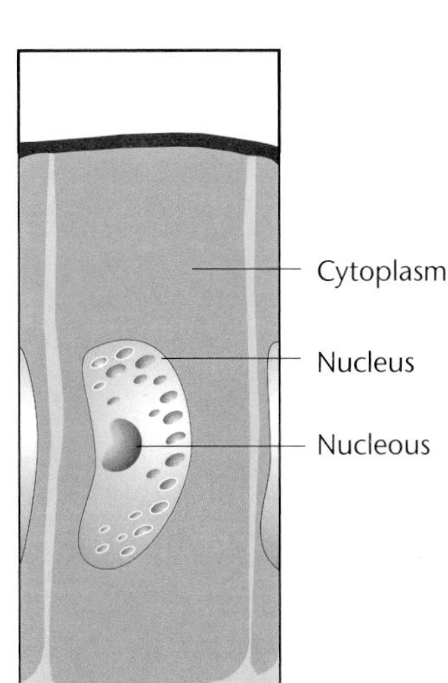

science lessons. It is also known as a compound microscope because it has two lenses. The eyepiece lens and the objective lens combine to produce a greater magnification than would be possible with only a single lens.

The total magnification is calculated by multiplying the magnifications of the two lenses. For example, if the eyepiece lens has a magnification of × 10 and the objective lens has a magnification of × 50, the total magnification of the microscope is × 500.

2 Why are there no light microscopes with a magnification of x 2000?

3 Explain the difference between magnification and resolution.

Electron microscopes

The transmission electron microscope

The detailed ultrastructure of plant and animal cells was revealed in the 1950s when the **transmission electron microscope (TEM)** was first used. A specimen for the electron microscope has to be specially prepared; a simple smear of cells on a microscope slide would not work. Very thin slices of the specimen are cut, preserved and stained. The specimen is then placed in a chamber inside the electron microscope, which is sealed and the air is sucked out to produce a vacuum. Electromagnets focus a beam of electrons that passes through the specimen and onto a viewing screen (Fig. 3). A modern electron microscope can magnify objects up to 500 000 times.

The development of the electron microscope has had a huge impact on biology. It makes it possible to see the details of cell organelles and has led to the discovery of new organelles. Look at the differences between the photographs (Fig. 4) of the animal cell taken with the light microscope and the electron microscope .

The electron microscope reveals that cells contain many membranes. The cytoplasm is surrounded by the cell surface membrane or plasma membrane. Other organelles, such as mitochondria, have an outer and an inner membrane. A network of membranes, the **endoplasmic reticulum**, spreads throughout the cytoplasm.

Fig. 3 The electron microscope

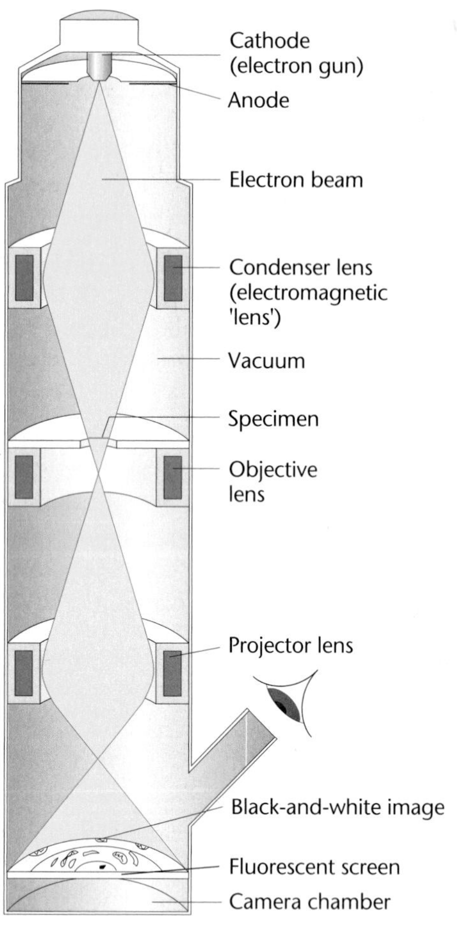

4 The wavelength of the beam of electrons that is used in an electron microscope is approximately 0.005 nm. What is the theoretical limit of the electron microscope's resolving power?

5 List the organelles that can be seen:

a with the light microscope

b with the electron microscope only.

Fig. 4 A generalised animal cell

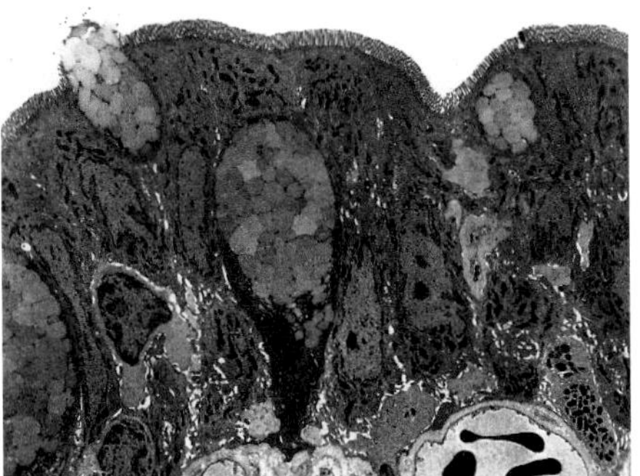

An electron micrograph of epithelial cells from the small intestine.

Magnification × 2000.

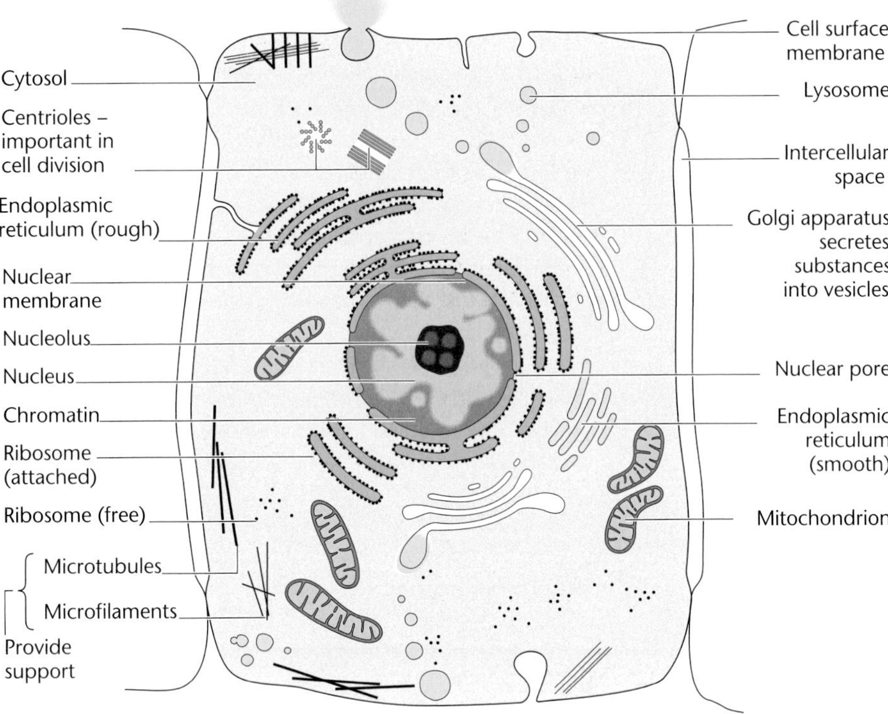

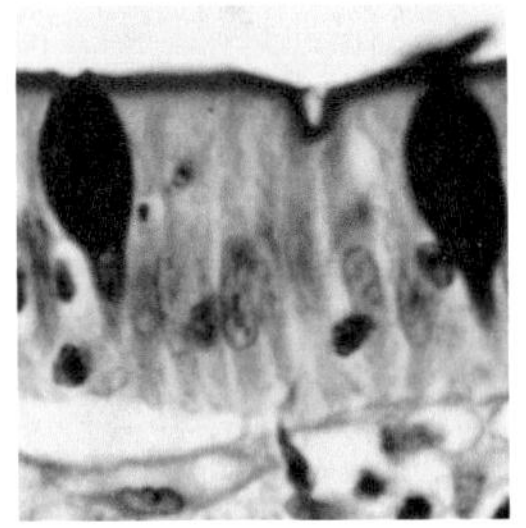

A light micrograph of epithelial cells from the small intestine.

Magnification × 1000.

The scanning electron microscope

The **scanning electron microscope (SEM)** is most often used to magnify the surface features of an organism or a cell. The electron beam is produced in the same way as the beam in the transmission electron microscope. But the beam is then focused by one or two condenser lenses into a beam with a very fine focal spot, sized 1–5 nm. The beam passes through pairs of scanning coils in the objective lens, which deflect the beam horizontally and vertically so that it scans over a rectangular area of the sample surface. The electrons in the scanning beam cause secondary electrons to be emitted from the surface of the sample. The secondary electrons are detected by a device which produces a signal that can be viewed and saved as a digital image.

The resolution of the SEM is not high enough to image down to the atomic scale, as is possible in the TEM. However, the SEM can:

- image a comparatively large area of the specimen
- image bulk materials (not just thin films or foils).

The resolution of an SEM ranges from less than 1 nm to 20 nm, depending on the instrument. In general, SEM images are much easier to interpret than TEM images.

Scanning electron micrographs show three-dimensional features of structures.

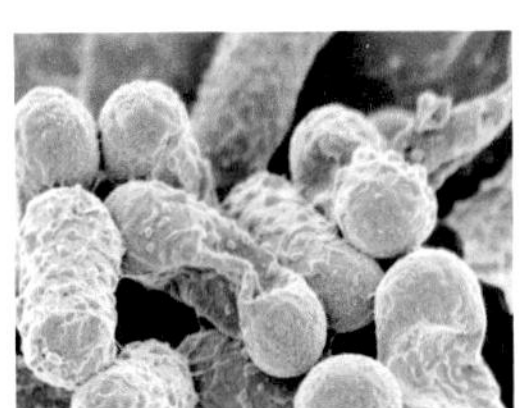

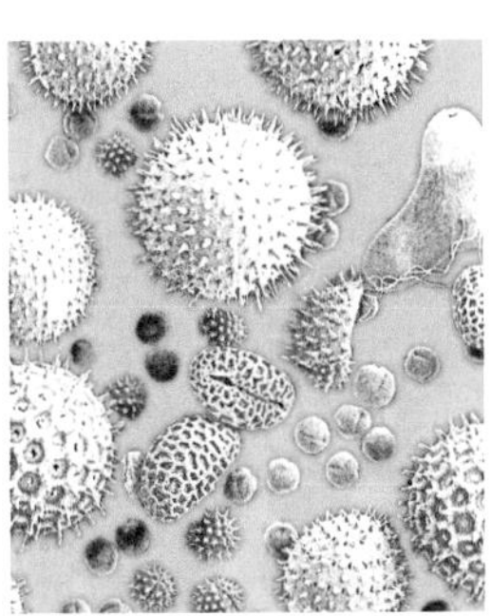

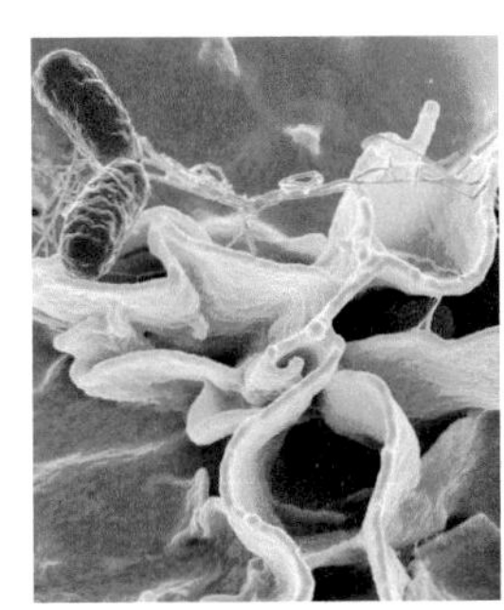

6 What advantages has

a the TEM over the SEM?

b the SEM over the TEM?

Cell fractionation and ultracentrifugation

It is easy to see that the availability of the electron microscope was a major help to scientists in working out the structure of cell organelles. But just looking at the organelles in preserved sections of tissue cannot reveal much about their function. In fact, people had found out a great deal about what organelles do, particularly about chloroplasts and mitochondria and the sequence of reactions in photosynthesis and respiration, before they had access to electron micrographs.

They investigated individual organelles by breaking down cells and then separating out organelles such as mitochondria, ribosomes and membranes. This is known as **cell fractionation**. To do this successfully, the organelles had to remain intact, with all internal enzymes and structures in place. The technique that made this possible is called **ultracentrifugation**.

As Fig. 5 shows, if a mixture of different sizes is spun at high speed in a centrifuge, the larger particles tend to accumulate at the bottom of the tube. Since the organelles vary in size, this principle can be used to separate them. Cells are broken up in a homogeniser, a device rather like a kitchen blender that breaks down the outer membrane of the cells but leaves the organelles intact. The liquid used in the homogeniser is ice cold to reduce the rate of enzyme activity and has the same concentration of solutes as the organelles to prevent shrinkage or bursting due to osmosis.

The homogenate is then spun at relatively slow speed. The nuclei, the largest organelles in the cells, collect at the bottom of the tube. The suspension above is known as the supernatant. This supernatant is then spun at a higher speed. This time the mitochondria separate out. To separate out the ribosomes the process is repeated at a higher centrifuge speed.

In 1937, Hans Krebs isolated mitochondria from animal cells and worked out the reactions in the final stages of respiration. One series of these reactions has been named the Krebs cycle in his honour.

Fig. 5 Differential centrifugation

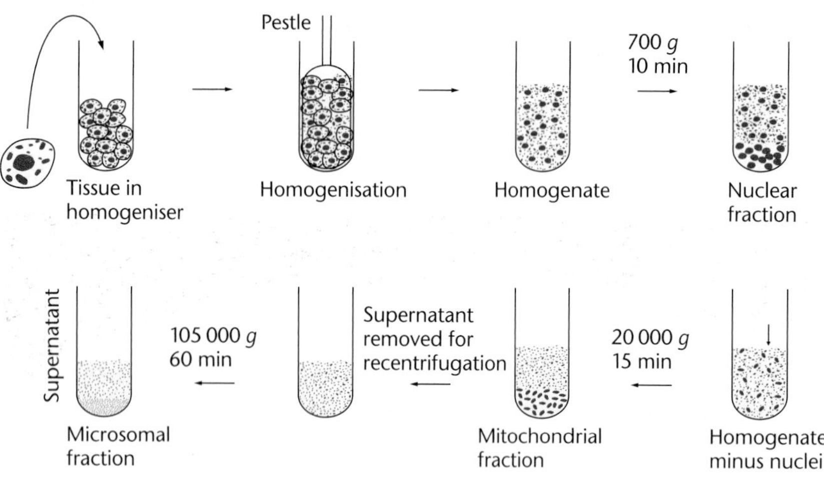

7 Why are the cells homogenised

a in ice-cold water?

b in a solution with the same concentration of solute?

8

a Place the following organelles in order of size with the largest first: ribosomes, mitochondria, nuclei.

b Draw a centrifuge tube to show the order of settling of these three after they have been centrifuged.

2.2 Major cell organelles

Both animal and plant cells have a **plasma** or **cell surface membrane**, a nucleus, mitochondria, endoplasmic reticulum, ribosomes, a **Golgi apparatus** and **lysosomes**. We look at all of these organelles in more detail in this section. In addition, plant cells have a cell wall and chloroplasts. These are described in Chapter 11.

Membranes

The word membrane means 'very thin layer.' It is important that you understand the difference between a cell membrane and a body membrane such as the alveolar membrane in the lungs or the corneal membrane covering the front of your eye. These membranes are made from a thin layer of whole cells.

The photograph in Fig. 6 is a high-power transmission electron micrograph of the plasma membrane. You can see the membrane as two dark bands separated by a clear central area. It looks like this because the membrane is not a single layer; it consists of two layers of **phospholipid** molecules. Proteins are embedded in each thin layer of phospholipids; some proteins span the membrane from one side to the other, others appear on one face of the membrane only. The phospholipid and protein molecules fit together to form a continuous pattern like the tiles in a mosaic. However, unlike mosaic tiles, the molecules that make up a membrane are not fixed in place; the position of the proteins can change from moment to moment. For this reason, cell membranes are said to have a fluid-mosaic structure. The fluid-mosaic model of cell membrane structure was first put forward in 1972, as a result of work with the transmission electron microscope.

The protein molecules that form part the membrane may occur only in the upper or lower layer of lipid molecules. Some protein molecules span the membrane completely. These proteins are often important in transport of substances across the membrane (see page 46). Proteins that protrude from the outside of the surface membrane are often involved in recognition, or they may act as receptors for **hormones**.

Fig. 6 The fluid-mosaic structure of cell membranes

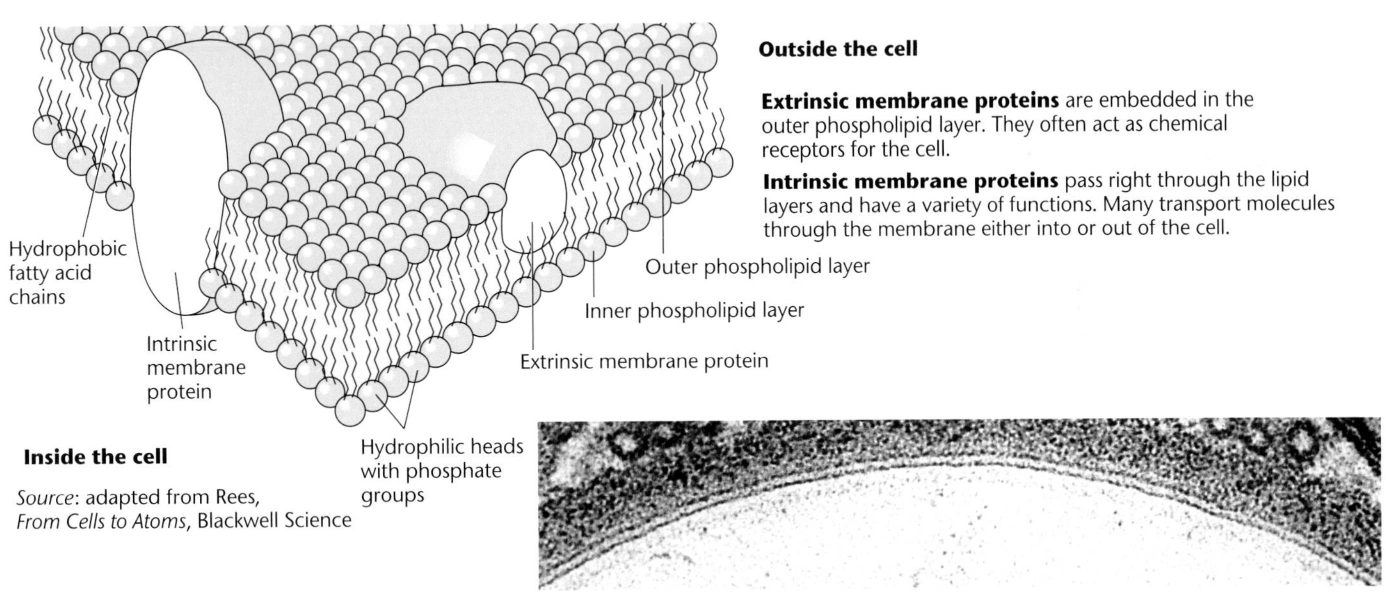

Outside the cell
Extrinsic membrane proteins are embedded in the outer phospholipid layer. They often act as chemical receptors for the cell.
Intrinsic membrane proteins pass right through the lipid layers and have a variety of functions. Many transport molecules through the membrane either into or out of the cell.
Hydrophobic fatty acid chains
Outer phospholipid layer
Inner phospholipid layer
Intrinsic membrane protein
Extrinsic membrane protein
Hydrophilic heads with phosphate groups
Inside the cell

Source: adapted from Rees, *From Cells to Atoms*, Blackwell Science

Transmission electron micrograph of a cell surface membrane. Magnification × 190 000.

The image of the cell surface membrane is quite fuzzy at the magnification of the photograph in Fig. 6.

9

a Estimate the thickness of the membrane in nanometres.

b How could you make this estimate more reliable?

10 This thickness is approximately twice the length of a phospholipid. Estimate the length of a lipid molecule in nanometres.

Lipids

Phospholipids are a type of lipid. Learning more about this important group of compounds will help you to understand membranes better.

In general, lipids contain the elements carbon, hydrogen and oxygen. Fats and oils, and other major types of lipid, are important energy stores in animals and plants. Lipids are large molecules that consist of **triglycerides**. These form when a condensation reaction joins three fatty acid molecules and one molecule of **glycerol**.

A glycerol molecule contains a chain of three carbon atoms with attached hydrogen atoms and **hydroxyl groups (OH)**. The glycerol molecule is common to all triglycerides so the properties of different fats depend on the structure of the fatty acids that are linked to it.

A single fatty acid molecule contains an **acid group (COOH)** attached to a hydrocarbon chain, a chain of carbon atoms with attached hydrogen atoms. We usually denote the hydrocarbon chain with the letter 'R', giving a general formula for a fatty acid as R–COOH. The hydrocarbon chain itself can contain up to 24 carbon atoms and forms a long 'tail' to the molecule. If every carbon atom in the chain is joined by single C–C bond, we say the fatty acid is **saturated** (Fig. 8a). If there is at least one double C=C bond somewhere in the hydrocarbon chain, we say the fatty acid is **unsaturated** (Fig. 8b). A chain with many double C=C bonds is polyunsaturated (Fig. 8c). Most animal fats are saturated while most plant fats are unsaturated.

Lipids do not dissolve in water. Most do not even mix with water very well and tend to form a layer on top of it. If stirred into water, lipid molecules group together to form droplets.

The acid group that is part of a fatty acid molecule ionises in water to form H^+ and COO^- ions. This end of a fatty acid molecule is therefore attracted to water, but the long carbon tail at the other end repels water. Molecules that repel water are said to be **hydrophobic**; those that are attracted to water are said to be **hydrophilic**.

Fig. 7 The formation of a triglyceride

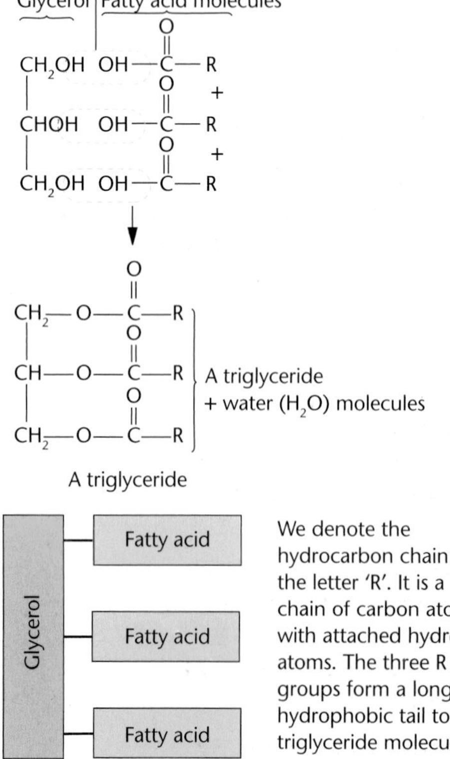

Fig. 8 Molecular structure of saturated and unsaturated fats

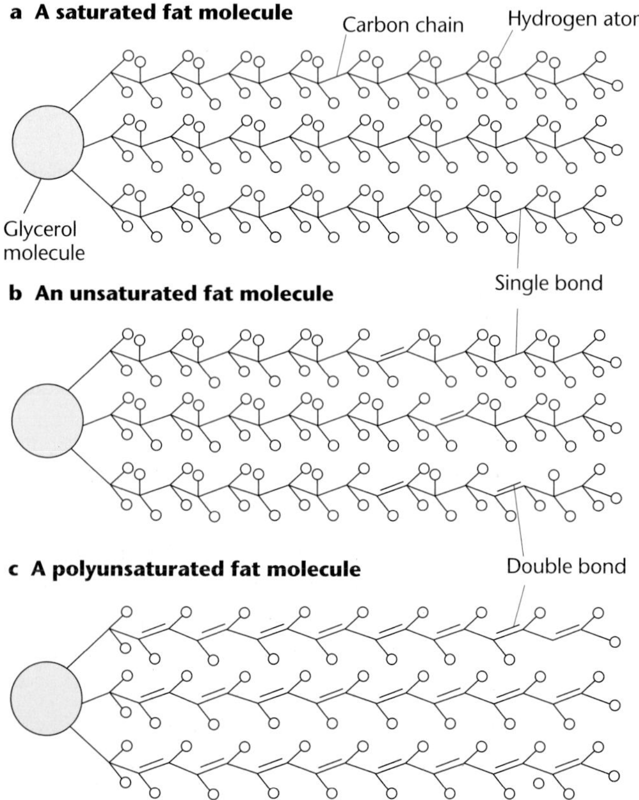

Phospholipids and membrane phospholipids are similar in structure to triglycerides, but one of the three fatty acids is replaced by a negatively charged phosphate group (Fig. 9a).

Like fatty acids, phospholipids do not dissolve in water. The tail of a phospholipid molecule is hydrophobic but the 'head' of the molecule is hydrophilic, it 'loves' water. So, when phospholipids are placed in water, they prefer to sit on the surface with their heads down and tails up, or they form spheres. These spheres can be either single-layered or they have the same double-layered structure as cell membranes (Fig. 9b). In both cases, the hydrophilic heads get to be near the water while the hydrophobic tails stay well away from it.

When phospholipids form a cell membrane, the layer of polar heads make up the outer surfaces of the membrane and the hydrophobic tails are enclosed inside. This arrangement forms a 'skin' around the cell. Similar membranes surround individual internal cell organelles.

The emulsion test for a lipid

- Shake the sample with ethanol in a test tube.
- Allow to settle.
- Pour the liquid into cold water in another test tube.
- A cloudy-white emulsion indicates the presence of lipid.

Fig. 9 Phospholipids

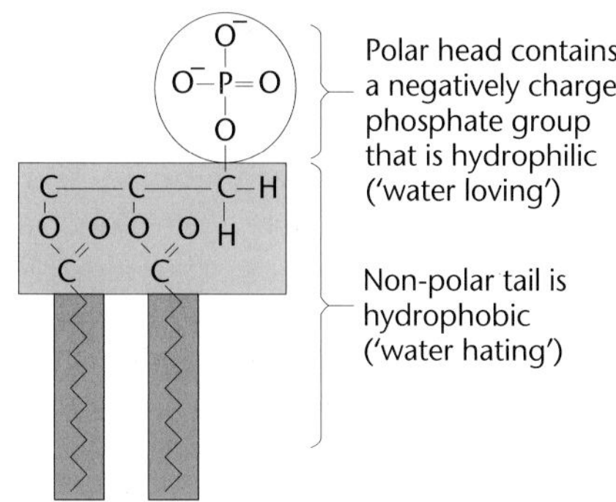

a The structure of a phospholipid. The phosphate gives the molecule a polar head and a non-polar tail.

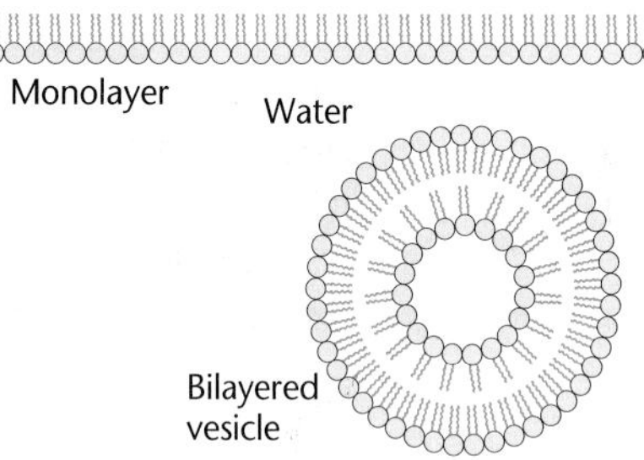

b In water, the hydrophilic heads of the polar phospholipids face outwards while the hydrophobic tails point inwards. Double-layered vesicles form in the water and a phospholipid monolayer covers the surface.

11 What are the two components of a lipid molecule?

12 What is the difference between a lipid and a phospholipid?

13 What is the difference between a saturated fatty acid and an unsaturated fatty acid?

key facts

- All cells are surrounded by a cell membrane and many organelles also have their own membranes.
- Cell membranes consist of a double layer of phospholipid molecules with embedded proteins. Some proteins are present in only one of the phospholipid layers; others span the membrane.
- The proteins are free to move within the phospholipid layer. Membranes are said to have a fluid-mosaic structure.
- Phospholipids are a type of lipid; lipids contain carbon, hydrogen and oxygen.
- Fats and oils are also lipids; their basic unit is a triglyceride. This forms when one glycerol and three fatty acids are joined by a condensation reaction.
- The properties of fats and oils depend on the nature of their fatty acids. In saturated fats, fatty acid chains contain only single C–C bonds. The fatty acids of unsaturated fats contain at least one double C=C bond.
- Phospholipids are similar to triglycerides but one of the fatty acids is replaced by a phosphate group.

The nucleus

The nucleus, shown in the diagram and photograph in Fig. 10, is the largest organelle in the cell. It is usually spherical and has a diameter of about 10 μm. Most of the cell's **deoxyribonucleic acid (DNA)** is in the nucleus. This nucleic acid contains all the information required to make a new copy of the cell and to control the cell's activities. In a dividing cell, the DNA molecules are condensed into **chromosomes** that become clearly visible. At other times the nucleus has a grainy appearance because the DNA molecules extend throughout the nucleus as **chromatin**. Nuclei have one or more nucleoli that are visible in electron micrographs as darkly stained spheres. **Nucleoli** produce the **ribonucleic acid (RNA)** needed to make ribosomes, the organelles that produce proteins using an RNA template. The nucleus is bound by a nuclear membrane, which is a **phospholipid bilayer**. The membrane contains pores large enough to allow big molecules such as messenger RNA to pass out of the nucleus and into the cytoplasm where they go to the ribosomes to be 'read' to produce proteins.

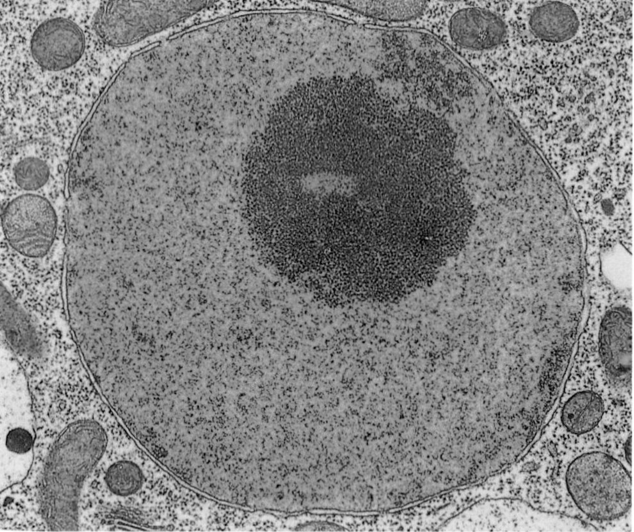

A false-colour electron micrograph of a cell nucleus.

Magnification × 16 500.

14 Suggest why the nuclear membrane has large pores and is so intimately associated with the rough endoplasmic reticulum.

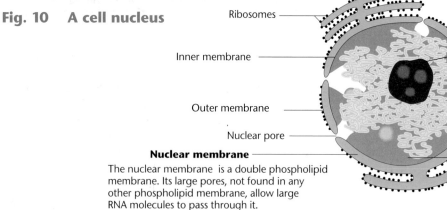

Fig. 10 A cell nucleus

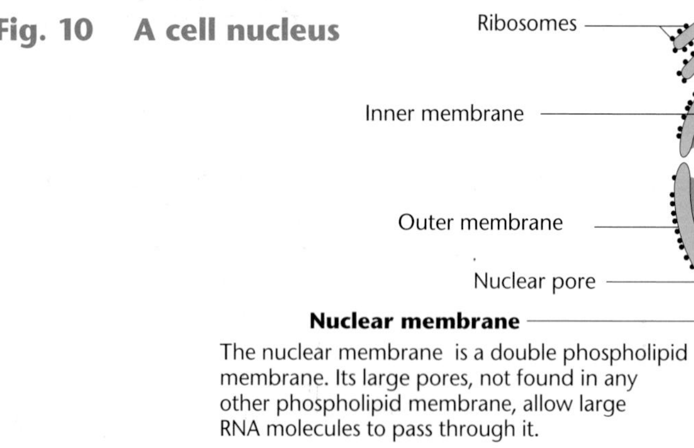
Endoplasmic reticulum
Ribosomes
Nucleolus
The nucleolus produces the RNA that is used to make ribosomes.
Inner membrane
Outer membrane
Nuclear pore
Nuclear membrane
The nuclear membrane is a double phospholipid membrane. Its large pores, not found in any other phospholipid membrane, allow large RNA molecules to pass through it.
Chromatin
When a cell is not dividing, the DNA is in the form of chromatin, which stains darkly.

key facts

- The nucleus contains the cell's DNA, the information the cell needs to divide and to control its activities.
- In a non-dividing cell, DNA molecules are extended as chromatin; when the cell divides, the DNA molecules condense into visible chromosomes.
- The nucleus has its own membrane that has pores to allow outward traffic of RNA.
- Nucleoli in the nucleus make ribosomal RNA.

Mitochondria

Mitochondria are found in all living plant and animal cells. They are the sites of **aerobic respiration**, the biochemical process that oxidises glucose to release energy.

Each mitochondrion (Fig. 11) has two phospholipid membranes: an outer membrane that surrounds the entire organelle and a highly folded inner membrane.

The matrix – the central fluid-filled space – contains the free enzymes that catalyse reactions in the early stages of respiration. The **cristae** that result from the intricate folds of the inner phospholipid membrane have a large internal surface area and hold many of the enzymes involved in the final stages of respiration in place. At the cristae, the transfer of energy to a molecule called **adenosine triphosphate (ATP)** occurs. ATP is the 'energy currency' of the cell – its energy is used to contract muscles, to build up large molecules from smaller units and to power active transport (see page 47).

15 Look at Fig. 11. Use information from the diagram to suggest the advantage that having a highly folded inner membrane gives to the mitochondrion.

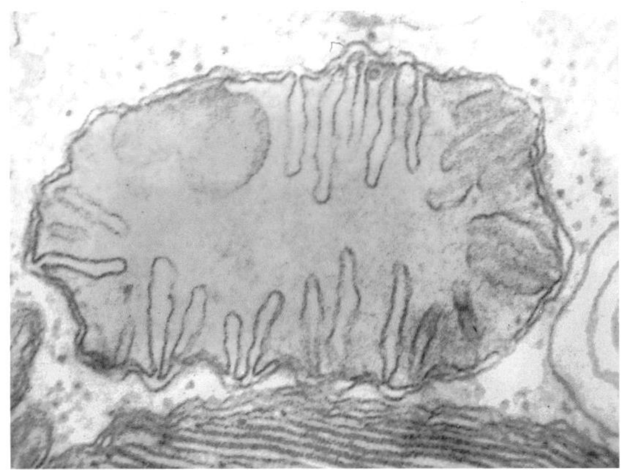

A false-colour electron micrograph of a section through a mitochondrion, showing the folded cristae.

Magnification × 144 000.

Fig. 11 The mitochondrion

Matrix
The matrix contains many of the enzymes that control the early stages of respiration.

DNA
The DNA threads contain the information required for the mitochondria to replicate.

Membrane-bound enzyme molecules

Inner membrane

Intermembrane space

Ribosome

Cristae
The inner phospholipid membrane is highly folded to form cristae. Most of the enzymes that control energy transfer reactions in the later stages of respiration are attached to the cristae.

Outer membrane
The outer membrane controls the passage of materials into and out of the mitochondrion.

Respiration

Aerobic respiration is a complex process involving many stages. The overall process releases 36 ATP molecules by oxidising one molecule of glucose.

Since a complete glucose molecule is too big to enter the mitochondrion, it is first split into two molecules of pyruvate, a smaller three-carbon compound. Enzymes in the central fluid-filled matrix control the further breakdown of pyruvate into carbon dioxide and hydrogen in a series of reactions known as the Krebs cycle. Finally, enzymes and carrier molecules attached to the cristae combine hydrogen with oxygen to produce water, and release ATP. Some ATP is produced as a result of earlier stages, but the majority results from the final stage.

The ATP is made available for all the processes in the cell that need energy. Cells that require large amounts of energy, such as secretory, nerve and muscle cells, contain large numbers of mitochondria.

key facts

- Mitochondria are the site of aerobic respiration.
- Mitochondria have a double phospholipid membrane.
- The fluid in the inner matrix contains many of the enzymes involved in the early stages of respiration.
- The highly folded inner membrane bears many of the enzymes involved in the formation of ATP.
- ATP is a source of energy for cells.

Endoplasmic reticulum

The **endoplasmic reticulum (ER)** is a series of thin, intricate channels. It exists in the space created by folds in the phospholipid membrane and is continuous with the nuclear membrane (see Fig. 10 and Fig. 12).

The narrow fluid-filled space between these membranes acts as a transport system to move materials through the cell. The folded membrane of the ER creates a large area within the cell.

Ribosomes and the ER

Much of the outside surface of the ER is dotted with ribosomes; this gives the membrane a grainy appearance and also its name, rough ER. Smooth ER has no attached ribosomes.

Ribosomes are small, dense organelles, about 20 nm in diameter. They carry out protein synthesis. When a ribosome binds to a length of messenger RNA, it uses information encoded in the RNA to assemble amino acids in the correct order to form a specific protein. Free ribosomes, ones that float around in the cytoplasm, produce proteins for use inside the cell. Ribosomes attached to the ER produce proteins destined for export from the cell for use elsewhere.

16 The electron micrograph in Fig. 12 does not seem to include many of the features of the ER that are shown in the three-dimensional diagram. Suggest a reason for this.

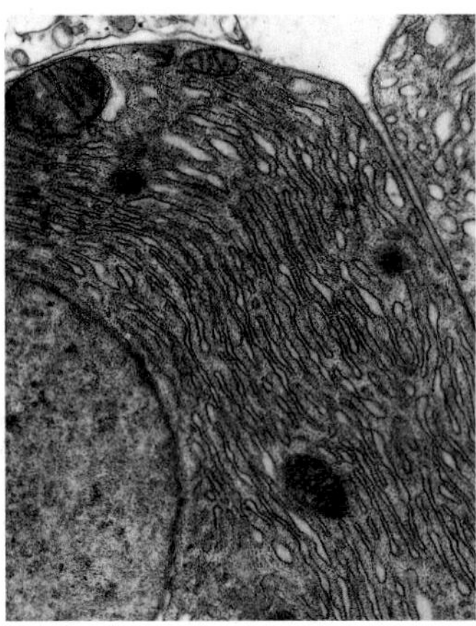

An electron micrograph of endoplasmic reticulum and ribosomes.

Magnification × 7500.

Fig. 12 Endoplasmic reticulum and ribosomes

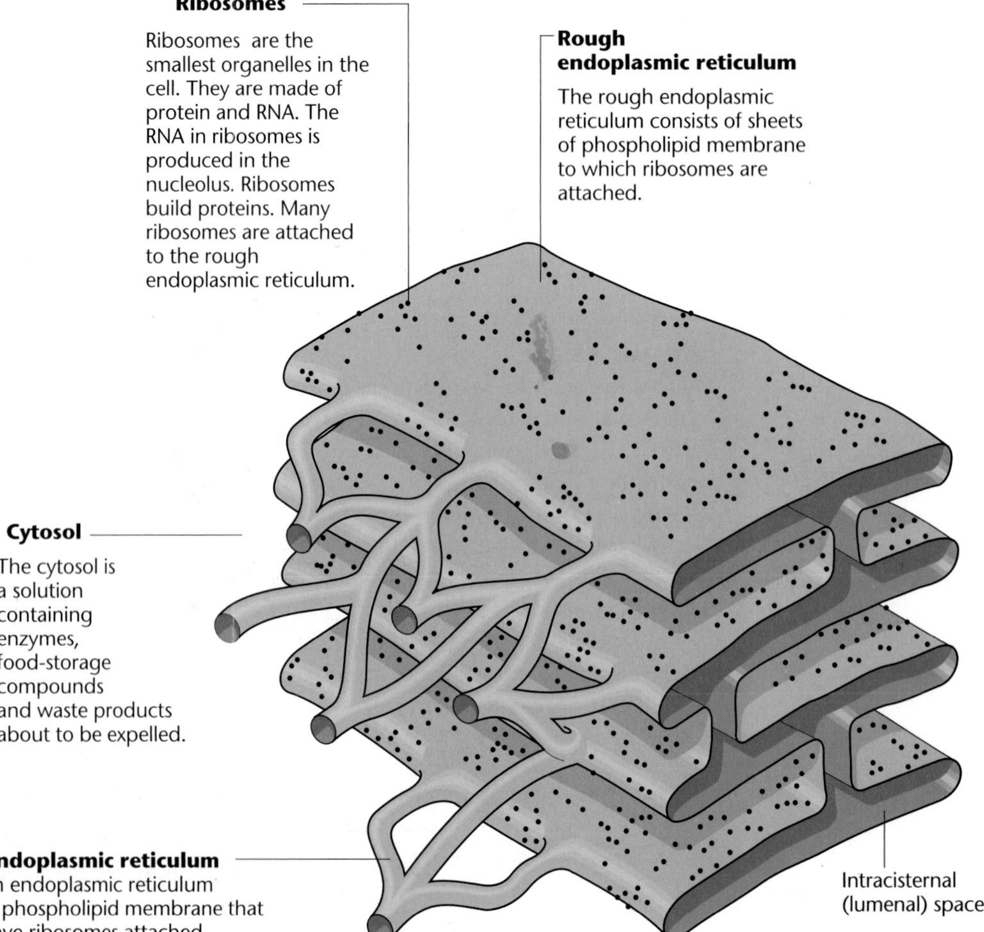

Ribosomes
Ribosomes are the smallest organelles in the cell. They are made of protein and RNA. The RNA in ribosomes is produced in the nucleolus. Ribosomes build proteins. Many ribosomes are attached to the rough endoplasmic reticulum.
Rough endoplasmic reticulum
The rough endoplasmic reticulum consists of sheets of phospholipid membrane to which ribosomes are attached.
Cytosol
The cytosol is a solution containing enzymes, food-storage compounds and waste products about to be expelled.
Smooth endoplasmic reticulum
The smooth endoplasmic reticulum is a tubular phospholipid membrane that does not have ribosomes attached.
Intracisternal (lumenal) space

Fig. 13 The Golgi apparatus

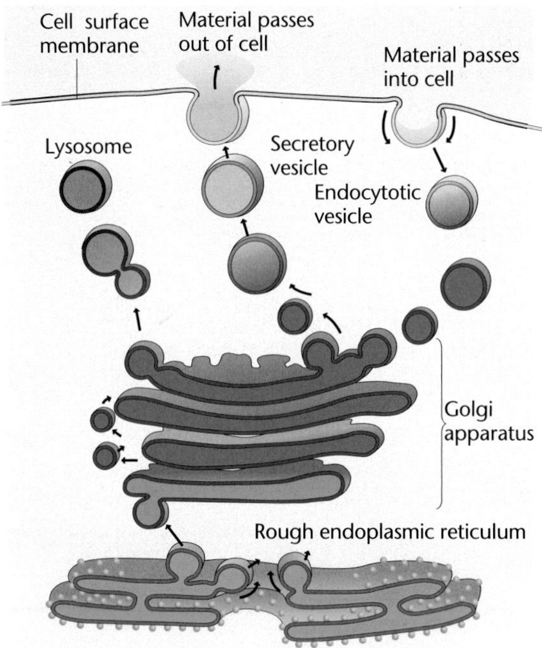

A summary of the functions of the Golgi apparatus. Material that enters the Golgi apparatus from the rough endoplasmic reticulum (ER) or from outside the cell is processed inside the flattened cavities. Processed material is packaged into vesicles. Secretory vesicles release their contents outside the cell; lysosomes take material to other organelles in the cell.

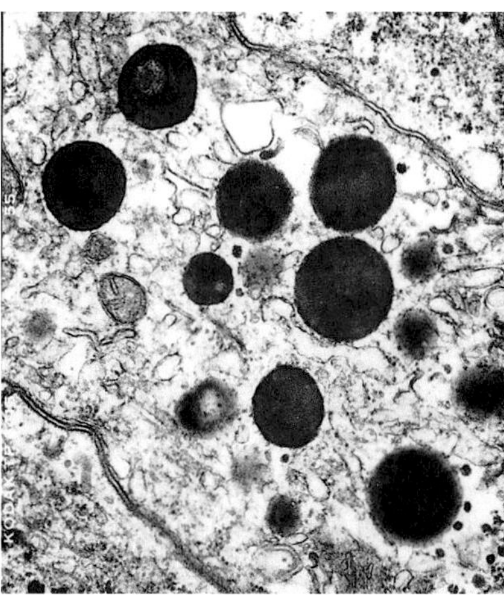

Lysosomes are vesicles that are filled with digestive enzymes. The lysosomes in this photograph have been stained to show up clearly.

Golgi apparatus and lysosomes

The Golgi apparatus is a group of flattened cavities (Fig. 13).

Its function is to take enzymes and other proteins that have been synthesised in the ER and to package them into membrane-bound vesicles. The appearance of the Golgi apparatus is constantly changing as material comes in on one side from the ER and is lost from the other as completed vesicles 'bud off'. Such vesicles transport materials to other parts of the cell, or fuse with the cell surface membrane, releasing their contents outside the cell.

Lysosomes (see photograph) are vesicles that contain digestive enzymes. These can destroy old or surplus organelles inside the cell or they can be used to break down material that has been taken into the cell by the process of endocytosis. Whole cells and tissues that are no longer required can be destroyed if cells nearby allow lysosomes to release their contents at the cell surface. The body uses this process to break down excess muscle in the uterus after birth, and to destroy milk-producing tissue in the breasts after a baby has been weaned.

17 Cells in the pancreas produce enzymes. Explain why these cells have large amounts of rough endoplasmic reticulum and Golgi bodies.

key facts

- The endoplasmic reticulum is a series of thin, intricate channels created by folds in phospholipid membranes. This narrow fluid-filled space acts as a transport system to move materials through the cell.
- Ribosomes are involved in protein synthesis.
- Endoplasmic reticulum with ribosomes attached is known as rough endoplasmic reticulum.
- The Golgi apparatus packages proteins into membrane-bound vesicles that transport materials to other parts of the cell.

examination questions

1

a The figure below shows the structure of a molecule of glycerol and a molecule of fatty acid.
Draw a diagram to show the structure of a triglyceride molecule. (2)

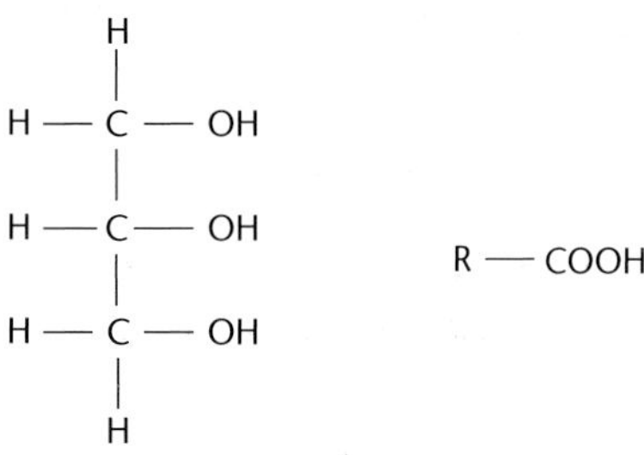

Glycerol Fatty acid

b Explain why triglycerides are **not** considered to be polymers. (1)

c The figure below shows two types of fat storage cell. Mammals living in cold conditions have more brown fat cells than mammals living in tropical conditions.

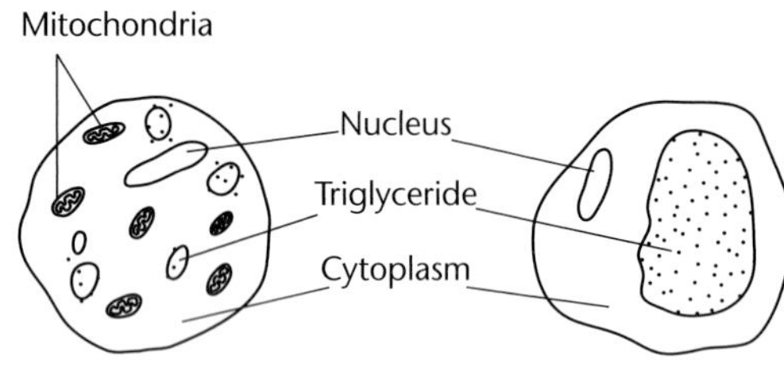

Brown fat cell White fat cell

Using evidence from the figure above to support your answer, suggest how the function of brown fat cells differs from that of white fat cells. (3)

Total 6

AQA, January 2004, Unit 1, Question 2

2 The diagram below shows a cross-section of a typical red blood cell. Red blood cells can carry both oxygen and carbon dioxide. Each cell has a thinner central area as shown in the diagram.

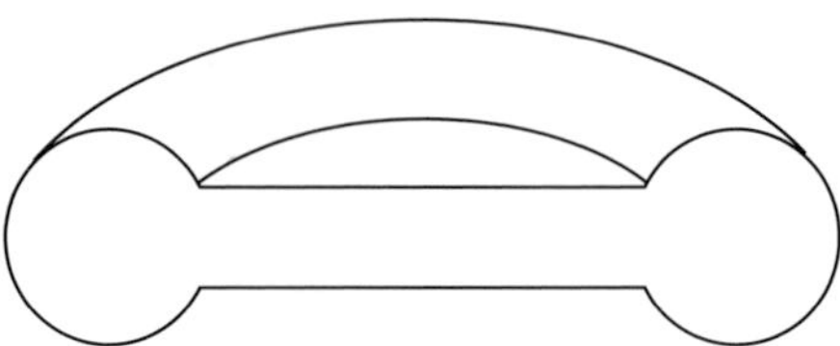

a Explain **one** advantage of the shape of red blood cells. (2)

b Red blood cells do not have a nucleus or rough endoplasmic reticulum. Give **one** function that red blood cells are therefore unable to carry out. (1)

c In an experiment, the phospholipids were extracted from the surface membrane of a single red blood cell. They were placed on the surface of water and allowed to spread out.

i Using O= to represent a single phospholipid molecule, draw on the diagram how you would expect **ten** phospholipid molecules to be arranged on the water surface. (2)

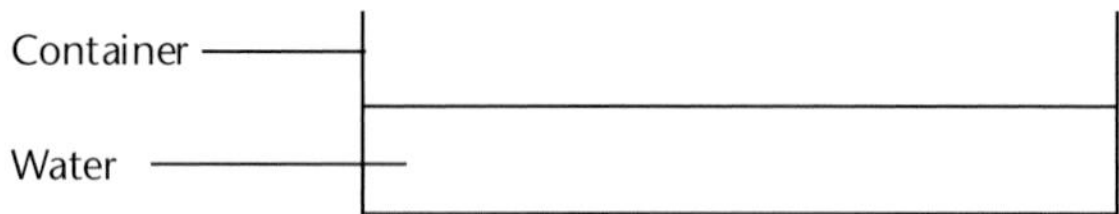

ii The area of water covered by the phospholipid molecules was calculated. This area was given an arbitrary value of 1. The surface area of an intact red blood cell was measured on the same scale. What would you expect the arbitrary value of the red blood cell surface area to be? Explain your answer. (2)

Total 7

AQA, January 2002, Unit 1, Question 5

3 The drawing shows an animal cell.

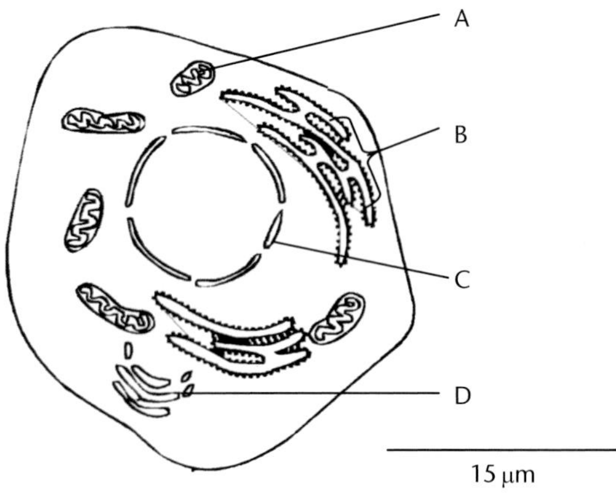

a Identify structures **A**, **B**, **C** and **D**. (4)

b Calculate the magnification of the drawing. Show your working. (2)

c The drawing was made from a photomicrograph. What type of microscope was used to produce the photomicrograph? Use information from the drawing to explain your choice. (2)

d Describe the functions of structures **B**, **C** and **D**. (3)

Total 11

4 Some animal cells were broken open, suspended in an ice-cold, isotonic solution and spun in an ultracentrifuge tube. Three fractions settled out as shown in the diagram.

A

B

C

a Explain why the broken down cells were placed in
  i isotonic solution
  ii ice-cold solution. (2)

b Explain why the fractions separated out. (2)

c The three fractions contained mitochondria nuclei and ribosomes.
In which fraction, A, B or C, would you expect to find each of the three organelles? (2)

Total 6

5 The diagram shows a mitochondrion.

a Describe in detail the structure of the part labeled A. (4)

b Name the parts labeled B and C. (2)

c Explain how the structure of the mitochondrion is related to its function. (3)

Total 9

how science works **assignment**

Magnification

Most biological illustrations have a scale shown, for example, × 60. This means that the object in the picture has been magnified 60 times. In exams you are often asked either to find the actual size of a specimen, or to calculate how many times it has been magnified.

You need to remember two formulae:

$$\text{Actual size} = \frac{\text{Image size}}{\text{Magnification}}$$

$$\text{Magnification} = \frac{\text{Image size}}{\text{Actual size}}$$

A1 The photograph below left shows a human egg covered with sperm. The egg is actually 0.1 mm across. Calculate:

a the magnification;

b the volume of the egg and a single sperm head in cubic micrometres;

c how many sperm heads would fit inside the egg.

A2 The photograph below shows bacteria magnified 9240 times. Calculate the actual mean length of the bacteria in micrometres.

A3 The photograph top right shows some pollen grains on the pistil of a flower. It has been magnified 420 times. Choose three of the pollen grains and calculate their actual mean width in millimetres.

A4 The ant in the photograph bottom right is magnified 25 times. Calculate the actual length of its head in milimetres, from the top of its head to its biting mouthparts.

A5 The photograph far right shows a section through part of an animal cell magnified 80 000 times. Calculate the length and width of the mitochondrion in nanometres.

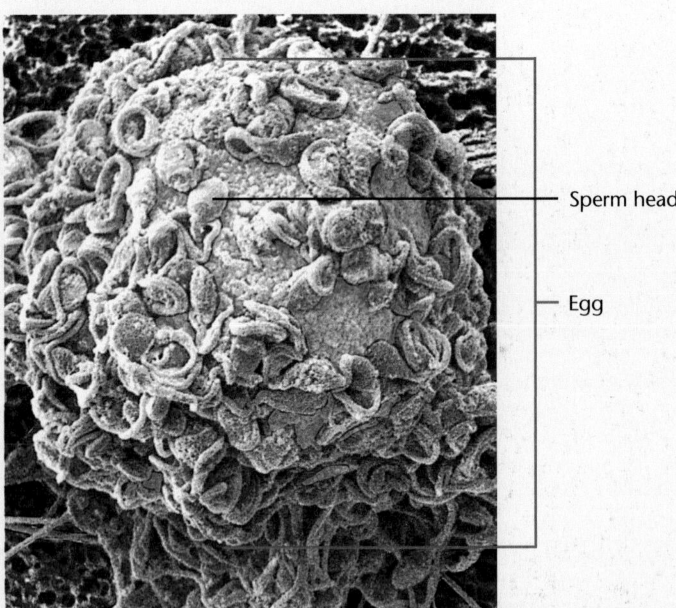

A human egg covered with sperm.
The egg is 0.1 mm across.

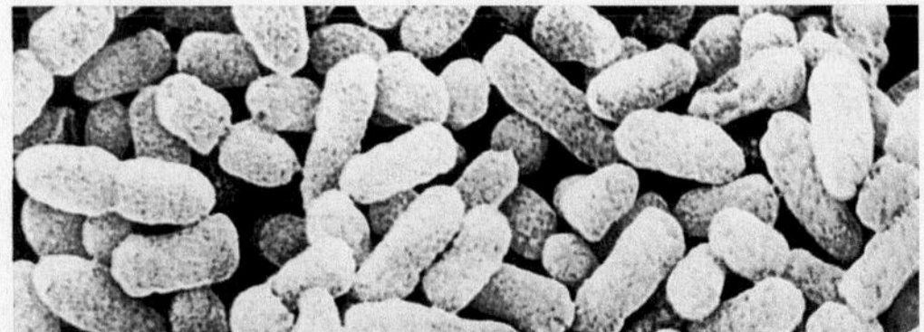

Bacteria magnified 9240 times

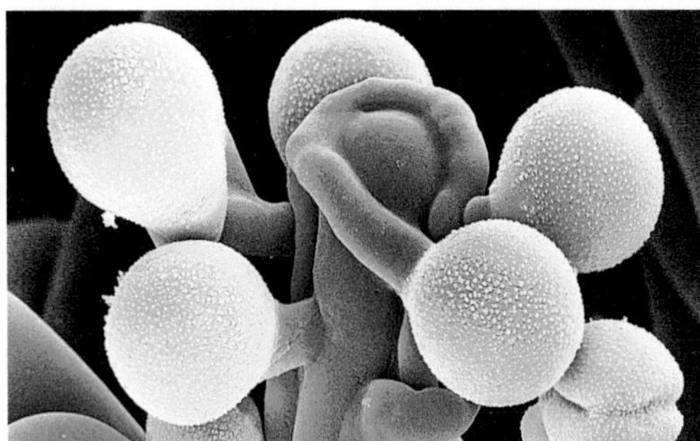

Pollen grains magnified 420 times

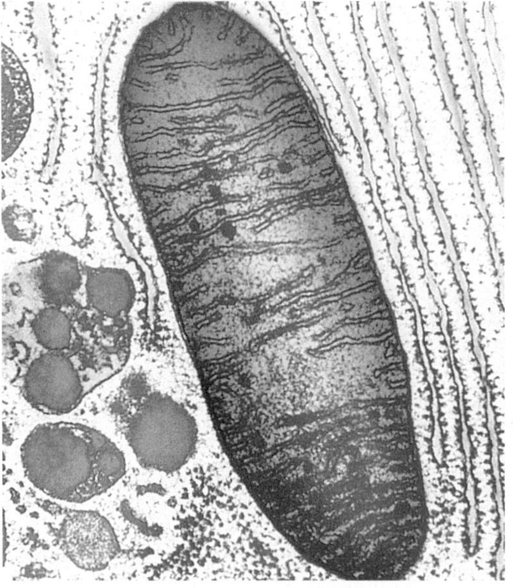

Section through part of an animal cell magnified 80 000 times.

An ant magnified 25 times

3 Cells and their environment

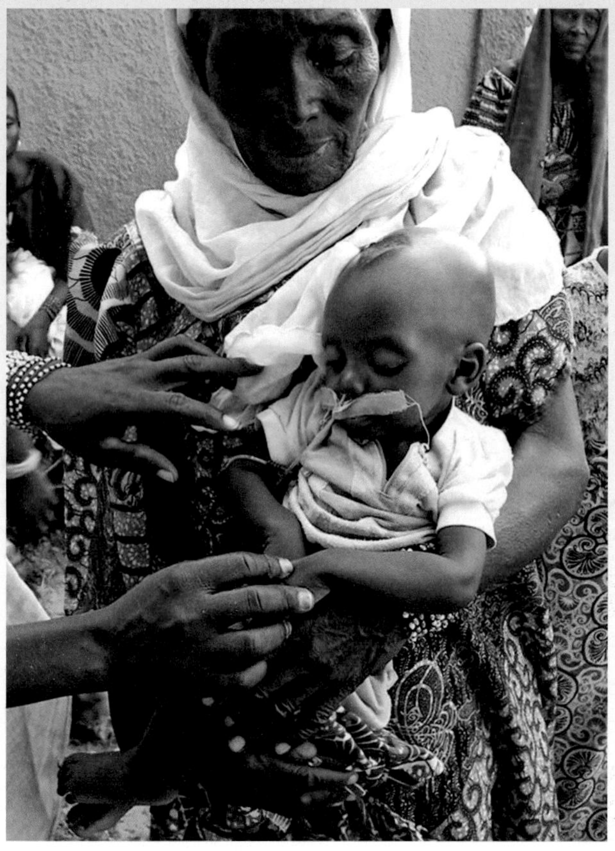

Oral rehydration therapy (ORT) is crucial for the treatment of infants with general diarrhoea and cholera.

About five million children die each year from **diarrhoea**. Most of these children's lives could be saved by giving them a simple solution containing ions and glucose. This is known as **oral rehydration therapy (ORT)**.

In the normal healthy intestine, there is a continuous exchange of water through the intestinal wall – up to 20 litres of water is secreted and very nearly as much is reabsorbed every 24 hours – this mechanism allows the absorption of soluble food into the bloodstream. When we have diarrhoea this balance is upset and much more water is secreted than is reabsorbed, causing a net loss to the body which can be as high as several litres a day. In addition to water, sodium ions (Na^+) are also lost. If diarrhoea is not treated, rapid depletion of water and Na^+ occurs. Death occurs if more than 10% of the body's fluid is lost. Simply giving a **saline solution** (water plus Na^+) by mouth does not help because the normal mechanism by which Na^+ is absorbed by the healthy intestinal wall is impaired when we have diarrhoea. If the Na^+ is not absorbed, neither can the water be absorbed. In fact, excess Na^+ in the lumen of the intestine causes increased secretion of water and the diarrhoea worsens.

If glucose is added to a saline solution, a new mechanism comes into play. The glucose molecules are absorbed through the intestinal wall and Na^+ is carried through in conjunction by **co-transport**. Water follows by **osmosis**. The leading medical journal the *Lancet* described the discovery of this mechanism of co-transport of sodium and glucose as 'potentially the most important medical advance this century'.

In this chapter you will learn about the processes by which substances enter and leave cells.

3.1 A balancing act

Table 1

	Concentration/mmol dm^{-3}		
	Plasma	**Interstitial fluid**	**Fluid inside cells**
Sodium	140	144	10
Potassium	4	4	155
Calcium	2.5	2	1
Magnesium	1	1	15
Chloride	102	114	5
Hydrogen carbonate	27	30	10
Phosphate	1	1	50
Sulfate	0.5	0.5	10

For the body cells to work efficiently, the composition of the **blood plasma** and the **interstitial fluid** has to be kept within close limits. Table 1 shows the concentration of ions in plasma, interstitial fluid and the cells themselves.

Exercise has major effects on the composition of these body fluids. When we exercise we get hot and we sweat. The fluid lost by the skin evaporates and this has a cooling effect. That is good, but losing

1 Which ions are present in the greatest concentration:

a in the plasma?

b in the interstitial fluid?

c inside the cells?

large amounts of sweat leads to dehydration if we don't drink fluid to replace what has been lost. Drinking water helps, but is that all we lose? Sweat is formed from the blood plasma. After we have been sweating, our blood plasma has fewer mineral ions. Just as we need to replace the water lost in sweat, these lost minerals need to be replaced too. Sports drinks claim to do this because they contain mineral ions as well as water. The concentration of Na^+ and Cl^- ions in these drinks is very similar to the concentration of the ions in blood plasma. Solutions containing the same concentration of mineral ions are said to be **isotonic**.

Controlled studies have shown that athletes who drink small amounts of isotonic drinks when they are doing hard sustained exercise can exercise for longer than those who just drink water. In one controlled investigation, cyclists given isotonic drinks cycled as hard as they could for about 17 minutes longer than those given water. In a race, this could mean the difference between a gold medal and fourth place.

The optimum biconcave shape of the red blood cell provides the maximum surface area for the exchange of molecules. This biconcave shape depends on the concentration of ions in the blood plasma. A high ion concentration causes red blood cells to become star-shaped.

Magnification × 3000.

Fig. 1 Red blood cells and salt

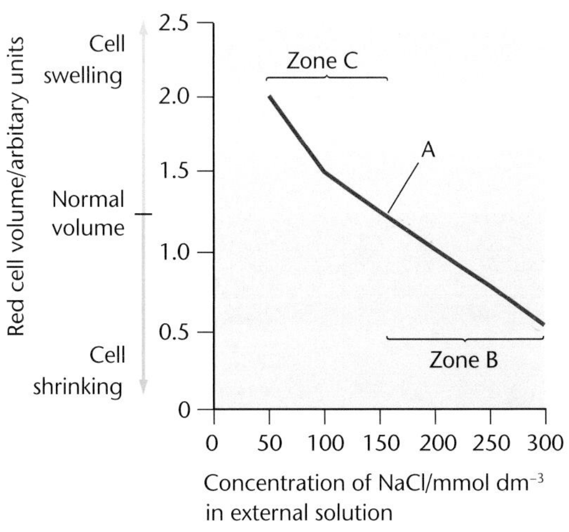

How do sports drinks work? One theory is that keeping the body's fluid and mineral ion levels constant allows the body to keep its blood plasma in balance. As long as the plasma and the inside of the red blood cells are isotonic, the cells are fine. They are normal volume and there is no net water gain or water loss (point A in Fig. 1). When the body loses a lot of sweat, the mineral ion concentration of the plasma can become slightly higher than the inside of the cells, and water leaves the cells to try to restore the balance. The volume of the blood cells decreases (Zone B in Fig. 1). In extreme cases, this can cause the red blood cells to become star-shaped. If an athlete drinks only water after exercise, the plasma can become less concentrated than the inside of the cells and water enters the cells. This causes them to swell (Zone C in Fig. 1).

In both cases, the changes in volume of the red cells, even if they are slight, affect the way the red cells work. To carry oxygen efficiently, the volume of the red blood cells needs to be kept as constant as possible. Drinking isotonic drinks during exercise makes this easier for the body to achieve.

2 List the variables that would have been controlled in the investigation into the effect of sport's drinks on cyclists.

3

a Look at Fig. 1. Describe the relationship between the volume of red blood cells and the concentration of sodium chloride (NaCl) in the external solution.

b Suggest one way in which the function of red blood cells might be affected if they were swollen (but not burst).

3.2 Moving substances in and out of cells

Red blood cells work better when their environment is ideal and this is also true for all the other cells in the body. Maintaining this ideal environment needs constant effort and depends largely on the way substances pass in and out of individual cells.

Cells are complex factories and they need to constantly import raw materials and get rid of waste. Molecules and ions of different sizes and electrical charges enter and leave all the time. Some of the exchange of materials occurs as a result of passive processes such as diffusion and osmosis, but not all movement of substances is free movement. This would make it impossible for a cell to maintain high concentrations of valuable substances that it needs to function efficiently. Cells must control the passage of substances through their membranes. The presence of special 'gateways' for individual substances and the use of energy-requiring transport mechanisms such as **active transport** enables a cell to keep what it needs. The same transport processes also occur inside the cell to segregate particular substances within organelles.

We will look at diffusion, osmosis, facilitated diffusion, active transport and co-transport in detail in the rest of this chapter.

3.3 Diffusion

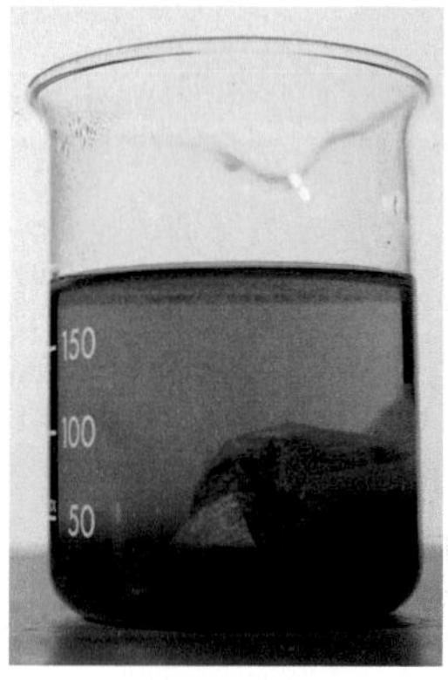

Illustrating diffusion with a tea bag: the flavour and colour from the tea inside the bag diffuse through the water.

All the particles in liquids and gases are in constant random motion. This motion results in a net movement of particles from a region of high concentration to a region of lower concentration (Fig. 2). This process is called **diffusion**.

In a mixture of gases, diffusion causes each gas to spread evenly through the space that the mixture occupies. In the same way, a soluble substance spreads through a liquid until it is evenly dispersed.

Particles of gas or solute can also diffuse through a membrane, as long as the membrane has pores that are larger than the particles (Fig. 3).

Fig. 2 The process of diffusion

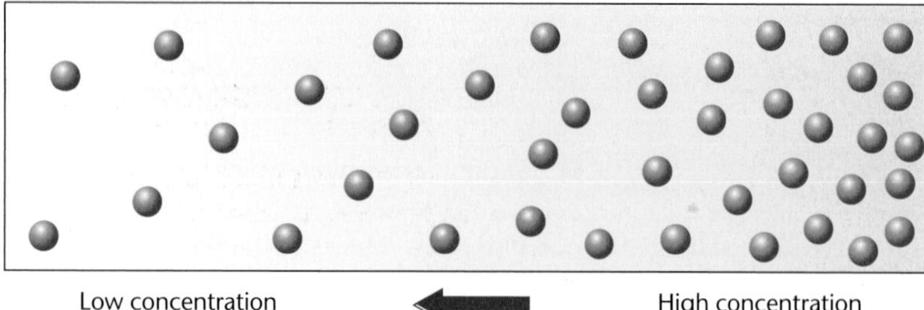

Low concentration

High concentration

Net movement of molecules

4 Look at the photograph of the tea bag in the beaker. Why is the solution darker near the tea bag?

Fig. 3 Diffusion through a membrane

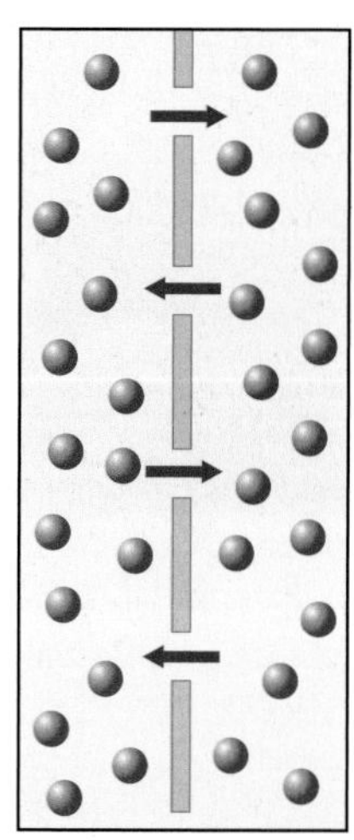

Pores in this membrane are wide enough to allow diffusion

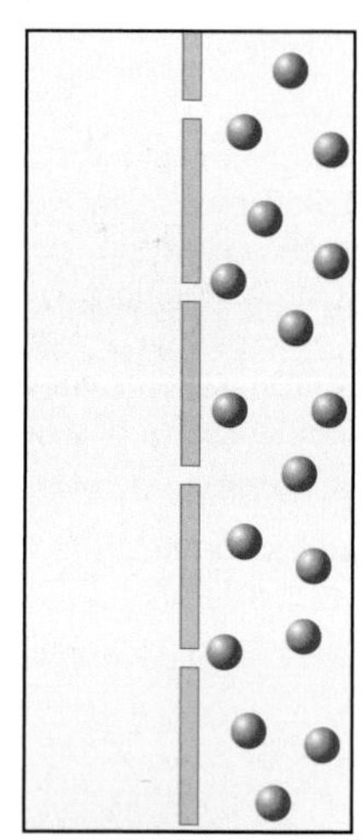

Pores in this membrane are too narrow to allow diffusion

Diffusion is a passive process. It happens without any energy input from the organism. The rate at which diffusion occurs depends on three factors:

- the difference in concentration between two areas, called the **concentration gradient**
- the distance between the areas
- the size of the molecules that are diffusing.

So the greater the concentration gradient and the smaller the particles, the quicker the net movement of molecules from the area of high concentration to the area of low concentration. Diffusion also happens faster when molecules need to move only microscopic distances than when they have to travel much larger distances. For example, it would take a small molecule such as oxygen at least 4 minutes to diffuse to the centre of a cell 1 mm in diameter. Some cells, such as those in the human gut and inside plant leaves, have special adaptations to allow diffusion to happen quickly. These adaptations reduce the distance over which diffusion occurs.

The overall rate at which a substance diffuses through a membrane also depends on the surface area in contact with the substance. The biconcave-disc shape of a red blood cell gives it a much greater surface area than if it were spherical, allowing the maximum amount of oxygen to diffuse into it. Look out for other examples of cells that maximise their surface area for efficient diffusion later in this chapter.

hsw

how science works

The upper diagram on the right shows a cross-section of a red blood cell. The diagram to the left of it shows a section through a spherical cell with the same volume as the red blood cell. In this exercise, we look at the efficiency of diffusion in a biconcave cell compared with a spherical cell.

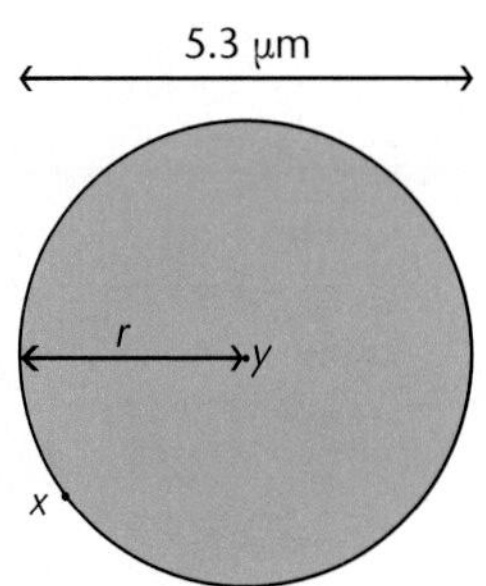

The surface area of a sphere is:

Area = $4\pi r^2$

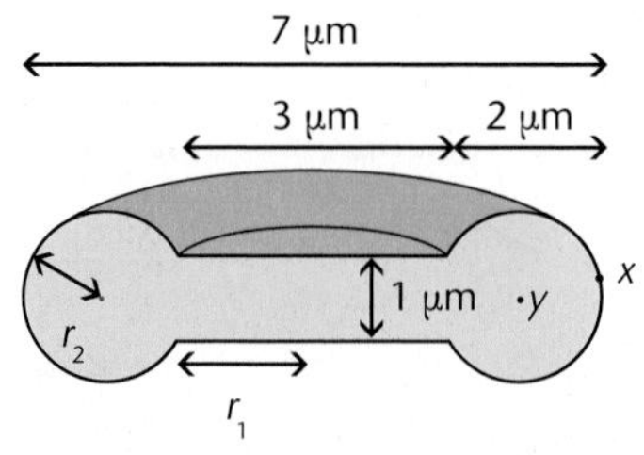

The equation for the surface area of a biconcave disc is very complicated. However, we can come up with an rough value by treating it as a disc:

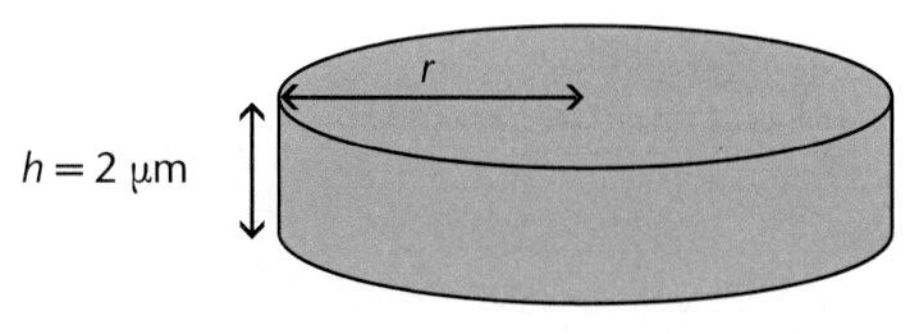

Area = $2(\pi r^2) + 2\pi rh$

Distance that molecules need to diffuse	Time required for sr molecules to diffuse
1 μm	0.4 milliseconds
10 μm	50 milliseconds
100 μm	5 seconds
1000 μm (1 mm)	8.3 minutes
10 000 μm (1 cm)	4 hours

5 Using the formulae given, estimate the surface area of each of the two cells.

6 Use the information in the table to calculate the speed particles diffuse across a distance of:

a 10 μm

b 1 nm.

7 How long would it take for a molecule of oxygen to diffuse from the point marked *x* to the point marked *y* on:

a the red blood cell?

b the spherical cell?

8 Assume that the concentration difference of oxygen between the inside and outside of the two cells is 1, and both cells have a membrane that is 10 nm thick. Use this equation:

$$\text{rate of diffusion} = \frac{\text{surface area} \times \text{concentration difference}}{\text{thickness of membrane}}$$

to calculate the rate of diffusion of oxygen that would occur in:

a the red blood cell

b the spherical cell.

9 Look at the change in volume of the red blood cell in the graph in Fig. 1. Explain how the rate of diffusion of oxygen into the cell would change if the red blood cell swells.

3.4 Osmosis

Not all substances can pass through the cell surface membrane; water molecules do but larger solute molecules do not. This means the membrane is **partially permeable**. The movement of water through a partially permeable membrane from a region of higher concentration of water molecules to a region of lower concentration of water molecules is called osmosis (Fig. 4).

Fig. 4 Osmosis

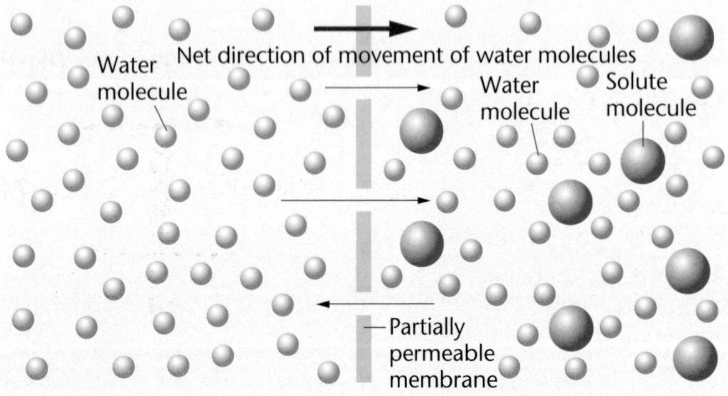

Water molecules pass through the membrane in both directions, but there is a net movement towards the region of lower concentration. The larger molecules in the cell cannot move outwards through the membrane to balance the inflow of water.

Fig. 5 It's not just the number of water molecules

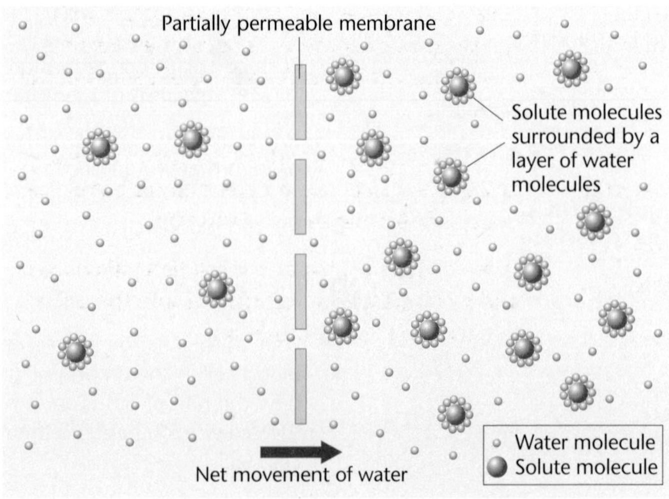

Osmosis depends on factors other than the differences in the number of water molecules (see Fig. 5). Solute molecules make weak chemical bonds with water molecules. Solutions with many solute molecules bind most of the water molecules. However, a solution with only a few solute molecules binds only a few water molecules. Water molecules bound to solute molecules move more slowly than free water molecules. So, as well as there being fewer water molecules, those that are present move more slowly than water molecules in pure water. This enhances osmosis by lowering the concentration of free water in the concentrated solution.

Fig. 6 explains the change in the volume of the red blood shown in Fig. 1 in terms of osmosis.

- Isotonic solutions have the same concentrations of water molecules, so the rate at which water molecules diffuse into and out of the cell is the same.

Fig. 6 Different solutions

External ion concentration 150 mmol dm^{-3}

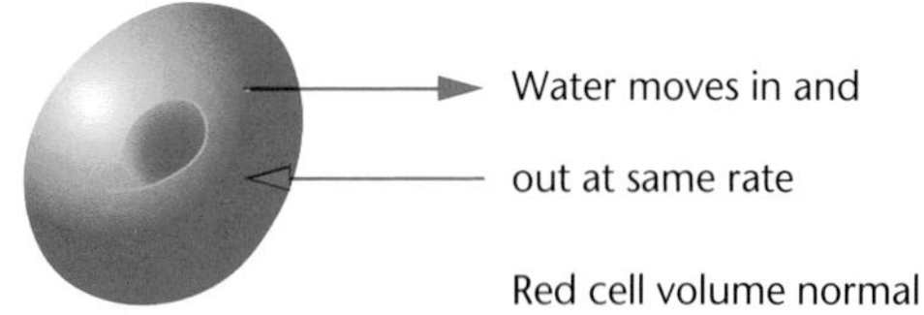

External ion concentration 100 mmol dm^{-3}

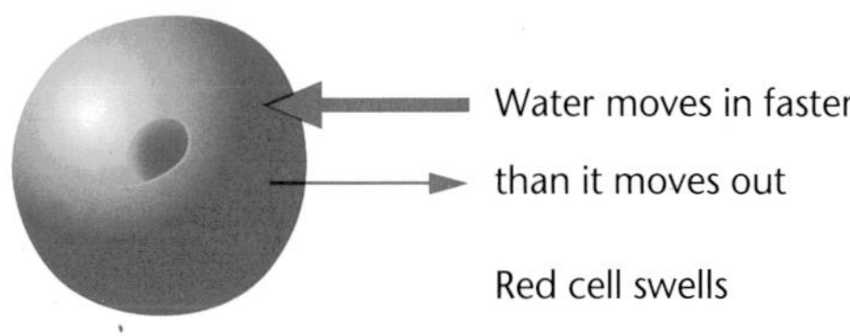

External ion concentration 300 mmol dm^{-3}

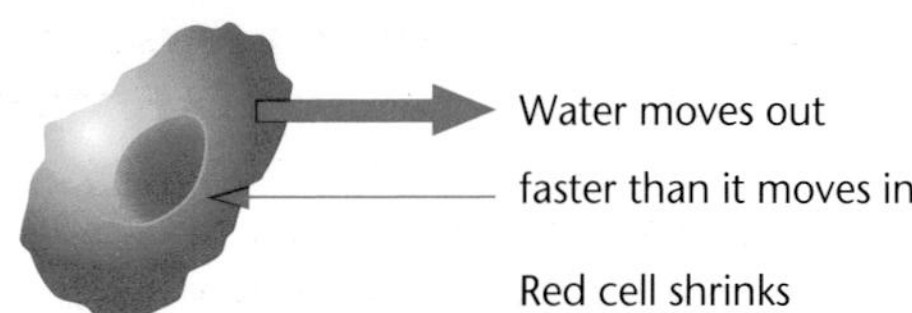

- **Hypertonic** solutions have a lower concentration of water molecules compared with the inside of a cell and so there is a net movement of water molecules out of the cell.
- **Hypotonic** solutions have a higher concentration of water molecules compared with the inside of a cell so there is a net movement of water molecules into the cell. This causes the cell to swell; a cell that is full of water but has not burst is said to be turgid.

Water potential

The ability of water molecules to move is known as their **water potential** (Fig. 7).

The symbol for water potential is the Greek letter ψ, pronounced 'sigh'.

The water potential of pure water at atmospheric pressure is given the value zero. You already know that water molecules that have formed bonds with solute molecules move around more slowly than free water molecules in pure water. Since the water molecules in solutions cannot move as easily as in pure water, solutions always have a water potential value that is negative (it is always less than zero). Solutions that have the lowest water potential have the largest negative values. A solution with a water potential of –200 kPa, for example, has a lower water potential than a solution with a water potential of –100 kPa. Water molecules always move towards a region of lower water potential – to where the water potential is relatively more negative, as shown in Fig. 7.

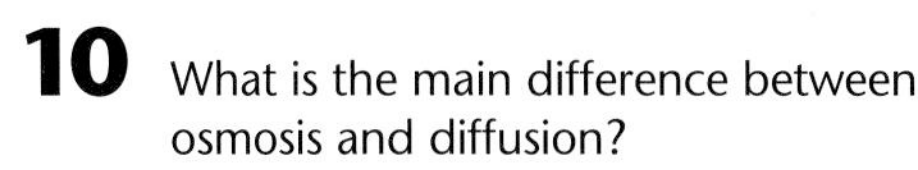

10
What is the main difference between osmosis and diffusion?

Fig. 7 Water potential

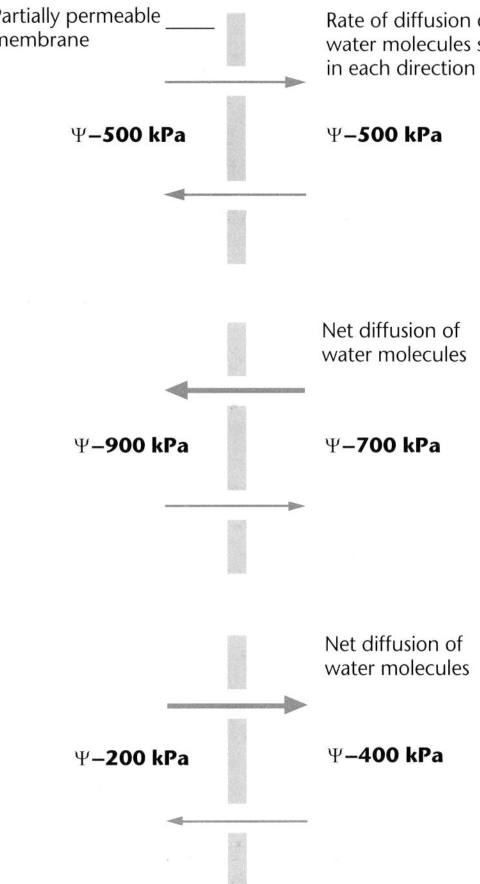

key facts

- Diffusion is the passive movement of molecules along a concentration gradient from a region of high concentration to a region of low concentration.
- The rate of diffusion can be increased by increasing the concentration gradient, increasing the surface area across which diffusion occurs, increasing the temperature or by decreasing the distance across which the molecules need to travel.
- Osmosis is a special type of diffusion.
- In osmosis there is a net diffusion of water molecules through a partially permeable membrane, from an area of high concentration of water molecules to an area of lower concentration of water molecules.
- A cell bathed in a solution that is isotonic with the cytoplasm can maintain a constant volume since water diffuses into and out of the cell at the same rate.
- A cell bathed in a hypertonic solution loses water and shrivels up because water diffuses out of the cell at a faster rate than it diffuses in.
- A cell bathed in a hypotonic solution gains water and swells because water diffuses into the cell at a faster rate than it diffuses out.

The ability of water molecules to move depends on the pressures that are acting on them. The unit of water potential, the kilopascal (kPa), is therefore a unit of physical pressure.

Water potential in a cell depends on two factors:

- the concentration of solutes inside the cell;
- the pressure exerted on the cell contents by the stretched plasma membrane or cell wall.

11 Look again at Fig. 5 on page 42. On which side of the membrane do the water molecules have the more negative water potential? Explain your answer.

A practice ISA on the effect of solute concentration on the rate of uptake of water by plant tissue can be found at www.collinseducation.co.uk/advancedscienceaqa

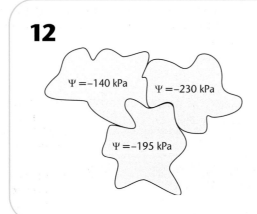

12

Ψ =–140 kPa

Ψ =–230 kPa

Ψ =–195 kPa

The diagram shows the water potential of three cells that are in contact.

a Copy the diagram. Use arrows to show the net direction of water movements between the three cells.

b Which cell has the most negative water potential, and which the least negative?

hsw how science works

Energetic dancing in a nightclub is like any other strenuous exercise – it makes you sweat and it can leave you dehydrated. Ecstasy is an illegal drug that is sometimes used by dancers. People have died after taking Ecstasy in a club and dancing all night. They didn't overdose on the drug; they seem to have overdosed on water. Ecstasy seems to induce repetitive behaviour in some people – this has occasionally led to people drinking 20 litres of water or smoking 100 cigarettes in the space of 3 hours.

The urge to drink water constantly may also have been encouraged by the popular belief that water is an antidote to Ecstasy (it isn't), or the drug may affect the way the brain controls the body's water balance. Whatever the reason, people have collapsed because the large amount of water they have drunk has diluted their blood so much that they have developed a condition called oedema. This happens when cells and tissues in the body absorb too much water and swell; if this happens in the brain it is really bad news. As brain tissue swells it squeezes against the skull and blood vessels and brain cells become squashed. If the centres in the brain that regulate breathing and the beating of the heart are damaged irreversibly, death is inevitable.

13 Use the concept of water potential to explain why drinking large amounts of pure water after sweating may lead to oedema in the brain.

14 Would it be a good idea for clubs to offer free isotonic drinks instead of water? What would you say to a club manager who asked for your advice?

3.5 Facilitated diffusion

In addition to mineral ions, most sports drinks contain small amounts of sugars for 'instant energy'. The market for these drinks is millions of pounds per year, so scientists have done a lot of research to find out which sugars are absorbed quickest, and what is the optimum concentration of sugars for the maximum rate of absorption. Some of their findings are surprising. Glucose, for example, is absorbed more quickly than fructose (even though both molecules have the same formula, $C_6H_{12}O_6$) and it is absorbed more quickly from solutions that contain sodium chloride in addition to water and glucose.

In the next section we see how the structure of cell membranes accounts for these observations.

When we drink an isotonic drink, the water, glucose molecules and mineral ions are absorbed by the cells that line the gut. From here they pass into the blood plasma, and then travel around the body to where they are needed. Most of the initial absorption takes place in the small intestine. The cells that line the small intestine are highly specialised to carry out their function (Fig. 8).

The cell surface is highly folded, which greatly increases the surface area of the cell. A greater surface area means that there is more cell membrane across which diffusion, osmosis, **facilitated diffusion** and active transport can take place.

Each of the three types of particle – water, glucose and mineral ions – is absorbed in a different way through the membrane of the intestine cells. To understand how, you need to revise the structure of the plasma membrane of these cells.

The bulk of a cell membrane is made up of a double layer of phospholipids (Fig. 6 on p 27). Molecules that do not dissolve in lipids can only diffuse across the phospholipid regions of the membrane if they are small enough to pass between the phospholipid molecules.

15 Explain how the folded cell surface membrane of intestinal **epithelial cells** helps them to carry out their function.

16 Water molecules can diffuse between the phospholipid molecules in the cell surface membrane. But they do so at a rate fifteen times slower than the rate that they diffuse in pure water. Suggest why.

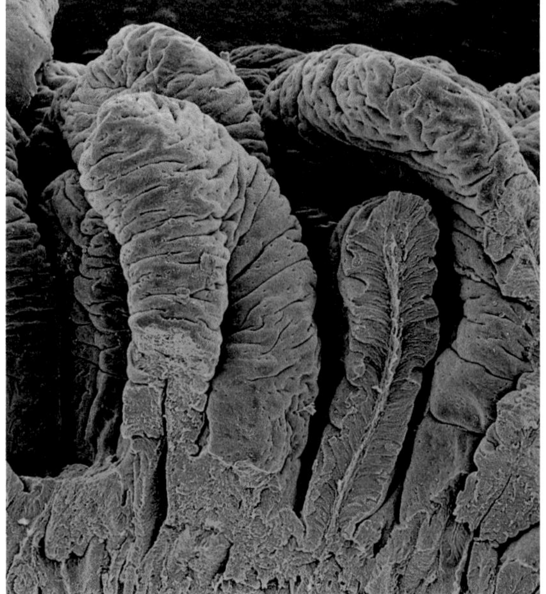

False-colour scanning electron micrograph of a section through the wall of the human intestine, showing many folded villi. These increase the surface area over which food molecules can be absorbed.

Magnification × 70 000.

Fig. 8 An intestinal epithelial cell

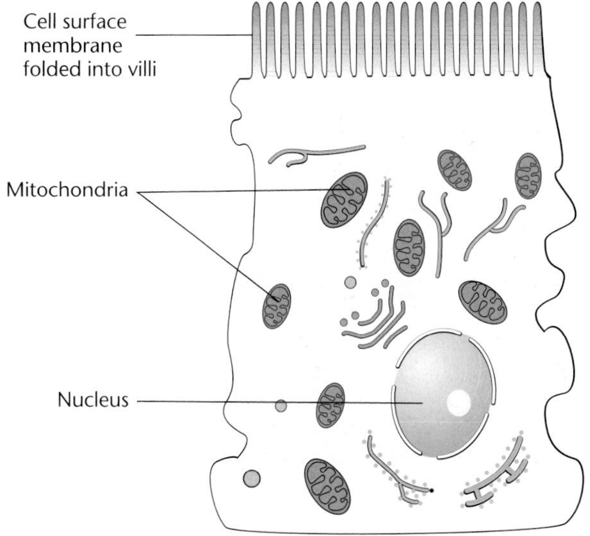

Larger lipid-insoluble molecules, such as glucose, need more help. They get it from intrinsic proteins within the phospholipid membrane. These protein molecules, called **carrier proteins**, provide transmembrane passages through which molecules such as glucose can diffuse (Fig. 9). This process is known as facilitated diffusion.

Specific carrier protein molecules transport each substance across the membrane. The proteins recognise the ions or molecules they interact with by their shape, rather like enzymes recognise substrate molecules. Like diffusion, facilitated diffusion only occurs down a concentration gradient and does not need an input of energy from the cell.

Fig. 9 Facilitated diffusion

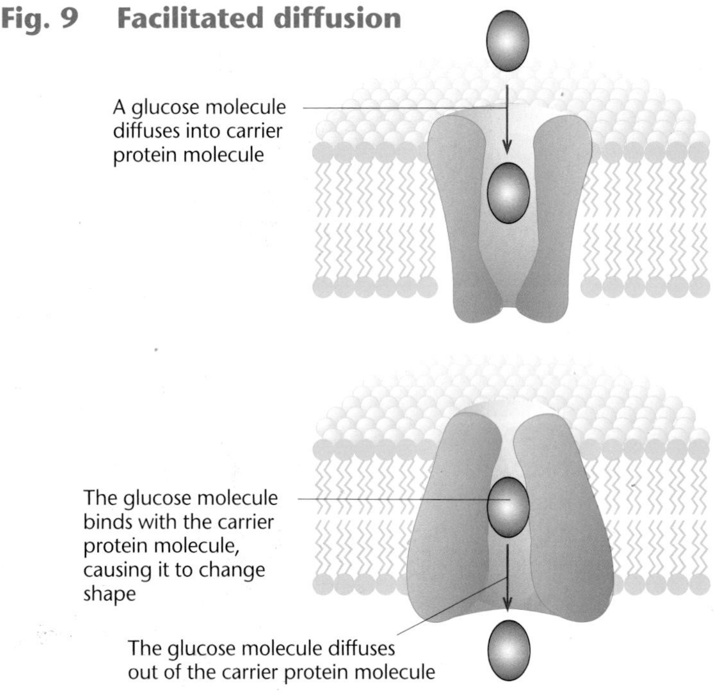
A glucose molecule diffuses into carrier protein molecule
The glucose molecule binds with the carrier protein molecule, causing it to change shape
The glucose molecule diffuses out of the carrier protein molecule

Fig. 11 Diffusion of water and glucose

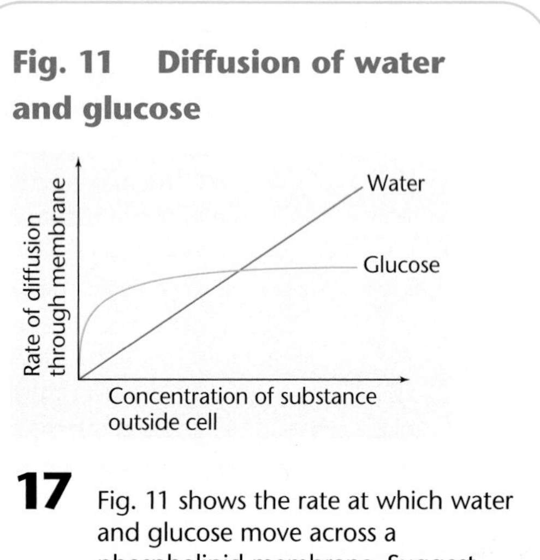
Rate of diffusion through membrane
Water
Glucose
Concentration of substance outside cell

17 Fig. 11 shows the rate at which water and glucose move across a phospholipid membrane. Suggest one reason for the difference between the two curves.

Fig. 10 Gated channel proteins

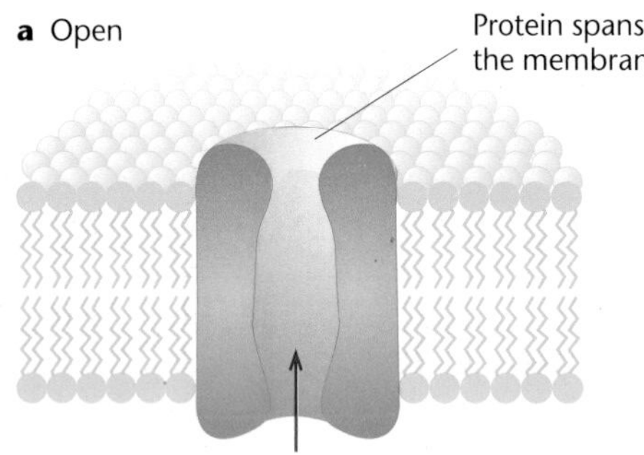
a Open
Protein spans the membrane
Ions such as Na+ pass through the protein pore
b Closed
Protein changes shape, so pore becomes too narrow to allow ions through

Mineral ions also move across membranes by facilitated diffusion but they travel along an **electrochemical gradient**. This is a gradient that occurs between two regions with different charges. So, for example, when there is a high concentration of positively charged sodium ions (Na^+) in one area and a high concentration of negatively charged chloride ions (Cl^-) in another area, the sodium ions will move from positive to negative and the chloride ions will move from negative to positive. Specific proteins called **channel proteins** allow ions to pass through the membrane. Channel proteins are selective – each one has a specific three-dimensional shape with a particular arrangement of electrical charges inside the channel and will only accept one type of ion. Channel proteins can be open or closed and so are called **gated channels** (Fig. 10). The mechanism that enables nerve impulses to be transmitted along nerve cells depends on gated ion channels that open and then close.

3.6 Active transport

Substances can pass in and out of cells along their concentration gradient in several different ways, none of which requires the input of energy. But what happens when a cell needs to move substances against a concentration gradient? In this situation the cell has to expend some energy. Some intrinsic protein molecules act as molecular pumps. They allow the cell to use **active transport** to accumulate glucose or ions against their concentration gradient (Fig. 12). Animal and plant cells that specialise in absorption usually have abundant mitochondria which provide the ATP needed to power active transport.

Fig. 12 Active transport

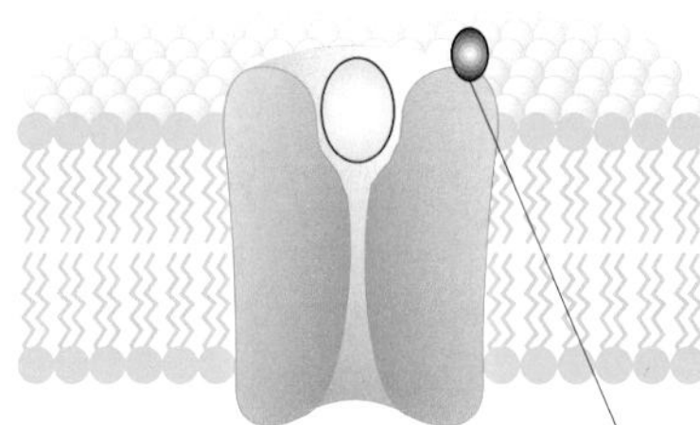

Phosphate group attached to carrier protein by energy from ATP activates protein to accept particle to be transported

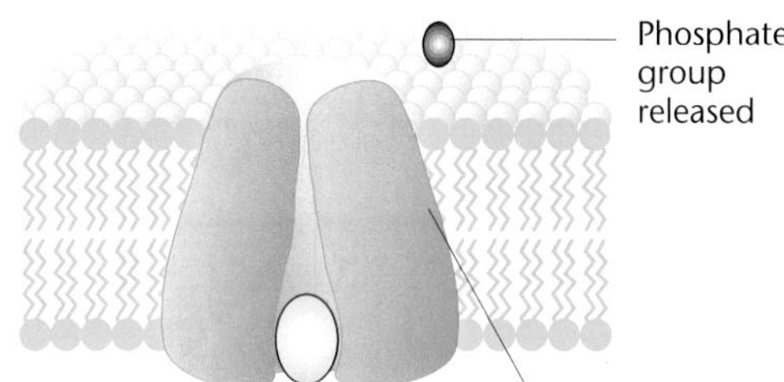

Energy attaching phosphate group to carrier protein used to change shape of carrier molecule – transporting particle across membrane

18 How do the many mitochondria in an intestinal epithelial cell help it to carry out its function?

3.7 Co-transport

When we exercise and sweat heavily, we lose water, the blood plasma becomes hypertonic and our performance suffers. Sports scientists now recommend that runners should take frequent drinks containing water, mineral ions and glucose during long-distance runs.

As well as replacing the sodium chloride lost in sweat, the sodium ions stimulate the rate of glucose uptake into the blood – so the athlete gets a quick 'glucose fix'. This happens because of symports – specialised intrinsic proteins in the cell surface membranes of epithelial cells of the small intestine (see Fig. 13 on page 48). The glucose–sodium symport transports glucose and sodium into the cell at the same time, but it only works when both substances are present. At the junction between the epithelial cell and a blood capillary, glucose is transported into the blood by facilitated diffusion and sodium ions by active transport.

So, isotonic drinks with a little glucose in them certainly seem to be the best option for someone involved in stamina sports.

Several vital processes depend on active transport including:

- absorption of amino acids from the gut
- absorption of mineral ions by plant roots
- excretion of urea and hydrogen ions by the mammalian kidney
- exchange of sodium and potassium ions in nerve cells
- loading of sugar from the leaf into the phloem in plants.

The phloem in the leaves of a plant is rich in sugars because of the high rate of active transport. Lower down the plant, sugars are converted to starch, and the phloem there contains less sugar. People who grow the sugar palm exploit this difference in sugar concentration. The sugar is 'tapped' by cutting off the flowers at the top of the plant, just before they are fully formed. When this is done, over 10 litres of rich sugary sap can be collected from the phloem of just one plant. The sap contains 10% sugar and it is boiled to make molasses or refined sugar.

19

a Why do sports scientists recommend that the maximum concentration of sodium in the drink should be 200 mg per litre?

b Why do runners need sugar?

20 At the junction between the epithelial cell and a blood capillary, glucose is transported into the blood by facilitated diffusion and sodium ions by active transport. Explain why different processes are required to transport these two substances from the epithelial cell into the blood.

21 Explain the importance of adding salt to the sports glucose drink.

Fig. 13 Glucose–sodium symports in the small intestine

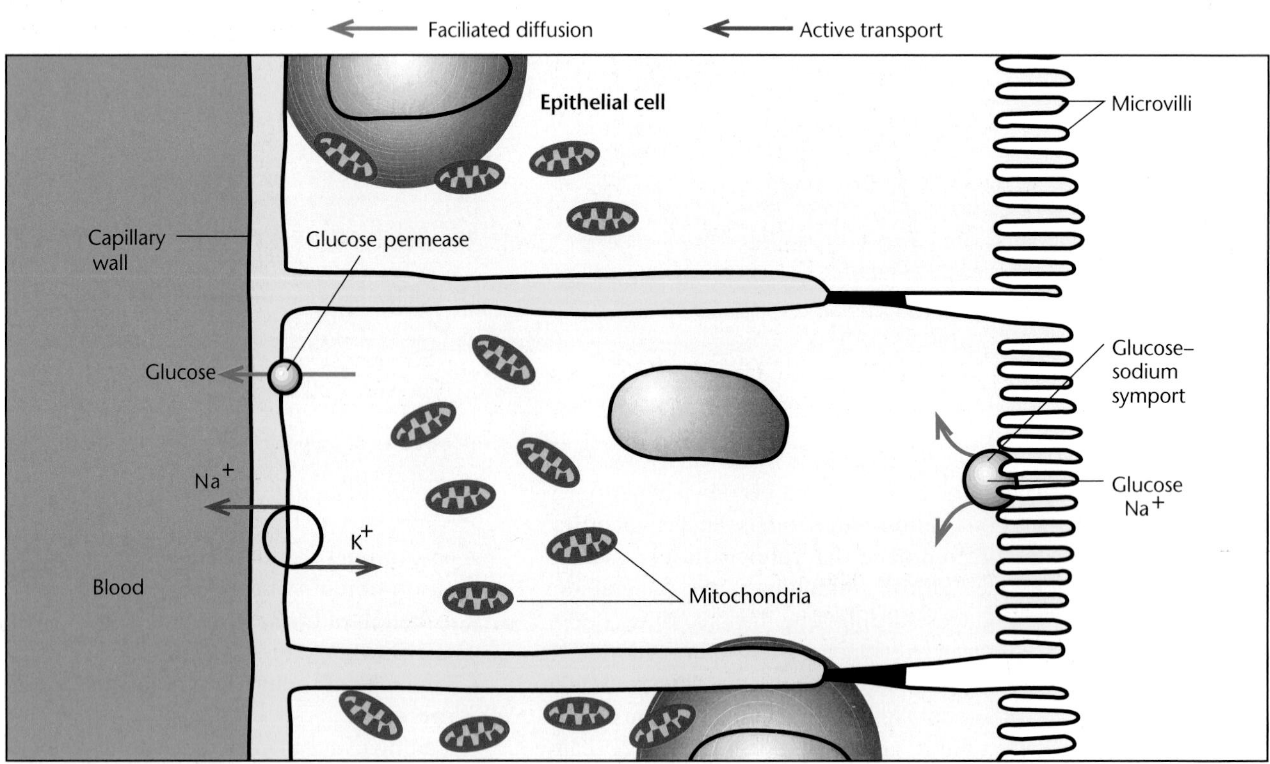

key facts

- Facilitated diffusion is the movement of water-soluble molecules through specific intrinsic carrier protein molecules or channel protein molecules in the plasma membrane. It does not require energy.
- Active transport moves ions and molecules across a membrane against their concentration gradient. It requires energy.
- In co-transport, glucose and sodium ions are moved together from the intestinal lumen into the blood by a combination of facilitated diffusion and active transport. This system is known as a symport.
- Cells that are specialised for absorption have a large surface area and often have many mitochondria.

how science works

The kidneys are the organs in the body that remove waste from the blood. If the kidneys fail, these wastes build up in the body. Today, kidney failure can be treated quite successfully using dialysis. Dialysis depends on a membrane that is differentially permeable; it allows small particles to pass through but it retains large particles and blood cells. The membrane can be part of an artificial kidney machine or the blood can be dialysed using the peritoneal membrane in the abdomen. This method is known as peritoneal dialysis (PD).

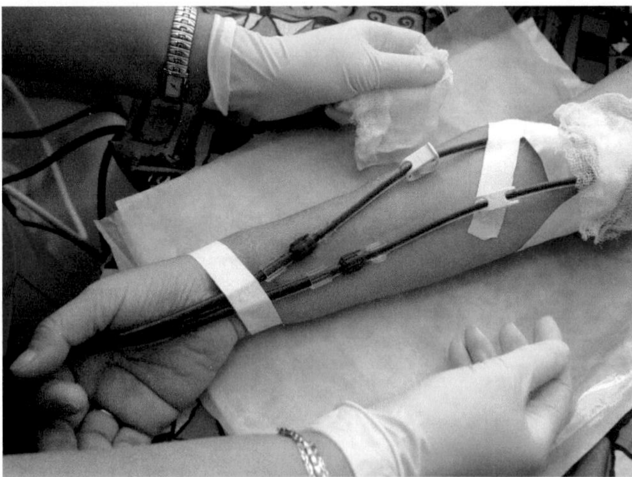

During haemodialysis, the patient is linked up to a kidney dialysis machine. For several hours, blood flows from the forearm into the machine. The other tube is removing waste from the blood, together with excess salt and water.

Peritoneal dialysis

The peritoneal cavity in the abdomen is lined by the thin peritoneal membrane, which surrounds the intestines and other internal organs. When someone has PD, his or her abdomen is filled with dialysis fluid via a catheter. The catheter is a special tube that is left permanently in place. A short operation is needed to put the catheter in place. The body's waste substances leave the blood, pass through the peritoneal membrane and mix with the dialysis fluid. After a time, this fluid is drained from the body and discarded. This 'flushing' process is repeated between three and five times a day. PD enables the person having treatment to live normally whilst it is going on. Dialysis by machine involves many hours sitting in hospital, attached to a large piece of medical apparatus. There are two types of PD, continuous and intermittent.

Continuous peritoneal dialysis (CPD)

In CPD about 2 litres of dialysis fluid is in the peritoneal cavity at any one time, so the blood is cleaned constantly. The fluid is changed regularly. There are two ways of doing this. In continuous ambulatory PD (CAPD), the dialysis fluid is drained four times a day and the peritoneal cavity is refilled with fresh fluid. This process takes about 45 minutes and most people space the sessions to suit them, perhaps early in the morning, at lunchtime, in the late afternoon and just before bed. In continuous cycling PD the change of fluid is done by an automatic cycler machine that works overnight.

About two litres of dialysis fluid are left in the peritoneal cavity for the day and not changed until the next night.

Intermittent peritoneal dialysis (IPD)

IPD usually requires a trip to hospital and most patients with kidney failure need between 36 and 44 hours of IPD each week. In IPD, the dialysis fluid is left in the peritoneal cavity for a short time and then drained out. One complete cycle takes about one hour. Because the dialysis fluid in IPD is not changed as frequently as in CPD, the person must take particular care with their diet and fluid intake to avoid a build up of food wastes and water.

22 Dialysis fluid is similar in concentration to the blood plasma in a person with healthy kidneys.

a What substances would you expect to find in fresh dialysis fluid?

b Would you expect that fresh dialysis fluid to contain urea? Explain your answer.

c Explain why excess water will pass into the dialysis fluid.

23 Describe how dialysis fluid is introduced into the body then drained from the body.

24

a Explain the difference between continuous peritoneal dialysis (CPD) and intermittent peritoneal dialysis (IPD).

b Summarise the advantages and disadvantages of CPD compared with IPD.

3.8 Cholera

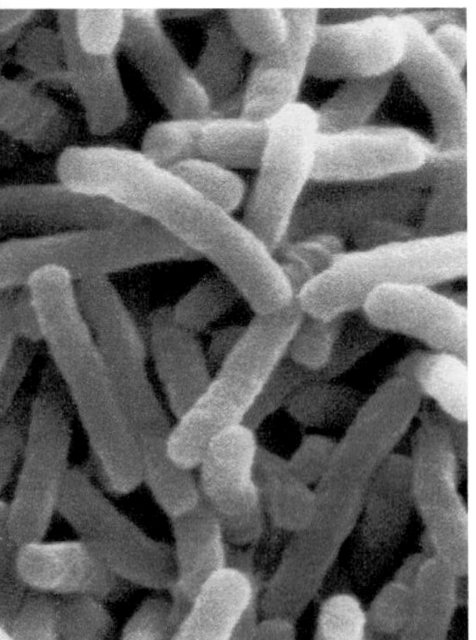

Scanning electron micrograph during early infection with *Vibrio cholerae*: the curved bacteria are adhering to the epithelial surface.

Prokaryotic cells

The most serious type of diarrhoea is caused by the bacterium *Vibrio cholerae*, which are transmitted to humans by ingesting contaminated water or food.

Bacterial cells are fundamentally different from the cells that make up our bodies. Human body cells are **eukaryotic** cells; they have a nucleus containing DNA and a collection of organelles that have specific functions. Bacterial cells are prokaryotic (meaning 'before the nucleus'). **Prokaryotic cells** do not have a nucleus. Instead they have a large piece of circular DNA and many smaller circular rings of DNA called **plasmids**. The cell wall of prokaryotes contains carbohydrates called **polysaccharides**, but these are different from the **cellulose** that makes up plant cell walls.

Many prokaryotes also secrete a protective **capsule** outside the cell wall. Some prokaryotes have a long whip-like structure called a **flagellum**, which enables them to move.

Fig. 14 Bacteria

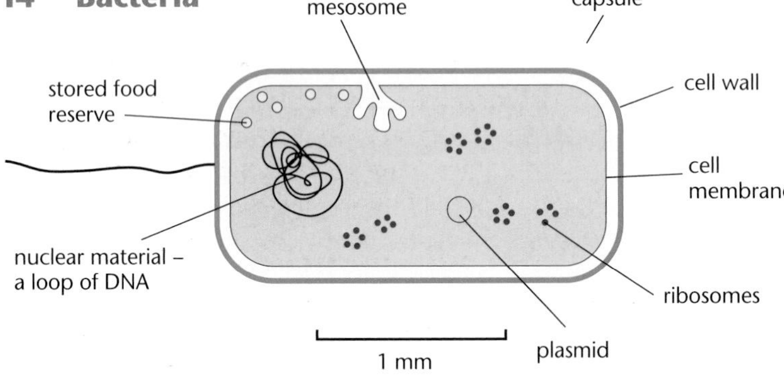

25 List the differences between a prokaryotic cell and a eukaryotic cell.

Cholera infection

Most persons infected with cholera do not become ill, although the bacterium is present in their faeces for 7–14 days. When illness does occur, more than 90% of cases are mild or moderately severe and are difficult to distinguish from other types of acute diarrhoea. Fewer than 10% of people infected develop typical cholera with severe dehydration.

V. cholerae produces cholera toxin whose action on the epithelium lining of the small intestine causes the massive diarrhoea of the disease. In its most severe forms, cholera is one of the most rapidly fatal illnesses known. A child may die within 2–3 hours if no treatment is provided. In adults the disease progresses from the first diarrhoeal attack to shock in 12 hours, followed by death in 18 hours if no treatment is available. In its later stages the diarrhoea looks like 'rice water'.

Most of the *V. cholerae* bacteria in contaminated water are killed by the acidic conditions in the human stomach. The surviving bacteria produce a flagellum to propel themselves through the thick mucus that lines the small intestine to get to the intestinal wall where they can thrive. They then start producing the toxic proteins that give the infected person diarrhoea.

Cholera toxin causes the intestinal epithelial cells to pump out chloride ions into the intestinal **lumen**. The electrochemical gradient caused by this prevents sodium ions entering the intestinal epithelial cells, resulting in a high concentration of sodium ions and chloride ions in the intestinal lumen. This makes the water potential of the contents of the lumen more negative, causing water to move out of the intestinal cells into the lumen by osmosis. About 6 litres of water per day enter the intestinal lumen in this way and is lost as diarrhoea, causing the body to become dehydrated.

Rehydration therapy

Table 2 shows the concentration of ions in the diarrhoea of someone with cholera.

The quickest way to rehydrate a person with cholera is to give an intravenous solution with the same concentrations of ions as blood plasma. This

Table 2

	Concentration of ions in diarrhoea/mmol dm^{-3}			
	Na^+	K^+	Cl^-	HCO_3^-
adults	140	13	104	44
children (under 5 yrs)	101	27	92	32

needs to be carried out by trained medical personnel, and there are risks associated with the procedure. Oral rehydration therapy (ORT) solutions were developed so that parents could give their children the life-saving fluid by mouth.

The first solutions contained the same concentrations of ions found in the children's diarrhoea (given in Table 2) and 8% glucose to nourish the child. Unfortunately these solutions caused a condition called hypernatraemia (excess sodium in the blood) because the ORT fluid had a more negative water potential than the child's blood, causing water to leave the blood by osmosis ('osmotic diarrhoea').

26 In these investigations:

a What were the independent and dependent variables?

b What was the control variable?

c What further controls should have been considered?

27 Why does water 'follow by osmosis'?

ORT trials on children

A series of trials was then carried out on children to solve this problem. Children with the same degree of diarrhoea were given ORT fluids with different amounts of glucose and ions and their effects on the degree of diarrhoea compared. The apparently obvious answer was to assume that the sodium concentration in the oral rehydration fluid used was too high and to reduce it (even to as low as 25 or 30 mmol dm^{-3}). Unfortunately, this was not the answer – it was the excess glucose that caused osmotic diarrhoea. The less obvious but correct answer was to reduce the glucose content. Scientists now recognise that a 1:1 ratio in terms of molarity provides the optimum concentrations of sodium ions and glucose for co-transport. There is rapid absorption of sodium ions ad glucose. Water follows by osmosis.

ORT fluids now usually contain 90 mmol dm^{-3} of Na^+ and 111 mmol dm^{-3} glucose.

Ethics of trialling ORT solutions

Medical progress involves trying out new procedures on human patients. An international code of practice governs such trials.

- The patients should be volunteers.
- Patients should be made fully aware of the risks involved before they agree to take part (give consent).
- Researchers should take all possible precautions to minimise risks to the volunteers, and there should be no chance of permanent damage from the treatment.
- Only competent experienced professionals should carry out research.
- All results should enter the public domain so that successful treatments can be made widely available.

Some of the trials of ORT solutions were carried out on children during severe cholera epidemics. Some researchers ignored parts of the code of practice in the interests of saving the lives of those children.

28 In your opinion, which parts of the code of practice should never be ignored?

key facts

- Cholera is caused by the *Vibrio cholerae* bacteria.
- Toxins secreted by the bacteria cause intestinal cells to secrete chloride ions into the lumen of the intestine.
- Sodium ions follow down an electrochemical gradient.
- Water follows by osmosis, causing diarrhoea.
- Cholera is treated mainly by oral replacement therapy (ORT).
- ORT fluids contain carefully calculated amounts of ions and glucose.
- Sodium ions and glucose are absorbed from the fluid by co-transport; water follows by osmosis.
- Prokaryotic cells:
  - do not have a nucleus
  - have a large piece of circular DNA and many smaller circular rings of DNA called plasmids
  - have a cell wall
  - may secrete a protective capsule outside the cell wall
  - may have a long whip-like structure called a flagellum which enables them to move.

examination questions

1 In an experiment ten cylinders cut from a potato were placed in each of sucrose solutions of different concentrations for one hour. The length of each cylinder was measured before and after immersion in sucrose solution. The graph shows the results of the experiment.

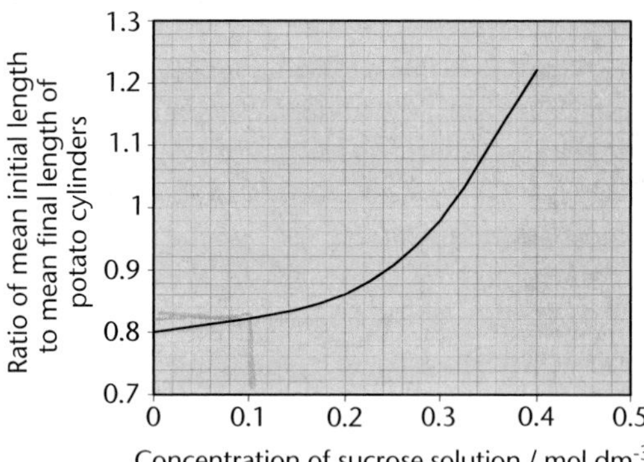

a For this experiment:
- **i** name the independent variable; (1)
- **ii** name the dependent variable; (1)
- **iii** name **two** variables that should be controlled. (1)

b Why were 10 potato cylinders placed in each solution? (1)

c In which concentration of sucrose solution did the potato cylinders remain the same length? (1)

d Explain in terms of water potential the result for
- **i** the potato cylinders in 0.1 mol dm^{-3} sucrose solution; (2)
- **ii** the potato cylinders in 0.4 mol dm^{-3} sucrose solution; (2)

Total 9

2 Read the report from UNICEF about ORS treatment.

For more than 25 years UNICEF and WHO have recommended a single formulation of glucose-based ORS treat dehydration from diarrhoea. This product, which provides a solution containing 90 mEq/l of sodium with a total osmolarity of 311 mOsm/l, has proven effective in worldwide use. However, this ORS solution does not reduce stool output or duration of diarrhoea. For this reason the current ORS might have had less than optimal acceptance by mothers and health workers, preferring a treatment that causes diarrhoea to stop. During the past 20 years numerous studies have been undertaken to develop an 'improved' ORS. The goal was a product that would be at least as safe and effective as standard ORS for preventing or treating dehydration from all types of diarrhoea but which, in addition, would reduce stool output.

Scientists working on ORS treatments use osmolarity rather than water potential to describe solutions. The tables show the composition of the standard ORS solution and the new reduced osmolarity solution.

	Standard ORS solution (mmol/l)	Reduced Osmolarity ORS solution (mmol/l)
Glucose	111	111
Sodium	90	50
Chloride	80	40
Potassium	20	20
Citrate	10	30
Osmolarity	311	251

a Why is ORS solution necessary in the treatment of cholera? (2)

b **i** Apart from osmolarity, what is the main difference between standard ORS solution and reduced osmolarity ORS solution? (1)
- **ii** Explain in terms of osmosis why reduced osmolarity solution ORS solution reduces stool output. (2)

Many children suffering from diarrhoea refuse to ORS because of its strong salty taste. Many parents and health workers flavor ORS with the child's favorite juice. Scientists investigated the effects of flavoring ORS on ion content and osmolarity. They also compared the palatability of various solutions with commercially flavored ORS.

The osmolality, sodium, potassium, chloride and glucose content of ORS solutions after flavoring with varying concentrations of apple juice, orange juice or orangeade was determined. Two of the solutions were offered to 30 children to assess palatability.

In further investigations, scientists found that additions to ORS of apple juice, orange juice or orangeade caused a decrease of sodium (–30 to –53 mmol/L) and chloride (–27 to –47 mmol/l) content, whereas osmolarity increased to greater than 311 mmol/l. The majority of children also preferred the commercially flavored ORS.

c **i** What ethical issues should the scientists have considered before carrying out the above investigation? (3)
- **ii** Did the investigation produce valid results? Explain the reason for your answer. (3)
- **iii** Explain why adding orange juice increased the osmolarity and decreased the ion content of the ORS solution. (2)
- **iv** Explain why adding orange juice reduces the effectiveness of an ORS solution. (2)
- **v** What advice would you give to a mother whose child dislikes ORS solution? (2)

Total 17

3 The diagram shows part of a cell surface membrane.

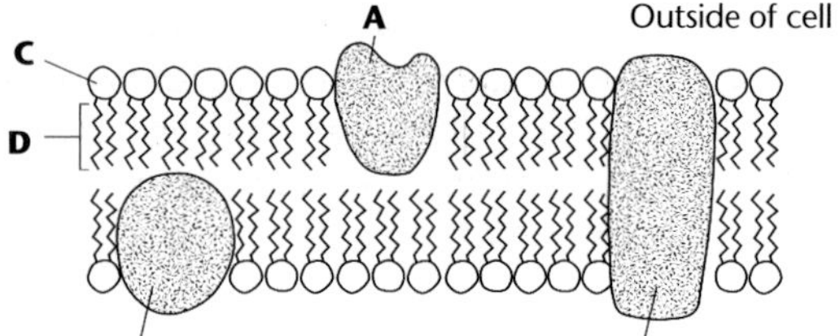

a Complete the table by writing the letter from the diagram which refers to each part of the membrane. (2)

Part of membrane	**Letter**
Channel protein	
Contains only the elements carbon and hydrogen	

b Explain why the structure of a membrane is described as *fluid-mosaic*. (2)

c When pieces of carrot are placed in water, chloride ions are released from the cell vacuoles. Identical pieces of carrot were placed in water at different temperatures. The concentration of chloride ions in the water was measured after a set period of time.
The graph shows the results.

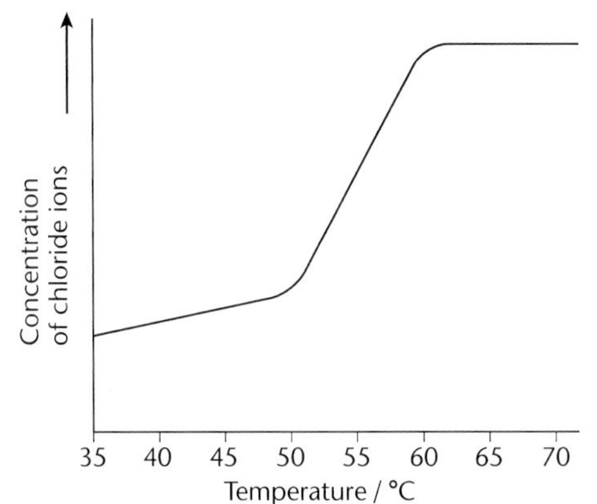

Describe and explain the shape of the curve. (3)

Total 7

AQA, January 2005, Unit 1, Question 4

4 The diagram shows a mitochondrion.

a i Name the part labelled X. (1)

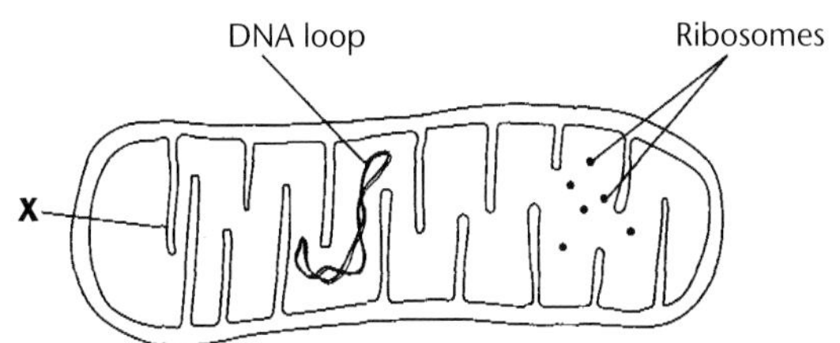

ii A human liver cell contains several hundred mitochondria. A cell from a plant root has only a small number.
Suggest an explanation for this difference. (2)

iii Mitochondria contain some DNA and ribosomes.
Suggest the function of these. (2)

b Mitochondria may be separated from homogenised cells by differential centrifugation. During this process the cells must be kept in an isotonic solution.
Explain why. (2)

c Ribosomes in bacterial cells differ from those in the cytoplasm of eukaryotic cells. When centrifuged at high speed, the eukaryotic cell ribosomes sediment more rapidly than bacterial ribosomes.
Explain what this tells you about the difference between bacterial and eukaryotic ribosomes. (1)

Total 8

AQA, January 2001, Unit 1, Question 5

5 The diagram shows a bacterium.

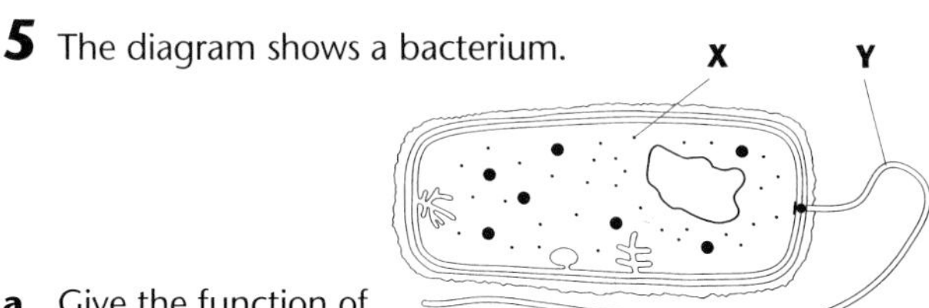

a Give the function of

i organelle **X**;

ii organelle **Y**. (2)

b i Give **two** ways in which the structure of this bacterium is similar to the structure of a cell lining the human small intestine. (2)

ii Give **two** ways in which the structure of this bacterium differs from the structure of a cell lining the human small intestine. (2)

Total 6

AQA, June 2005, Unit 7, Question 2

6 Cholera is a water-borne disease caused by the intestinal pathogen, *Vibrio cholerae*. The pathogen produces an exotoxin which acts specifically on the epithelial cells of the small intestine causing changes in membrane permeability. Individuals with cholera suffer from severe diarrhoea which may result in death.

a Suggest **two** precautions which could be used to prevent the transmission of cholera. (1)

b Explain how the effects of diarrhoea on the body can be treated. (2)

c i What is an exotoxin? (1)

ii Suggest why the cholera exotoxin is specific to the epithelial cells of the small intestine. (2)

d The cholera exotoxin affects the movement of ions through the intestinal wall. It causes the loss of chloride ions from the blood into the lumen of the small intestine. This prevents the movement of sodium ions from the lumen of the small intestine into the blood.

i Describe how sodium ions normally enter the blood from the cells of the intestinal wall against a concentration gradient. (2)

ii Use the information provided to explain why individuals with cholera have diarrhoea. (2)

Total 10

AQA, January 2004, Unit 7, Question 6

The discovery of the source of cholera

A doctor called John Snow investigated the cause of cholera in the eighteenth century, before the discovery of bacteria. Read this abridged version of Snow's account of an outbreak of cholera in Soho, London.

The most terrible outbreak of cholera which ever occurred in this kingdom, is probably that which took place in Broad Street, Golden Square, and the adjoining streets, a few weeks ago. Within two hundred and fifty yards of the spot where Cambridge Street joins Broad Street, there were upwards of five hundred fatal attacks of cholera in ten days. The mortality in this limited area probably equals any that was ever caused in this country, even by the plague; and it was much more sudden, as the greater number of cases terminated in a few hours. The mortality would undoubtedly have been much greater had it not been for the flight of the population. There were a few cases of cholera in the neighborhood of Broad Street, Golden Square, in the latter part of August; and the so-called outbreak, which commenced in the night between the 31st August and the 1st September, was, as in all similar instances, only a violent increase of the malady. As soon as I became acquainted with the situation and extent of this irruption of cholera, I suspected some contamination of the water of the much-frequented street pump in Broad Street, near the end of Cambridge Street; but on examining the water, on the evening of the 3rd September, I found so little impurity in it of an organic nature, that I hesitated to come to a conclusion. Further inquiry, however, showed me that there was no other circumstance or agent common to the circumscribed locality in which this sudden increase of cholera occurred, and not extending beyond it, except the water of the above mentioned pump. I found, moreover, that the water varied, during the next two days, in the amount of organic impurity, visible to the naked eye, on close inspection, in the form of small white, flocculent particles; and I concluded that, at the commencement of the outbreak, it might possibly have been still more impure. I requested permission, therefore, to take a list, at the General Register Office, of the deaths from cholera, registered during the week ending 2nd September, in the sub districts of Golden Square, Berwick Street, and St. Ann's, Soho, which was kindly granted. Eighty-nine deaths from cholera were registered, during the week, in the three sub-districts. Of these, only six occurred in the first four days of the week; four occurred on Thursday, the 31st August; and the remaining seventy-nine on Friday and Saturday. I considered, therefore, that the outbreak commenced on the Thursday; and I made inquiry, in detail, respecting the eighty-three deaths registered as having taken place during the last three days of the week.

On proceeding to the spot, I found that nearly all the deaths had taken place within a short distance of the pump. There were only ten deaths in houses situated decidedly nearer to another street pump. In five of these cases the families of the deceased persons informed me that they always went to the pump in Broad Street, as they preferred the water to that of the pump which was nearer. In three other cases, the deceased were children who went to school near the pump in Broad Street. Two of them were known to drink the water; and the parents of the third think it probable that he did so. The other two deaths, beyond the district which this pump supplies, represent only the amount of mortality from cholera that was occurring before the irruption took place.

The result of the inquiry then was, that there had been no particular outbreak or increase of cholera, in this part of London, except among the persons who were in the habit of drinking the water of the above-mentioned pump-well.

I had an interview with the Board of Guardians of St. James's parish, on the evening of Thursday, 7th September, and represented the above circumstances to them. In consequence of what I said, the handle of the pump was removed on the following day.

The additional facts that I have been able to ascertain are in accordance with those above related; and as regards the small number of those attacked, who were believed not to have drank the water from Broad Street pump, it must be obvious that there are various ways in which the deceased persons may have taken it without the knowledge of their friends. The water was used for mixing with spirits in all the public houses around. It was used likewise at dining rooms and coffee shops. The keeper of a coffee shop in the neighbourhood, which was frequented by mechanics, and where the pump water was supplied at dinner time, informed me (on 6th September) that she was already aware of nine of her customers who were dead. The pump water was also sold in various little shops, with a teaspoonful of effervescing powder in it, under the name of sherbet; and it may have been distributed in various other ways with which I am unacquainted. The pump was frequented much more than is usual, even for a London pump in a populous neighborhood.

There are certain circumstances bearing on the subject of this outbreak of cholera which require to be mentioned. The Workhouse in Poland Street is more than three-fourths surrounded by houses in which deaths from cholera occurred, yet out of five hundred and thirty-five inmates only five died of cholera, the other deaths which took place being those of persons admitted after they were attacked. The workhouse has a pump well on the premises, in addition to the supply from the Grand Junction Water Works, and the inmates are never sent to Broad Street for water. If the mortality in the workhouse had been equal to that in the streets immediately surrounding it on three sides, upwards of one hundred persons would have died.

There is a Brewery in Broad Street, near to the pump, and on perceiving that no brewer's men were registered as having died of cholera, I called on Mr Huggins, the proprietor. He informed me that there were above seventy workmen employed in the brewery, and that none of them had suffered from cholera, – at least in a severe form, – only two having been indisposed, and that not seriously, at the time the disease prevailed. The men are allowed a certain quantity of malt liquor, and Mr Huggins believes they do not drink water at all; and he is quite certain that the workmen never obtained water from the pump in the street. There is a deep well in the brewery, in addition to the New River water.

At the percussion-cap manufactory, 37 Broad Street, where, I understand, about two hundred workpeople were employed, two tubs were kept on the premises always supplied with water from the pump in the street, for those to drink who wished; and eighteen of these workpeople died of cholera at their own homes, sixteen men and two women.

A1 What type of research did Snow use in his investigation?

A2 Describe the evidence that led Snow to associate cholera with sewage contamination of water.

4 Food and digestion

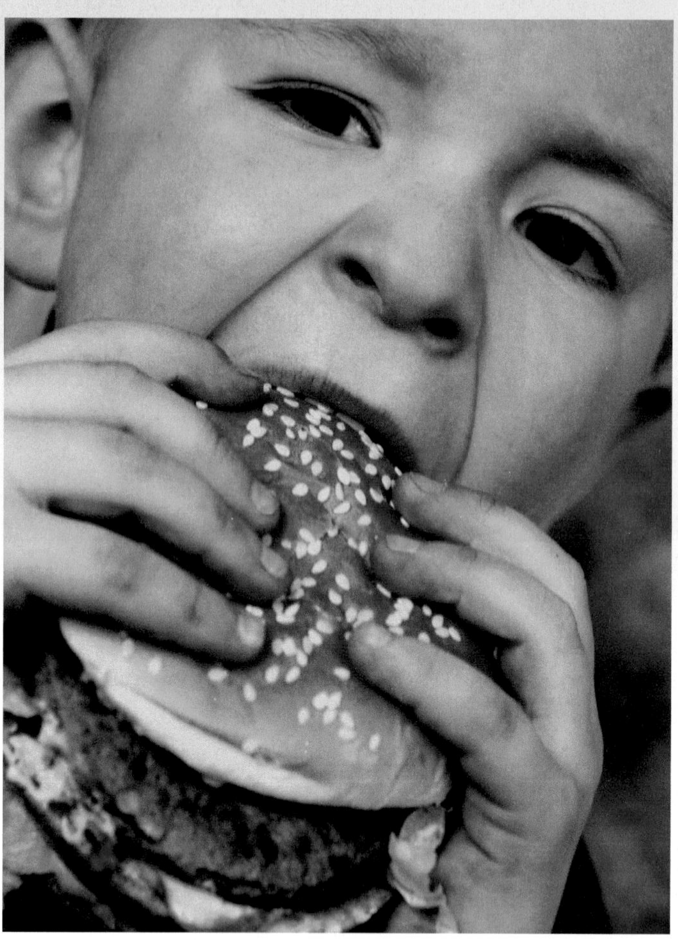

Good health depends on good digestion. Overeating and eating the wrong types of food can lead to poor or inadequate digestion, which can make you feel bloated or lethargic. Some companies claim that taking supplements that contain digestive enzymes can help you to digest your food better. Of course, you could choose to eat less or to eat healthier food...

Your body actually contains about 10 times more bacterial cells than body cells! The term 'the microbiome' has been coined to describe the dense ecology of the 500-plus species of bacteria living your gut. Scientists are beginning to realise that this huge population of bacteria makes significant contributions to our metabolic and physiological process and therefore to our wellbeing. It is now thought that the microbiome plays key roles in how we develop, from the moment we leave the womb. For example, if you do not absorb nutrients from your food too well, perhaps something went wrong when bacteria were controlling the way in which the lining of your gut developed. The microbiome may affect your risk of developing allergies by influencing the sensitivity of the immune system. So maybe that's why you get hay fever. Were the bacteria involved in shaping the connections between the brain and the adrenal glands? Is that why some people get digestive problems when they are stressed?

The greatest hype surrounds the notion that certain species of bacteria may play a role in weight gain, and in particular how effectively two particular species of bacteria work together. Perhaps getting this combination right is a revolutionary way to lose weight?

Many manufacturers claim that taking digestive enzymes as a food supplement can solve the problems associated with overeating. Food supplements that contain the bacteria found in the intestines are called 'probiotics' and the market for such products is huge – millions of people take these supplements daily. But will adding extra enzymes make our digestive systems more efficient, or is it just advertising hype? Do probiotic supplements improve our health, or do they simply give us an excuse to eat too much of the wrong types of food?

We probably know less about the digestive system than about any other major body system.

This chapter will introduce you to the digestive system.

4.1 Food and digestion

Every day in wealthy countries such as the UK, people eat large quantities of food. The food that we swallow is not in a form that can be used directly by the body; steak, which is muscle, cannot simply be added to our biceps to make them grow. So what happens to it?

Chewing breaks up food into smaller pieces and the mechanical churning action of the stomach continues this process. However, most food molecules are too large to be absorbed into the bloodstream. They have to be broken down into smaller molecules that are then able to pass into blood capillaries in the wall of the gut and then from the blood into the cells where they are to be used. Digestion depends on enzymes, which catalyse the reactions that break down food molecules, allowing the reactions to take place rapidly and at body temperature. In this chapter we look at the enzymes that work on food as it passes through the gut.

4.2 Enzymes

A huge number of chemical reactions take place inside every living cell. Some release energy, some synthesise new substances and others break down waste products. **Enzymes** enable these reactions to happen and, most importantly, enzymes make it possible for the reactions to happen quickly enough at body temperature. Most chemical reactions need an input of energy to set them going and normally this is supplied as heat. Enzymes catalyse reactions; they reduce the amount of energy that is needed to get a reaction going but they are not themselves changed at the end of the reaction. This makes them very valuable substances, not only in living organisms, but also for human use. For centuries, enzymes from yeast have been used to brew alcoholic drinks. Today, adding enzymes to detergents to digest food stains is commonplace, and in the future, many more industrial processes are likely to make use of enzymes.

Enzymes are proteins. The properties of proteins help to explain how enzymes work and how they supply the energy needed to get a reaction going. The structure of enzyme molecules does, however, make them sensitive to environmental conditions. Changes in temperature or pH can disrupt enzyme action. In humans, severe over-heating or getting so cold that core body temperature falls can upset the delicate balance of enzyme-controlled reactions in the body's cells and can be fatal.

4.3 Enzymes and chemical reactions

The energy in food can be released by combustion but a large input of energy is needed to get the reaction going. Once the sugar starts burning, the combustion reaction occurs very rapidly.

Sugar left in a bowl does not react with oxygen in the air. Sugar only reacts if heated strongly, and even then it is difficult to start it burning. In living cells, however, sugar is continuously broken down by respiration to release energy. This is just one of many chemical processes going on in a cell and this constant chemical activity is known as the cell's **metabolism**.

In respiration, glucose is broken down in a series of reactions called a metabolic pathway. This is quite different from burning, where the energy release is rapid and uncontrolled. A metabolic pathway releases energy in small steps. Enzymes enable metabolic reactions to occur in cells at normal environmental temperatures. They also determine when, where and how fast those reactions take place.

The substances that react together in any reaction catalysed by an enzyme are called the **substrates** of that reaction.

Activation energy

Why does sugar react with oxygen only when it is heated? Before it can react, the bonds linking the atoms in the molecules must be broken, and this requires energy. The heat from a match or a Bunsen burner can provide this. The energy gives the molecules of the substrates more kinetic energy. As a result, the molecules of glucose vibrate more rapidly in the solid sugar and the molecules of oxygen move faster in the air nearby. Some of these molecules collide, but does that always lead to a reaction? Normally the electrons that orbit an atom or molecule repel other electrons because both are negatively charged, but by making the molecules move faster, this repulsion can be overcome and the orbits can overlap. The bonds that hold together the molecules of the substrates break and new bonds form. The minimum amount of energy needed to do this and so set a reaction going is called the **activation energy** (Fig. 1 overleaf).

Once the bonds have been broken, new bonds can be formed. In the case of burning sugar, molecules of carbon dioxide and water are made. These have much less energy stored in their molecules, so the excess energy is released as heat. This keeps the reaction going since it provides energy to activate other substrate molecules. However, this sudden release of heat energy would be bad news inside a living organism; reactions here need to be controlled.

Fig. 1 Activation energy

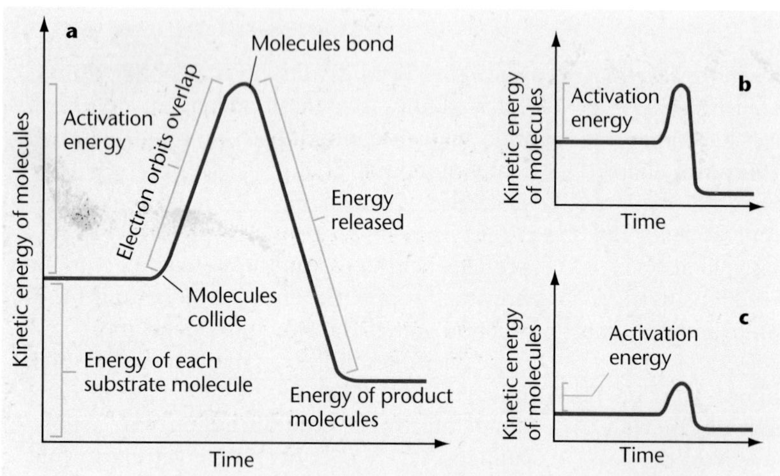

1 Would the substrate molecules in Fig. 1b react? Explain your answer.

2 The substrate molecules in Fig. 1c have less energy than the substrate molecules in Fig. 1b, but they still react. Explain why.

Lowering activation energy

Enzymes act as catalysts, lowering the activation energy by forming a complex with the substrate. This is described in more detail on page 63. Reactions catalysed by enzymes usually occur in stages, each with a lower activation energy than the 'total' reaction (Fig. 2). The cell can supply this much smaller amount of energy from ATP produced in respiration. This allows many reactions to take place easily at normal body temperature.

Fig. 2 Lowering activation energy

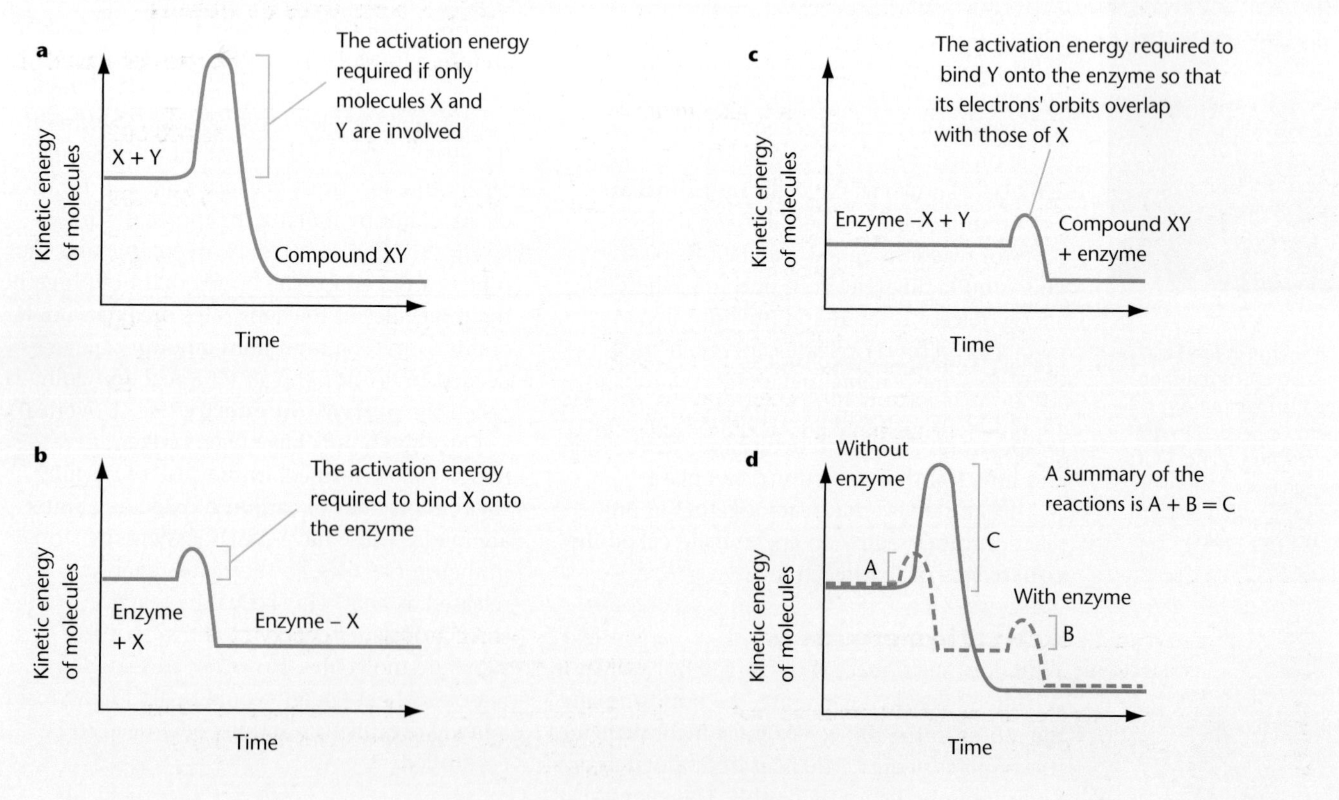

Fig. 3 The difference an enzyme makes

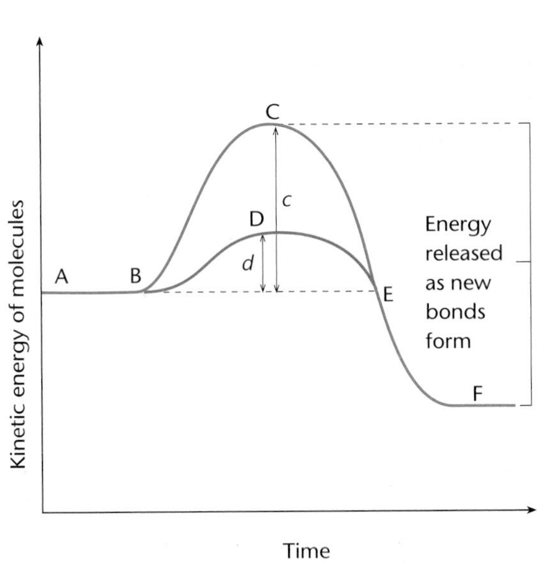

3 The graph in Fig. 3 shows the activation energies for the same reaction with and without an enzyme. Using the letters on the graph, indicate:

a the curve that shows the enzyme-controlled reaction

b the activation energy required when no enzyme is present

c the difference in activation energy with and without the enzyme.

Speedy enzymes

It is hard to appreciate how fast enzymes can work. Hydrogen peroxide is produced as a waste product in some cells but is highly toxic; so rapid removal is essential. The enzyme catalase breaks down hydrogen peroxide into water and oxygen. A single molecule of catalase can break down about 100 000 molecules of hydrogen peroxide per second, that's in less time than it takes to say 'one hundred thousand'.

Table 1 shows the **turnover number** of catalase and some other enzymes. This is the number of substrate molecules that can be acted upon by a single molecule of the enzyme in one minute.

Table 1 Enzymes in action

Enzyme	Turnover number
carbonic anhydrase	36 000 000
catalase	5 600 000
β-galactosidase	12 000
chymotrypsin	6 000
lysosome	60

key facts

- Enzymes are catalysts. They increase the rate of chemical reactions.
- Enzymes regulate the metabolic processes that occur in living cells.
- The energy needed to start a reaction is the activation energy.
- Enzymes lower the activation energy, so reactions can take place at the temperatures inside living organisms.

4.4 Enzymes are proteins

Enzymes are proteins. Proteins are important compounds in living organisms – not just as enzymes but also in many other ways. In Chapter 2 you learnt about the role of proteins in cell membranes. Proteins are also major structural components in tissues such as muscle, cartilage and bone, making them vital for growth. Haemoglobin, insulin and antibodies are all proteins and, since there are over 10 000 different proteins in the human body, you will come across many other examples.

The primary structure of proteins

How can proteins have such a range of different functions? Like polysaccharides, they are polymers. Protein monomers are amino acids and a protein molecule can contain hundreds or thousands of amino acid units. Unlike starch, which is made from only one type of monomer (glucose), proteins are made from 20 different naturally occurring amino acid monomer units. These can be assembled in any order, making an endless variety of protein structures possible. It is like having an alphabet of twenty letters to make up words hundreds of letters long.

All amino acids have an **amino group**, which is chemically basic. This means it reacts with and neutralises acids. They also have an acid group called the **carboxyl group**. Different amino acids have a different group attached to the central carbon atom. This group is referred to as the 'R' group (Table 2).

Table 2 Some amino acids

Amino acid	R group
Glycine	H
Alanine	CH_3
Cysteine	CH_2SH

Fig. 4 shows how two amino acids molecules join by a **condensation reaction**. This is the same type of reaction that joins the sugar monomers in polysaccharides and involves removal of water. The two amino acids link together by a peptide bond. The substance produced is a **dipeptide**. Notice how the acid group links up with the amino group. The result is that the dipeptide molecule has an acid group at one end and a basic amino group at the other. The long chains that eventually form the protein are made by adding more amino acids. Each protein has its own unique sequence of amino acids, known as the **primary structure** of the protein.

Fig. 4 Amino acids and dipeptides

A molecule of amino acid can be represented like this.

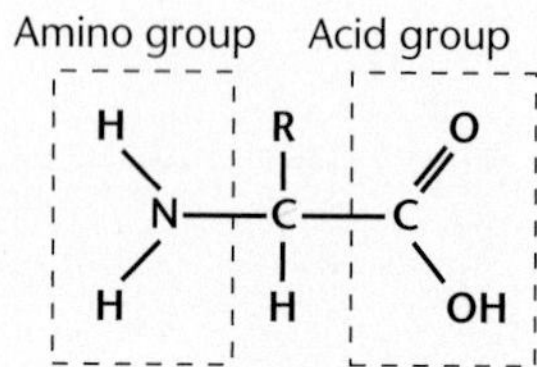

Two amino acid molecules join by a condensation reaction to form a dipeptide.

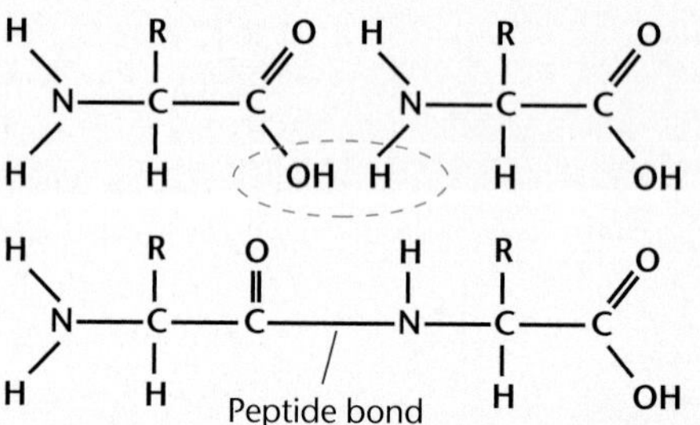

4 Look at Fig. 4 and answer the following questions.

a Which four elements are present in every amino acid?

b Which of these elements is not present in a carbohydrate or lipid?

c What is the formula of the amino group?

d What is the formula of the acid group?

e When the carboxyl group ionises in solution, which two ions are formed?

5 Draw the structural formula of a molecule of each of the three amino acids in Table 2.

6 Write an equation to show the formation of a dipeptide by condensation of alanine and glycine.

7 Draw the backbone chain of carbon and nitrogen atoms formed when three amino acid molecules condense. Label the peptide bonds.

Secondary, tertiary and quaternary structure

Many proteins consist of several short chains of amino acids. These chains are **polypeptides** and they can form coils or pleats. The coiling or pleating of a polypeptide chain is known as the **secondary structure** of a protein. The polypeptide chains often fold and are joined by weak chemical bonds that give a complex three-dimensional shape. This is the **tertiary structure** of a protein (Fig. 5).

Fig. 5 Primary, secondary and tertiary protein structure

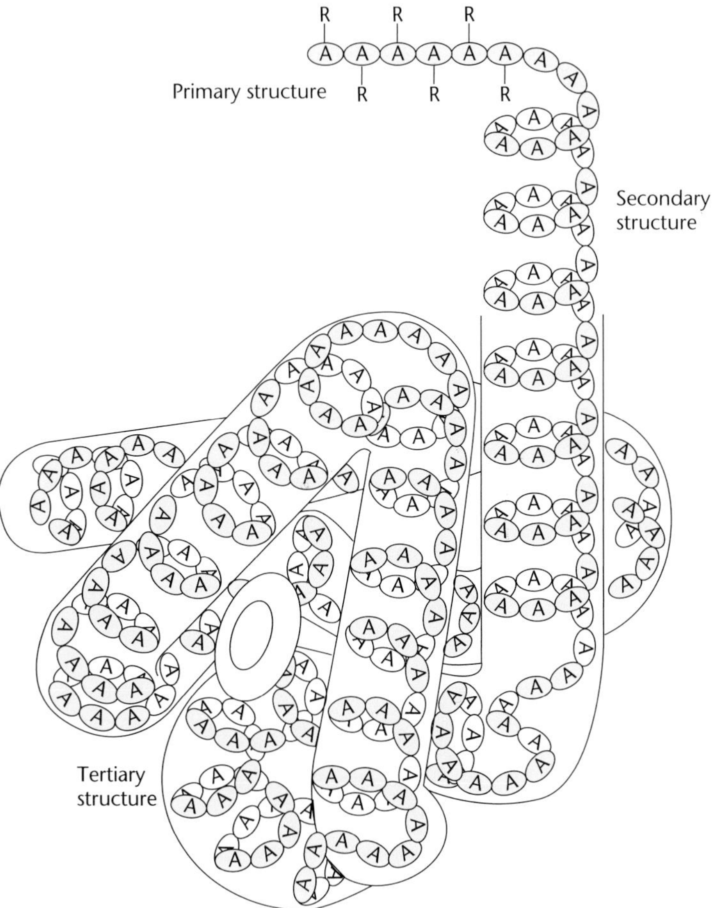

Some proteins, such as haemoglobin, have a fourth level of organisation, in which polypeptide chains and non-protein groups combine to give a **quaternary** structure (Fig. 6).

Fig. 6 Quaternary structure of haemoglobin

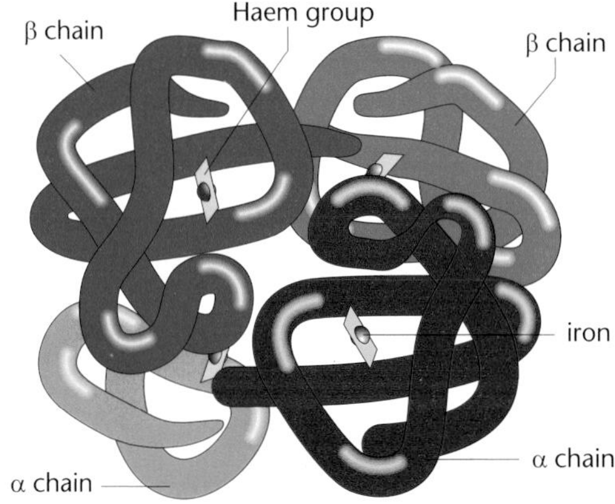

Haemoglobin has four polypeptide chains, each associated with an iron-containing non-protein part, called haem. Its shape, together with the iron in the haem, gives haemoglobin its special oxygen-absorbing properties.

The final shape of a protein molecule depends ultimately on its primary structure. The order of amino acids in a polypeptide chain determines how the molecule folds into its secondary and tertiary structure. Many of the connections that maintain the folds are hydrogen bonds, which occur where a hydrogen atom on one part of the polypeptide chain is attracted to an oxygen atom at another position on the chain (Fig. 7).

Fig. 7 Secondary protein structure

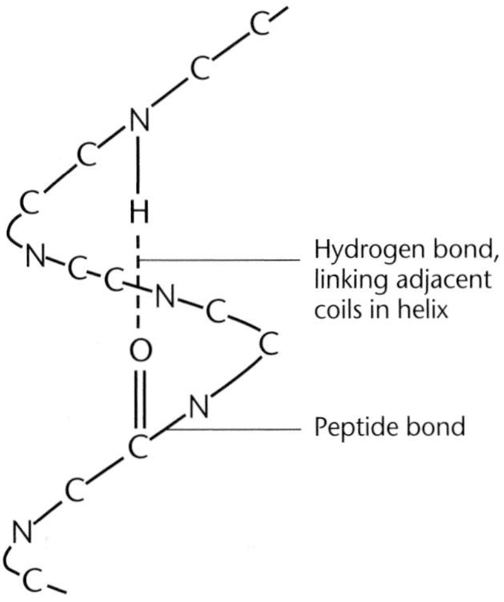

Other connections depend on the R groups. Two amino acids contain sulfur atoms, and these link together to form bonds that also determine the three-dimensional shape of a protein molecule.

There are two basic types of protein molecule: fibrous and globular.

Fibrous proteins

The structure of proteins relates to their function. In **fibrous proteins** such as keratin, which occurs in hair and nails, the chains of amino acids have a regular pattern of hydrogen bonds that cause them to coil into long helices that resemble thin springs. Three of these coils twist round each other, as in a rope. This tertiary structure makes the keratin strong and flexible and so well suited to its role.

Collagen, another fibrous structural protein, gives bone and cartilage their strength. Collagen molecules also have three twisted polypeptide chains, but these are more tightly bound than in keratin. This means that collagen can form more rigid structures. Often the molecules are grouped together to make almost rigid rods (Fig. 8).

This computer graphic shows a myoglobin molecule. Myoglobin is a globular protein that stores oxygen inside muscles. The shape is important because the haem that carries the oxygen must fit into exactly the right position. Notice the difference between the regular arrangement of collagen and the myoglobulin molecule shown here.

Fig. 8 Three-dimensional structure of fibrous proteins

Collagen fibres in the cornea of the eye are stacked in neat crosswise piles

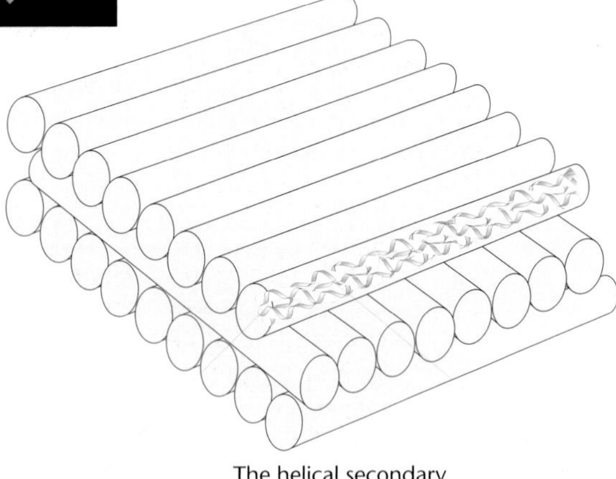

The helical secondary structure of collagen is typical of all fibrous proteins

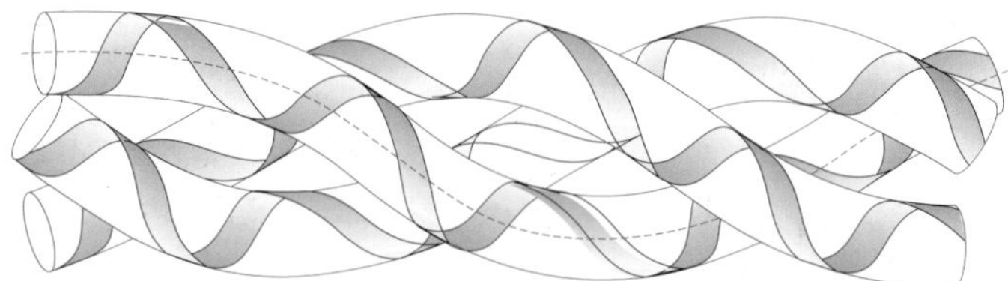

Polypeptide chain

8 In cartilage the collagen rods are arranged in many different directions. Suggest the advantage of this.

9 In the cornea collagen rods are stacked in neat piles. Suggest the advantage of this, bearing in mind that light enters the eye through the cornea.

Globular proteins

Enzymes are **globular proteins**, and their ability to function depends largely on their tertiary structure. Bonds between R groups at different points on the polypeptide chain make the molecule fold up into an almost spherical shape. This makes it easy for the molecules to move around inside cells or to be secreted through membranes. The exact three-dimensional shape of an enzyme molecule is very important. This is determined by the sequence of amino acids in the polypeptide chains that make up the enzyme (see Fig. 5) and so any change in the sequence can affect its function. If even one amino acid in the chain is changed, the 'correct' pattern of hydrogen bonds might not form to give the protein its proper shape and the enzyme may no longer interact with its substrates. The importance of the shape of the active site in an enzyme molecule is explained in the next section.

The biuret test for proteins

The substance is warmed gently in **biuret solution**. If protein is present a lilac/mauve colour will develop.

To answer these you may need to look back at the sections on cell membranes in Chapters 2 and 3.

10 Suggest why the proteins in cell membranes are globular rather than fibrous.

11 Explain why the carrier proteins used for facilitated diffusion have a variety of different shapes.

key facts

- Proteins are polymers of amino acids.
- Proteins contain nitrogen as well as carbon, hydrogen and oxygen.
- Amino acids have a basic amino group ($-NH_2$) and an acid carboxyl group ($-COOH$).
- Two amino acids combine by condensation to form a dipeptide.
- Coiling or pleating of the polypeptide chain produces the secondary structure.
- Folding produces the tertiary structure, and addition of other groups makes the quaternary structure.
- The precise shape of the protein molecules determines their function. Structural fibrous proteins have long twisted molecules; enzymes and carrier molecules are globular proteins with roughly spherical molecules.

4.5 How do enzymes work?

Enzyme molecules have a complex tertiary structure. The substrate molecules of the enzyme must be precisely the right shape to fit into part of the enzyme called the **active site**. The substrate molecules are attracted to the active site and form an **enzyme–substrate complex**. This complex exists for only a fraction of a second, during which time the products of the reaction form.

The traditional way of explaining how an enzyme and its substrates fit together was to use the analogy of a lock and key. The enzyme can be thought of as a lock into which the substrates, but no other molecules, fit. However biochemists now know that the active site changes shape as the substrate fits into it (Fig. 9). They have modified the **lock and key hypothesis** into the **induced fit hypothesis**.

Only the substrates for a reaction catalysed by a particular enzyme cause the changes in shape necessary for the enzyme to function.

When two substrate molecules are attracted to adjacent positions in the active site, the forces of attraction between them causes new bonds to form. Because the enzyme brings the two substrate molecules very close together, only small amounts of energy are needed for the two molecules to react, and the reaction needs much less energy than when uncatalysed. Once the product is formed it is released immediately, so the enzyme molecule is available for reuse.

Fig. 9 An enzyme in action

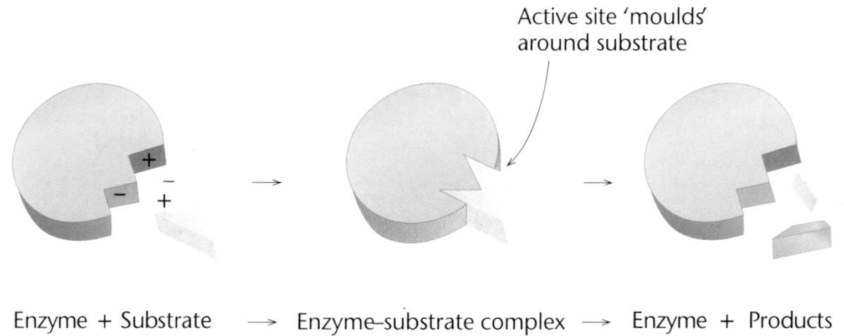

The induced fit hypothesis. Before substrate binding, the enzyme's active site is 'relaxed'. When the substrate binds, the active site is pulled into the correct shape by molecular interactions between the two molecules, and an enzyme–substrate complex forms. As the products fall away from the active site, the enzyme molecule becomes 'relaxed' again.

Enzymes are specific. A particular enzyme can catalyse only one chemical reaction. For example, catalase will only break down hydrogen peroxide. Some enzymes break only one type of bond – for example, peptidase breaks peptide bonds.

12 Use your knowledge of how enzymes work to explain why enzymes are specific.

13 hsw Explain why the 'induced fit' hypothesis of enzyme action is better than the 'lock and key' hypothesis.

key facts

- Enzyme molecules have a very specific shape, provided by the tertiary structure of the protein.
- The molecules of the reacting substances – the substrate – are attracted to a particular part of the enzyme molecule, called the active site. An enzyme–substrate complex is formed, the enzyme molecule is distorted and the substrates react.
- Activation energy is low in a reaction catalysed by an enzyme because little energy is needed to bring the two substrate molecules together.
- The products of the reaction are released rapidly, so the enzyme can be reused immediately.

Temperature and enzyme activity

Enzyme activity is affected by temperature (Fig. 10).

As the temperature rises, the rate of a chemical reaction normally increases. This is because the molecules gain kinetic energy, so they move faster, collide more often and the collisions are more likely to lead to a reaction. This also holds true for reactions controlled by enzymes because there is an increased chance of substrate molecules colliding with the active site. However, as temperatures rise, the atoms within the enzyme molecules also gain energy and vibrate so rapidly that the weak bonds that maintain the tertiary structure of the protein molecule break and the molecule can unravel. Since the shape of the active site in an enzyme molecule is crucial for it to work, any change in shape will inactivate the enzyme. Once broken, the hydrogen bonds do not re-form in their original positions. So, even when the temperature falls, the enzyme cannot regain its functional shape. We say the enzyme is denatured. Note that you should never say that enzymes are 'killed', since they were never 'alive' in the first place.

Most enzymes in the human body are denatured by temperatures above about 45°C. They work fastest just below this, at about 40°C; this is therefore their optimum temperature.

Human body temperature is about 37°C, but do not think that all the body's enzymes have an optimum at that temperature. Many work fastest at about 40°C. However, if our temperature does rise to this level, we feel ill and our metabolic pathways are no longer co-ordinated properly.

Many organisms have enzymes that work perfectly at much higher or lower temperatures. Some fish live in the Antarctic sea, where the water temperature never rises above 2°C. They are active at this temperature, but die if the water warms by more than a few degrees. Worms and bacteria can live near volcanic vents in the ocean and close to hot springs in Iceland, where the temperatures are close to boiling point. Many of the enzymes from these specialised organisms are proving useful in the manufacturing industry.

Fig. 10 Enzymes and temperature

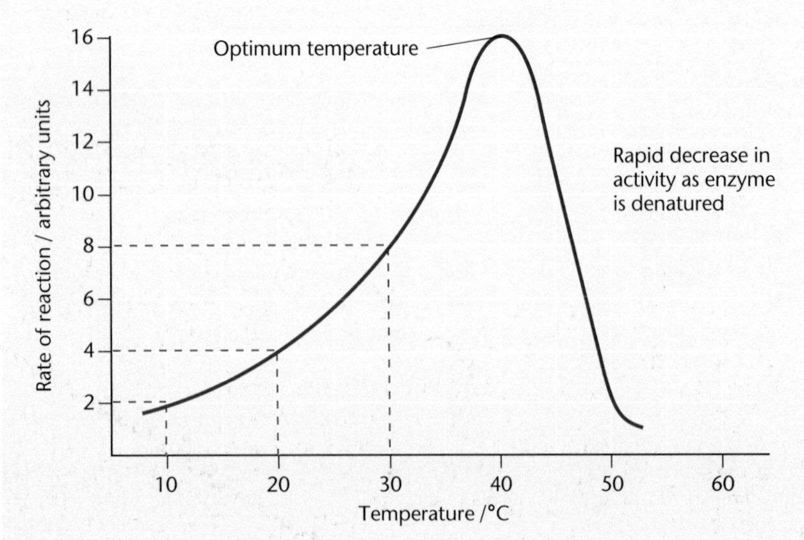

14 Explain why the rate of reaction is so slow below 10°C.

15 From Fig. 10, what are the rates of reaction at 10°C, 20°C and 30°C?

16 What do you notice about the effect of increasing the temperature by 10°C?

17 What would be the rate of reaction if some of the enzyme and its substrates were frozen and then warmed to 40°C?

Fig. 11 Effect of temperature on a protease

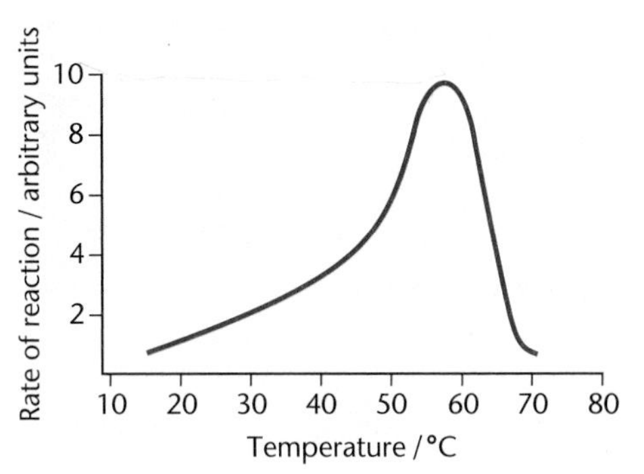

18 The graph in Fig. 11 shows the effect of temperature on a protease used in washing powders.

a What is the optimum temperature of this protease?

b Suggest the advantage of using a protease with this optimum temperature in a washing powder.

19 Proteases break down protein molecules into amino acids. Explain how proteases in washing powders help to remove protein stains such as blood or egg from clothes.

pH and enzyme activity

As well as having an optimum temperature, enzymes also have an optimum pH. Most enzymes are denatured in solutions that are strongly acidic or strongly alkaline. This is because the **hydrogen ions** (H^+) in an acid or the **hydroxyl ions** (OH^-) in an alkali are attracted to the charges on the amino acids in the polypeptide chains that make up the enzyme. They interact with the amino acids and disrupt the bonds that maintain the molecule's three-dimensional shape, destroying the active site of the enzyme. Once the original structure is lost, it cannot reform and the enzyme no longer binds to its substrates.

hsw how science works

Extremes of pH rarely occur inside the cells of living organisms. Most enzymes have an activity curve similar to curve A in the graph below, with an optimum somewhere near neutral pH. However, some enzymes work best at extreme pH values. For example, proteases in the acid environment of the stomach have an optimum pH of about 2 and are denatured above pH 6.

Curve B in the graph below shows the effect of pH on an enzyme that has been extracted from the marine shipworm, a mollusc that can bore into woodwork. This enzyme is a protease and is very good at removing the protein that accumulates on contact lenses. It has an advantage over other proteases used for this purpose because it remains active in the presence of the slightly alkaline solution of hydrogen peroxide that is used to disinfect lenses. A single solution that can cleanse and sterilise at the same time can thus be used in place of the conventional two solutions.

20 Sketch a graph to show the effect of pH on the activity of stomach protease.

21 Use the graph to explain why the protease obtained from the marine shipworm is better for cleaning contact lenses than a protease with a curve like curve A.

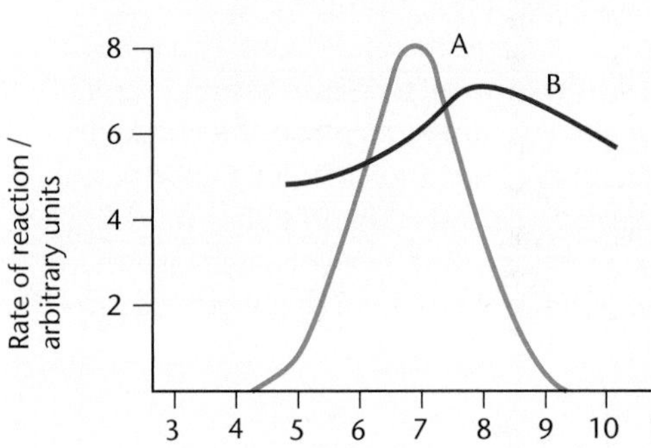

The marine shipworm *Spirorbis borgalis*

Concentration and enzyme activity

Enzymes work by forming enzyme–substrate complexes. The rate of reaction depends on the number of substrate molecules that bind with enzyme molecules in any given time. Clearly, the more enzyme molecules in the solution, the greater the chance of a substrate molecule finding an active site, and the faster the rate of reaction. Similarly, the more substrate molecules in a solution, the greater the chance that an enzyme–substrate complex will form, and the greater the rate of reaction.

These questions refer to Fig. 12.

22 Explain in terms of active sites why the curve in Fig. 12a flattens out.

23 What do the letters X and Y in Fig. 12b show?

24 What might increase the rate of the reaction at position Z in graph b?

25 Explain one set of conditions in which no increase would occur.

Fig. 12 Enzyme and substrate concentrations

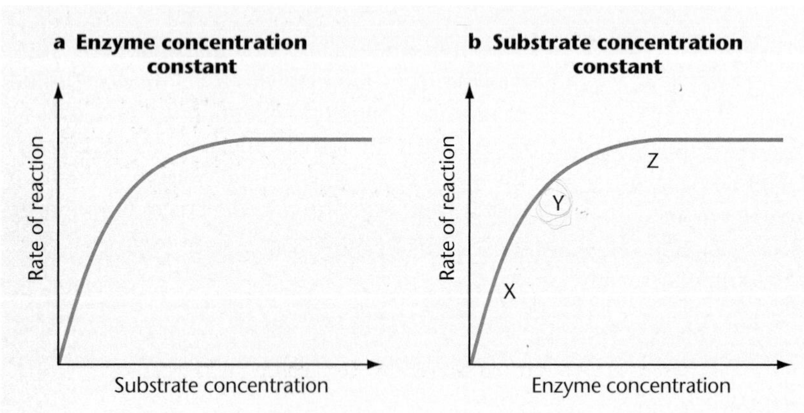

Fig. 13 Competitive and non-competitive inhibition

a Competitive inhibition

The usual substrate molecule can form a complex with the enzyme

A competitive inhibitor molecule can also form a complex with the enzyme, preventing the usual reaction from taking place. The substrate molecule and the inhibitor molecule compete for the active site

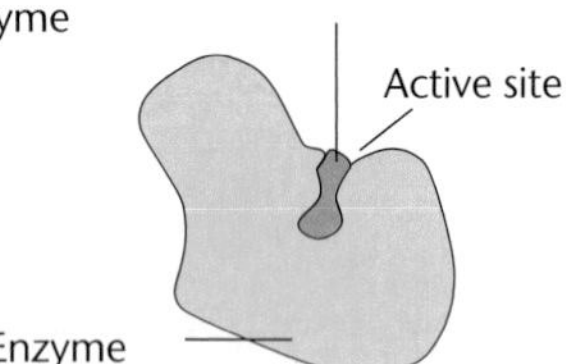

b Non-competitive inhibition

The substrate can form a complex with the enzyme

The substrate molecule is unable to form a complex with the enzyme

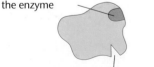

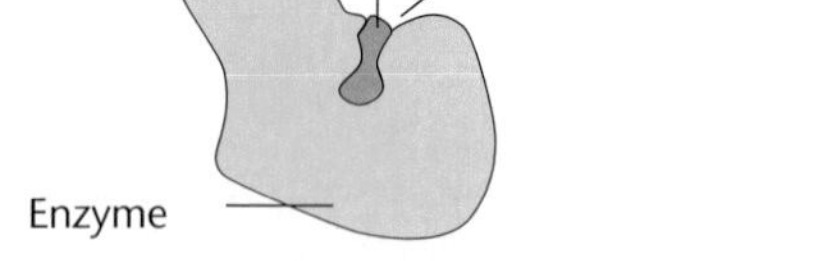

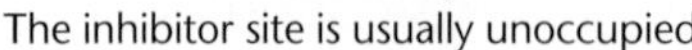

The inhibitor site is usually unoccupied

Competitive inhibition

Many enzymes have high substrate specificity. This means that only molecules of one particular substrate can attach to the active site. However, sometimes molecules with a very similar shape to the usual substrate attach to the active site. If such molecules are present in the same solution as the normal substrate they *compete* for the active site. They are called **competitive inhibitors** (Fig. 13a).

The molecules of the competitive inhibitor block the active site, so fewer molecules of the enzyme are available for the normal reaction and the reaction rate is reduced. The degree of inhibition depends on the relative concentrations of the substrate and the competitive inhibitor. The inhibitor does not attach permanently to the active site, nor does it damage the site. Therefore, if the concentration of the normal substrate is increased, its molecules compete more successfully for a place in the active site of the enzyme and the rate of the reaction increases.

Non-competitive inhibition

Some substances inhibit enzyme reactions in a different way. Their molecules do not attach to the active site, but to a different part of the enzyme. This alters the shape of the enzyme molecule so that the substrate can no longer bind to the active site. These substances are called **non-competitive inhibitors** (Fig. 13b). The shape of a non-competitive inhibitor molecule may be completely different from that of the usual substrate. The inhibitor may remain permanently attached to the enzyme, so the enzyme is effectively destroyed. Increasing the amount of substrate present does not help. Some non-competitive inhibitors can affect many enzymes. This is why heavy metal ions such as mercury, lead and arsenic are so poisonous; they prevent many of the body's metabolic reactions taking place. Others are more specific and they can be used as insecticides.

Fig. 14 Succinate and malonate

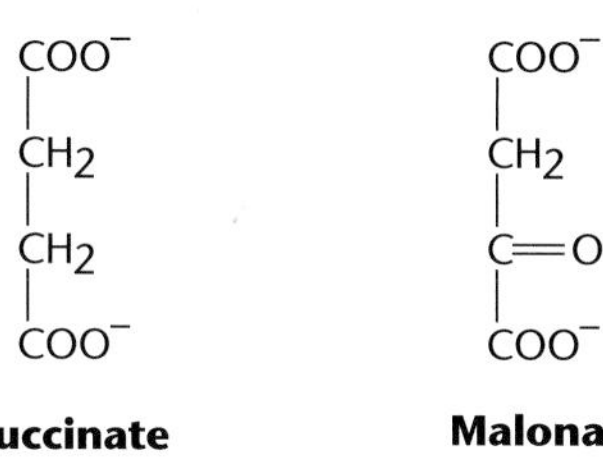

26 Fig. 14 shows the structure of succinate and malonate. Succinate dehydrogenase is an enzyme involved in the metabolic pathway of respiration. It removes hydrogen from succinate.

a Explain why malonate is a competitive inhibitor of succinate dehydrogenase.

b Molecules of non-competitive inhibitors are quite different from the molecules of the enzyme's substrate. Explain why this might be.

27 Some non-competitive inhibitors can be used as pesticides. What factors would have to be considered before using one as a pesticide for general use?

key facts

- Temperature, pH, inhibitors, enzyme concentration and substrate concentration are all factors that affect the rate of enzyme-controlled reactions.
- Rising temperature increases the rate of reaction, up to an optimum.
- High temperatures denature enzymes, thus destroying the active site.
- Most enzymes are denatured by strongly acid or alkaline conditions.
- Increasing the concentration of enzyme or substrate increases the rate of reaction, up to a maximum when all the active sites are occupied.
- Competitive inhibitors block the active site.
- Non-competitive inhibitors change the shape of the active site and prevent substrate binding.

4.6 Carbohydrates

All **carbohydrates** contain the elements carbon, hydrogen and oxygen. The hydrogen and oxygen atoms are always in the ratio 2:1, giving carbohydrates the general formula CH_2O.

There are three basic types of carbohydrate molecule: **monosaccharides**, **disaccharides** and **polysaccharides**. Monosaccharides are single sugars, small molecules that dissolve in water and taste sweet. These are the units from which all larger carbohydrates are made.

28 How is the composition of a carbohydrate different from that of a lipid?

Fig. 15 Glucose and fructose

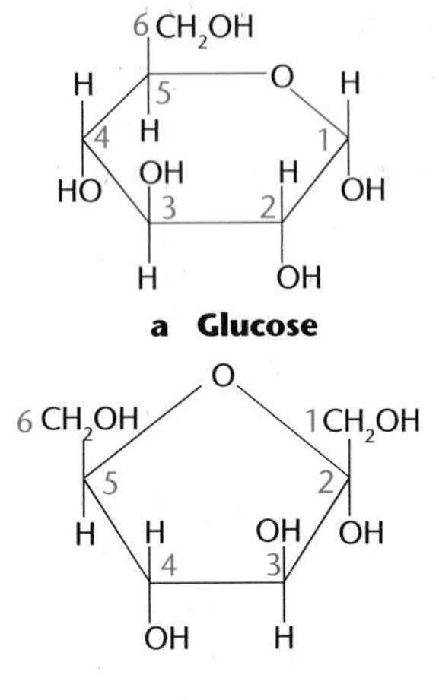

a Glucose

b Fructose

Glucose is a monosaccharide. It has the formula $C_6H_{12}O_6$; its structural formula is shown in Fig. 15a. Five of the carbon atoms (numbered 1–5 in the diagram) form a ring. Fructose, another monosaccharide is shown in Fig. 15b. Glucose and fructose are **isomers** – they have the same number of carbon, hydrogen and oxygen atoms but these are arranged differently. This gives the fructose molecule a different shape – and therefore different properties. Honey is a good source of fructose.

Disaccharides form when a condensation reaction joins two monosaccharides. Two glucose molecules combine to form maltose; glucose and fructose combine to form sucrose (Fig. 16). Sucrose is the most common sugar in plants and is also the sugar you probably put in your tea or coffee.

Glucose and galactose combine in the same way to form the disaccharide lactose. Lactose is commonly known as milk sugar.

Since sugars are small molecules that are soluble in water, they are easy to transport to different parts of an organism. The energy stored in their molecules can be used to power other essential chemical reactions. Sugars are also converted into glycoproteins, which are required for growth and repair of organs.

Polysaccharides are giant molecules made up from many single sugar molecules joined together by condensation reactions. Starch is a polysaccharide. Part of a starch molecule is shown in Fig. 17. Starch molecules are compact, coiled and branched, making them ideal 'energy' stores.

Fig. 16 The formation of maltose and sucrose

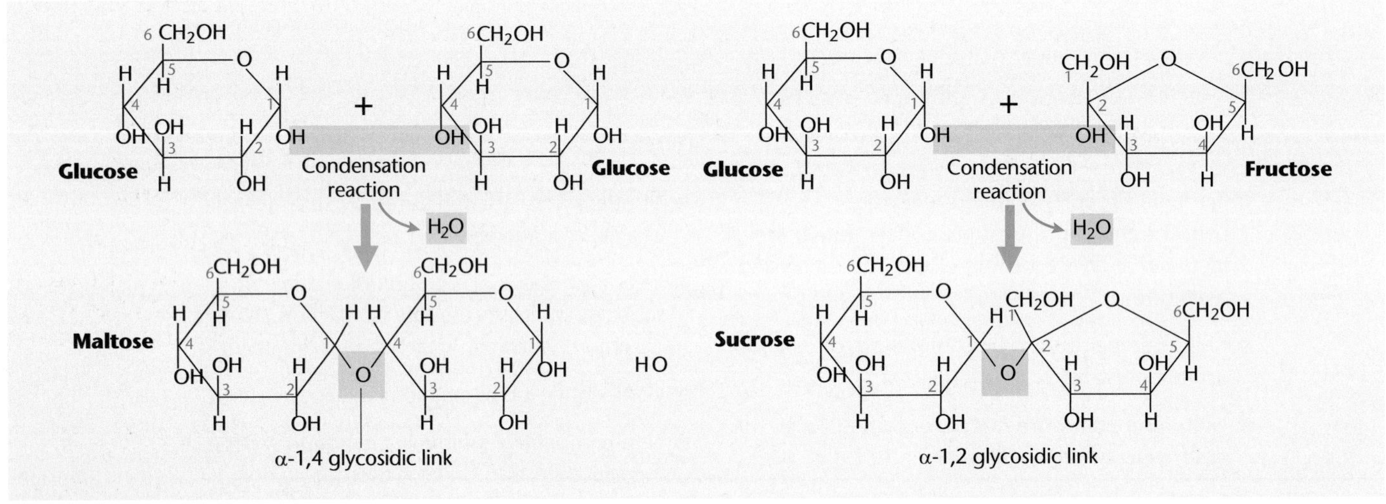

Fig. 17 Starch

Part of a starch molecule

Glucose monomers

Branches

Main chain

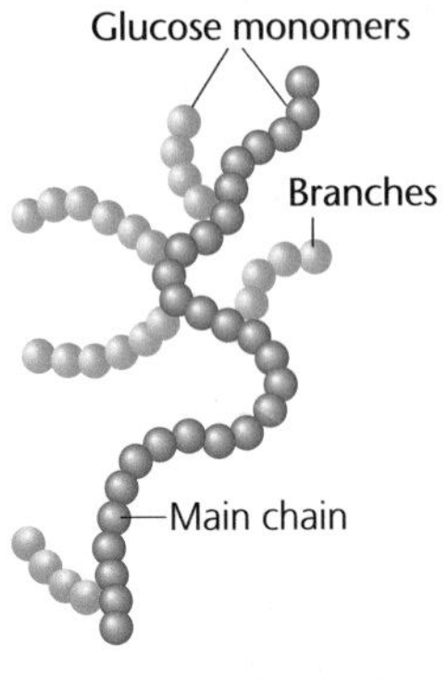

Monomers and polymers

'Mono' and 'poly' come from Greek words; mono means one, and poly many. These are used in many biological and chemical names.

Starch is a **polymer** – a molecule made up of repeating units, rather like links in a chain. The individual units are called **monomers**. The same monomer can be used to build different kinds of chains – long, short, straight, branched. For example, glucose is linked in different ways in the polymers starch and cellulose found in plants and the polymer **glycogen** in animals.

The glucose monomers in starch (called α-glucose) are joined by **α-glycosidic** bonds. These bonds produce twisted chains of monomers that form branched molecules. The coiled and branched chains of starch molecules (Fig. 17) give them a compact shape, which, together with their insolubility in water, makes them ideal 'energy' storage compounds. Starch is the major storage carbohydrate in plants and thus the major carbohydrate in our diet.

Biochemical tests for carbohydrates

To test for starch, add **iodine** or **potassium iodide solution** to the substance. A blue–black colour indicates the presence of starch.

Tests for sugars can distinguish two groups of sugars: the **reducing sugars** and the **non-reducing sugars**. Reducing sugars such as the monosaccharides glucose and fructose are readily oxidised. When this reaction occurs the sugars lose electrons to another substance, which is said to be *reduced*. Some disaccharides, notably sucrose, are not readily oxidised and so do not reduce other substances. For this reason they are termed non-reducing sugars.

The test for a reducing sugar is to add **Benedict's solution** to the substance, then heat the mixture in a water bath. An orange–red precipitate indicates the presence of a reducing sugar.

The test for a non-reducing sugar is to boil the substance with dilute acid, then neutralise by adding hydrogencarbonate. The mixture is then tested with Benedict's solution as above.

What is the difference between reducing and non-reducing sugars?

Look at the structure of glucose (Fig. 15a). You will see that the carbon atoms 1 and 5 are linked by an oxygen atom. In the presence of an oxidising agent – a substance that readily releases oxygen or takes up electrons – this link is broken. The –CHO group is oxidised to form the carboxyl acid group, –COOH. Reactions like this are called redox reactions (see Fig. 18). In any **redox reaction**, one substance is oxidised and loses electrons while another is reduced and gains electrons. The sugar acts as a reducing agent because it takes oxygen away from the oxidising agent. Such reactions are common in respiration and photosynthesis.

When two glucose molecules combine by a condensation reaction to form maltose, the glycosidic bond C–O–C, is formed between carbon atoms 1 and 4, as shown in Fig. 18. As you can see, there is still a carbon atom at one end of the molecule, which is free to form an exposed –CHO group. However, when glucose combines with fructose to make sucrose, carbon atom 1 of the glucose molecule becomes 'buried' inside the sucrose molecule, and is no longer free to produce a –CHO group. Sucrose is therefore a non-reducing sugar, whereas maltose, although a disaccharide, is still able to act as a reducing sugar.

Benedict's solution contains copper sulfate. In solution, copper sulfate has copper(II) ions, Cu^{2+}. Because glucose is a reducing sugar, it releases electrons to the copper(II) ions. These gain electrons, and copper(I) ions, Cu^{+}, are formed.

$$Cu^{2+} + e^{-} \rightarrow Cu^{+}$$

Copper(I) oxide is insoluble and produces a red precipitate. In the blue solution of copper sulfate, small quantities of precipitate can look yellow, or even green.

29 Explain why the Benedict's test works with maltose but not with sucrose.

30 Boiling a disaccharide with acid breaks it down into monosaccharides. Explain why Benedict's test gives a positive result after sucrose has been treated with acid.

31 For low concentrations of glucose, a positive Benedict's test ranges from green through yellow to orange and brick red. By comparing the colour of the sample solution with the colour of a standard solution, the glucose concentration of the sample can be estimated. Explain how you would prepare standard solutions for comparison.

Fig. 18 Redox reaction

A redox reaction

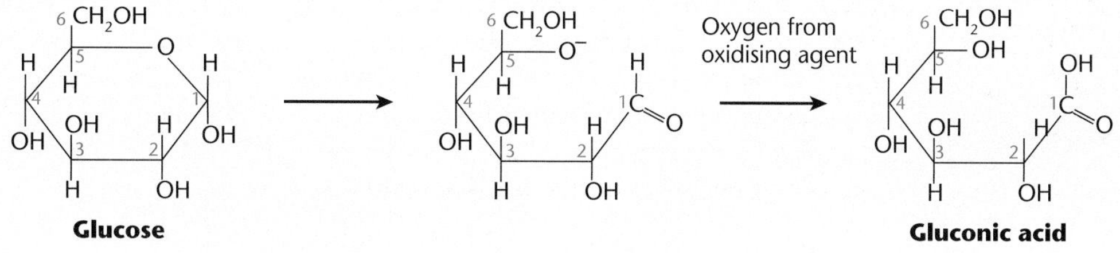

key facts

- The three classes of carbohydrate are the monosaccharides, the disaccharides and the polysaccharides.
- Monosaccharides (single sugars) combine to form di- or polysaccharides by condensation reactions.
- Monosaccharides and disaccharides are sugars – small, soluble, diffusible molecules that are easily transported around organisms. They can be used to release energy and to build other molecules.
- Polysaccharides (carbohydrate polymers) include starch.
- The properties of these polymers depend on the shape of their molecules. Starch molecules have compact, coiled and branched molecules, making them ideal 'energy' stores.

4.7 Digestion of carbohydrates

Food compounds such as starch, proteins and lipids are large, insoluble molecules that cannot be absorbed directly into the blood. They must first be digested into smaller molecules.

Digestion breaks down polymers into the monomers or smaller molecules of which they are made. The reaction that splits polymers is the reverse of the condensation reaction that joined them together. Condensation involves the removal of water; digestion involves the addition of water. The reaction involved in digestion is therefore termed **hydrolysis,** which means 'water splitting'. The hydrolysis reaction can be illustrated by looking at how the peptide bonds between the amino acids are broken when a protein is digested.

Fig. 19 Hydrolysis of a dipeptide

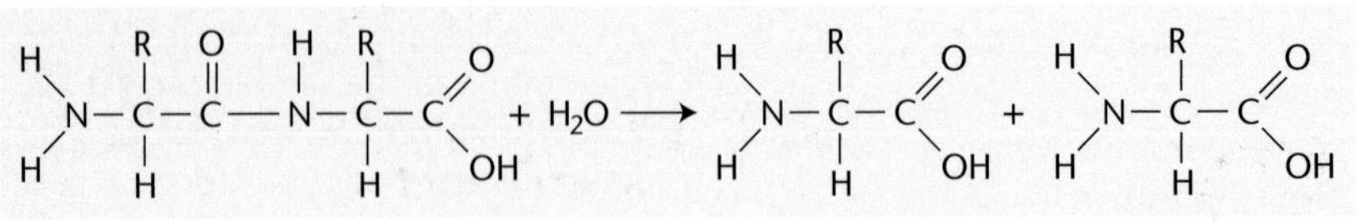

$$H_2N{-}CHR{-}CO{-}NH{-}CHR{-}COOH + H_2O \longrightarrow H_2N{-}CHR{-}COOH + H_2N{-}CHR{-}COOH$$

32 Draw a diagram to show the hydrolysis of the disaccharide, maltose. Refer to the diagram of condensation in Fig. 16 on page 68 of this chapter to help you.

The human gut

Digestion is catalysed by enzymes in all living organisms. Humans, like most other animals, secrete enzymes into the **lumen**, the central cavity of the gut where digestion takes place.

As food travels along the gut it passes through several different regions. Each region is adapted to carry out a specific function:

- to break large lumps of food into smaller lumps, a process called **mechanical digestion**
- to break down the food compounds by hydrolysis, a process called **chemical digestion**
- to absorb the soluble products of digestion;
- to get rid of undigested materials and other waste products.

Food enters the gut by the mouth. In the mouth cavity, chewing starts the process of mechanical digestion. Some chemical digestion also begins here since the saliva contains an enzyme that breaks down starch. Most of the time though, people don't chew their food for long enough for this to have much effect. The saliva softens and lubricates the food so that it can be swallowed easily. From the mouth, the food passes down the oesophagus into the stomach. The rest of the gut is in effect a long tube with a variable diameter. The muscles of the tube contract and relax in a regular and constant succession of waves. This process, called **peristalsis,** moves food steadily along through the gut.

Fig. 20 The digestive system

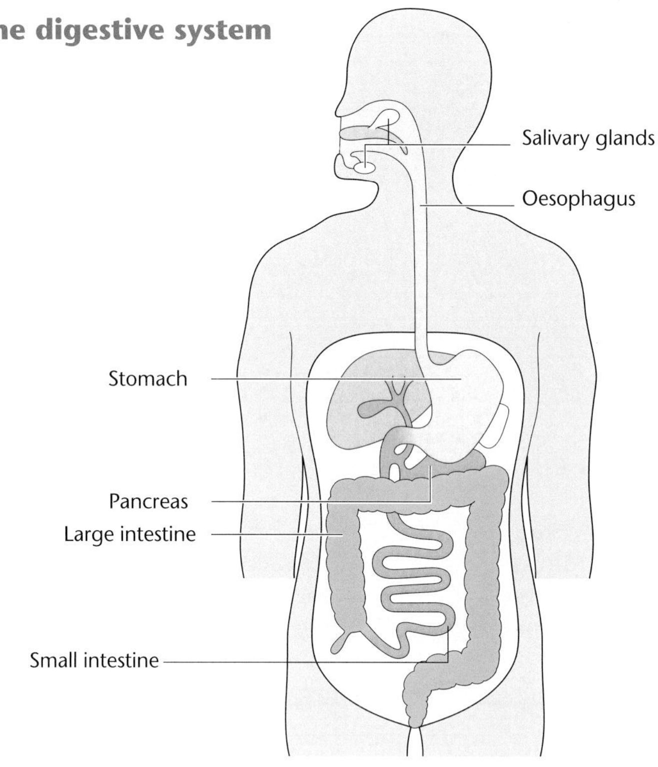

Digestive enzymes hydrolyse carbohydrates, lipids and proteins. Glands in the small intestine secrete digestive enzymes directly into the lumen of the gut. Other glands, such as the **salivary glands** and the **pancreas,** secrete their digestive enzymes through tubes called **ducts**. Whatever method is used to deliver the enzymes, the digestion is **extracellular**, because it happens in the lumen, outside the cells of the gut wall. In the small intestine, many of the enzymes secreted by cells in the gut wall stay attached to the cell membranes. Here digestion occurs very close to the surface of the gut wall rather than in the lumen. The enzymes are released into the lumen only when cells in the mucosa are broken down by the scouring effect of food passing along the intestine.

An electron micrograph showing the microvilli from an intestinal epithelial cell.

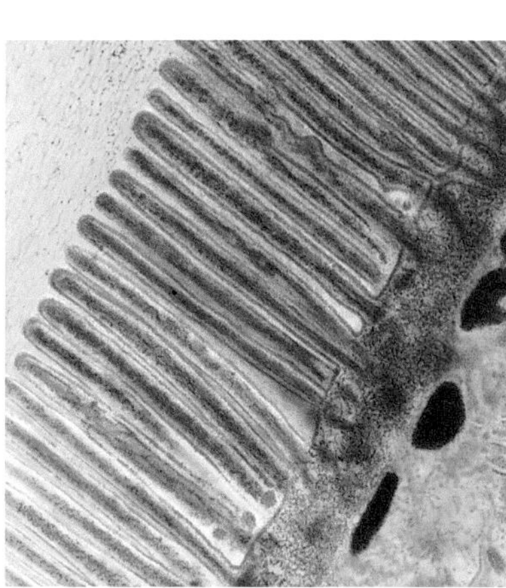

Starch is the main carbohydrate in our diet because it is such a common storage compound in plants. We eat small amounts of glycogen – animals use this polymer as their main carbohydrate storage compound, but they store only relatively small quantities in their liver and muscles. Another common carbohydrate polymer is cellulose, but humans do not have an enzyme to digest it. Cellulose therefore forms the bulk of the **fibre** in our diet, which is necessary to maintain a healthy colon and bowel.

The enzyme **amylase** breaks down starch into the disaccharide, **maltose**. Some digestion of starch occurs in the stomach as a result of the continuing action of the amylase that was added to food in the mouth. However, this tends to be short-lived as amylase is inactivated rapidly by the acid in the stomach. No further digestion of starch takes place until the food reaches the small intestine, where pancreatic amylase completes the process.

33 Explain how amylase is inactivated in the stomach.

The epithelial cells lining the small intestine have huge numbers of very thin, finger-like projections on their surface, called **microvilli** (Fig. 21).The membranes of these microvilli contain the enzymes that break down disaccharides into monosaccharides. Maltase, for example, digests maltose into glucose, which passes immediately into the cytoplasm of the

Fig. 21 An intestinal epithelial cell

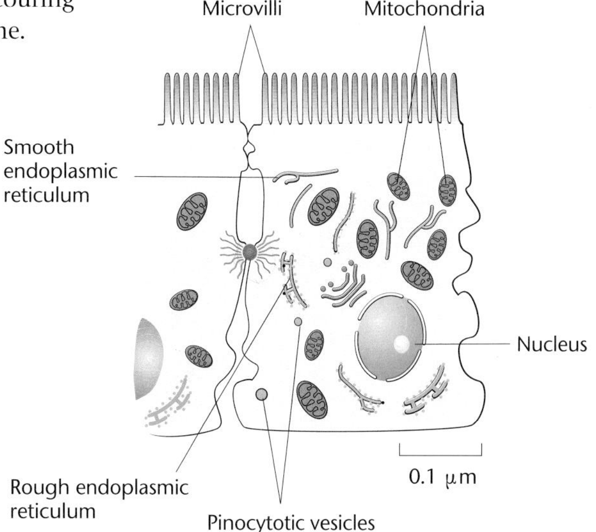

Fig. 22 Carbohydrate digestion

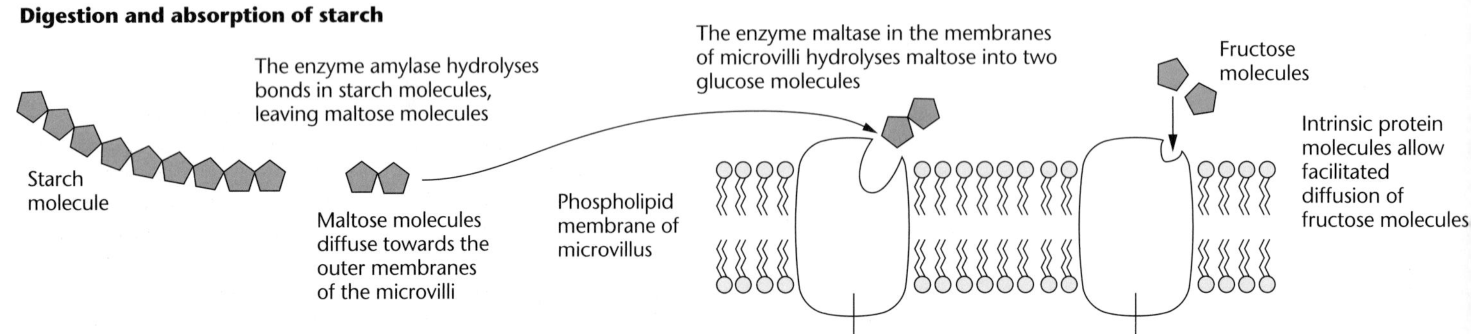
Digestion and absorption of starch
The enzyme amylase hydrolyses bonds in starch molecules, leaving maltose molecules
Starch molecule
Maltose molecules diffuse towards the outer membranes of the microvilli
Phospholipid membrane of microvillus
The enzyme maltase in the membranes of microvilli hydrolyses maltose into two glucose molecules
Fructose molecules
Intrinsic protein molecules allow facilitated diffusion of fructose molecules
Glucose molecules pass into the cytoplasm of the cells of the small intestine by active transport

nearby epithelial cells (Fig. 22). Carbohydrate digestion is complete by the time the gut contents have passed through the **small intestine**.

As digestion is completed, the products are absorbed from the small intestine into the blood or lymph systems. The duodenum and ileum are well adapted for absorption. The surface area through which the products of digestion can pass out of the lumen is huge – it covers about 350 m^2, over three times the floor area of the typical living room or bedroom. The large surface area is due to:

- the length of the gut, it is about 6 metres long and is coiled up inside the abdomen
- the large folds in the inner wall
- the huge numbers of projecting villi that cover these folds
- the extra surface area of the microvilli on the surface of the gut epithelial cells.

Simple sugars and amino acids enter the epithelial cells by active transport and facilitated diffusion. The carrier protein that transports glucose through the membrane also carries sodium ions. The passage of glucose is much more rapid when the concentration of sodium ions in the lumen is relatively high. This process is known as co-transport (see Chapter 3).

Lactose intolerance

Milk contains a disaccharide called **lactose**, or 'milk sugar'. Human milk is about 7% lactose and so is quite sweet. Babies are able to produce an enzyme called lactase in their intestine. Lactase digests lactose into two monosaccharides, glucose and **galactose**. Like maltase, lactase is bound to the membrane of the microvilli in the epithelial cells. In most mammals, the lactase degenerates when the baby is old enough to start eating solid food. In many humans, however, the enzyme persists and is still present in adults, although the amount produced often declines with age. In some people, most of the lactase disappears from the intestine at an early age. As a result they experience **lactose intolerance**. This condition is particularly common in people of oriental descent. They cannot digest lactose and this milk sugar passes intact from the ileum into the colon. Here, the excess sugar reduces the uptake of water by the body and tends to cause diarrhoea. Also bacteria feed on the lactose, producing acids, which irritate the bowel, causing excessive wind and abdominal pain. People who are lactose intolerant need to avoid milk and many dairy products, although natural yoghurt can usually be tolerated because the lactase produced by the bacteria in it pre-digest the lactose before it is eaten.

34 Explain why milk sugar is not digested by sucrase, which breaks down cane sugar.

35 What evidence suggests that lactose intolerance is inherited?

hsw

36 Use your knowledge of osmosis to explain why excess lactose in the colon causes diarrhoea.

how science works

Measuring enzyme activity

The activity of a digestive enzyme can be determined in two ways:

- by measuring the quantity of products
- by measuring how much substrate is used up in a given time.

One convenient technique for measuring the activity of the starch-digesting enzyme, amylase, is to use starch agar plates. Starch agar is made by adding starch to liquefied agar jelly. The molten starch agar is poured into a Petri dish and allowed to set. Samples to be tested are placed in cavities cut into the agar, or solid samples can be placed on the surface (Fig. 23).

Fig. 23 Starch agar plates

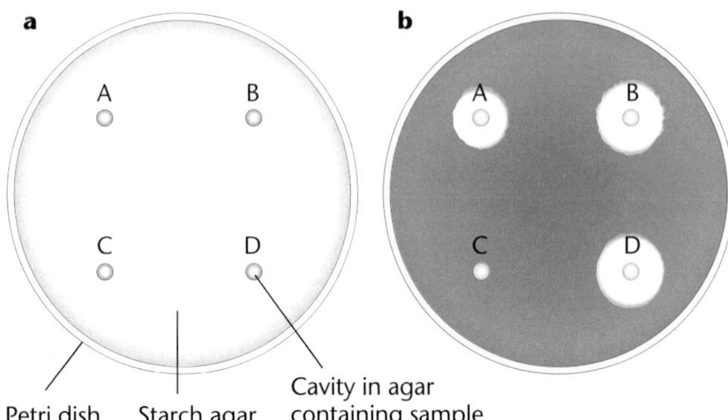

Cavity A contains the standard sample; B, C and D contain test samples

After several hours, the surface of the agar plate is covered with iodine solution. Areas that contain starch stain blue–black, whereas areas in which the starch has been digested remain clear. The size of a clear area can be used as a measure of the concentration of amylase in the sample. The larger the diameter of the clear area, the further the amylase must have diffused from the sample. This is because the higher the concentration in the sample, the steeper the diffusion gradient.

This technique, in which the quantity of a substance is found by comparing its activity with a standard sample, is called an **assay**. A similar technique can be used to measure the activity of other enzymes. For example, white protein powder can be suspended in the agar. Protein-digesting enzymes make the milky-white agar turn clear.

37 Look at the clear areas in Fig. 23.

a What can you conclude about the samples?

b What substance would you expect to find in the clear areas of agar?

38 A manufacturer wants to test three strains of a fungus as possible sources of amylase. Describe how starch agar plates could be used to find the strain that produces most amylase.

key facts

- Amylases break down starch into the disaccharide, maltose. Maltase then breaks maltose into glucose.
- The amylase produced by the salivary glands starts the digestion of starch in the mouth. However, most of the starch in food is digested in the small intestine.
- Enzymes in the microvilli of the intestinal epithelium complete the digestion of disaccharides to monosaccharides.
- Monosaccharides are absorbed into the blood capillaries in the villi of the small intestine.
- Monosaccharides enter the epithelial cells by active transport and facilitated diffusion. The process is speeded up by a process called co-transport involving sodium ions. The molecules are transferred to the blood capillaries by active transport.
- Lactose intolerance is caused by the inability to produce lactase to hydrolyse lactose in the diet. The lactose causes water to enter the lumen of the intestine via osmosis, causing diarrhoea.

examination questions

1 Liver contains the enzyme catalase. This enzyme catalyses the breakdown of hydrogen peroxide into water and oxygen.

$$2H_2O_2 \rightarrow 2H_2O + O_2$$

A 1 g piece of liver was dropped into a beaker containing 50 cm^3 of hydrogen peroxide. The loss of mass by the beaker and its contents was measured for the next 15 minutes. The graph shows the results of this experiment.

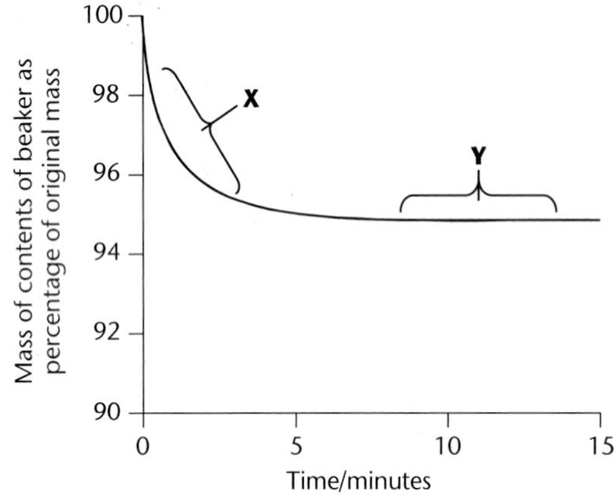

a Why was there a loss in mass during this experiment? (1)

b **i** Describe the most suitable control for this experiment. (2)

ii Explain why this control is necessary (2)

c Describe the relationship between the rate of this reaction and time. (2)

d Explain the reason for the shape of the curve at

i X

ii Y (4)

e Hydrogen peroxide decomposes slowly at room temperature, but very rapidly if catalase is added. Explain why in terms of activation energy. (2)

Total 13

2 The diagram shows part of the alimentary canal.

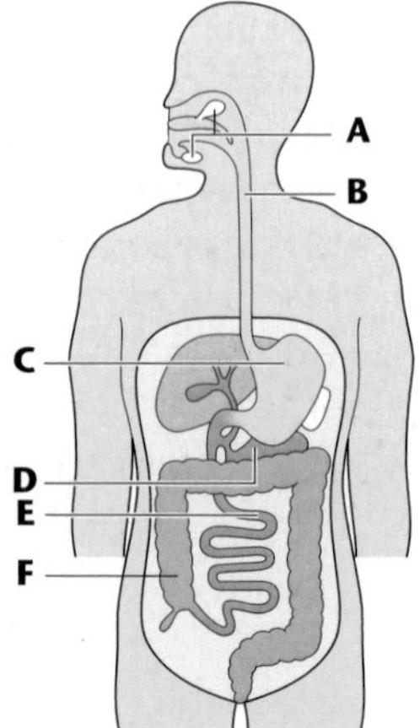

a Which of the organs, A-F

i produce amylase? (2)

ii produces maltase? (1)

b Explain how the products of carbohydrate digestion are absorbed into the blood. (3)

c **i** What is meant by *lactose intolerance*? (1)

ii One of the symptoms of lactose intolerance is diarrhoea.
Explain why. (3)

Total 10

3

a Describe the lock and key model for the action of an enzyme. (2)

b Explain why the induced fit model provides a better explanation of enzyme properties. (3)

c Explain the effect on enzyme action of

i a competitive inhibitor;

ii a non-competitive inhibitor. (4)

Total 9

4

a Describe a biochemical test to find out if a substance contains a protein. (2)

b The diagram shows the structural formulae of two amino acids.

```
        H    O                    H    O
        |    ||                   |    ||
H2N --- C --- C --- OH    H2N --- C --- C --- OH
        |                         |
        H                         CH2
                                  |
                                  SH
```

i Name **one** chemical element found in all amino acids, but **not** in monosaccharides. (1)

ii What type of chemical reaction occurs to form a dipeptide? (1)

iii Draw the structural formula of the dipeptide formed when these two amino acids combine. (1)

Total 5

AQA, January 2002, Unit 1, Question 1

5

a Many reactions take place in living cells at temperatures far lower than those required for the same reactions in a laboratory.
Explain how enzymes enable this to happen. (3)
An amylase enzyme converts starch to maltose syrup which is used in the brewing industry.

b Describe a biochemical test to identify

i starch; (2)

ii a reducing sugar such as maltose. (2)

c The graph shows the results of tests to determine the optimum temperature for the activity of this amylase.

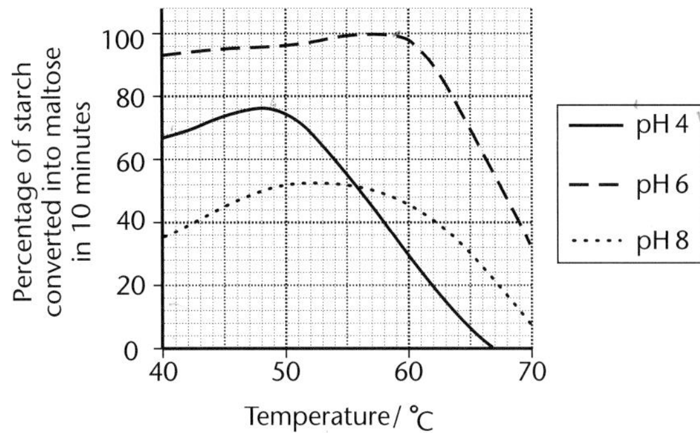

i Complete the table for the optimum temperature for the activity of amylase at each pH value. (1)

	pH		
	4	**6**	**8**
Optimum temperature / °C			

ii Describe and explain the effect of temperature on the rate of reaction of this enzyme at pH 4. (6)

Total 14

AQA, January 2003, Unit 1, Question 8

6 Sucrose is a disaccharide. It is formed from two monosaccharides **P** and **Q**.
The diagram shows the structure of molecules of sucrose and monosaccharide **P**.

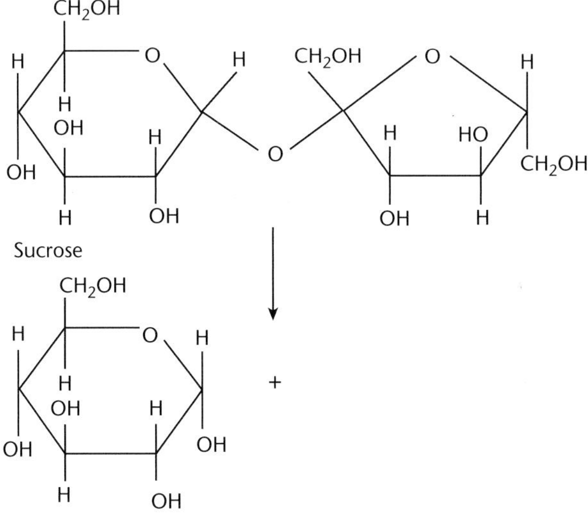

a i Name monosaccharide **Q** (1)

ii Draw the structure of a molecule of monosaccharide **Q** in the space above. (1)

b The enzyme sucrase catalyses the breakdown of sucrose into monosaccharides. What type of reaction is this breakdown? (1)

c The diagram shows apparatus used in breaking down sucrose. The enzyme sucrase is fixed to inert beads. Sucrose solution is then passed through the column.

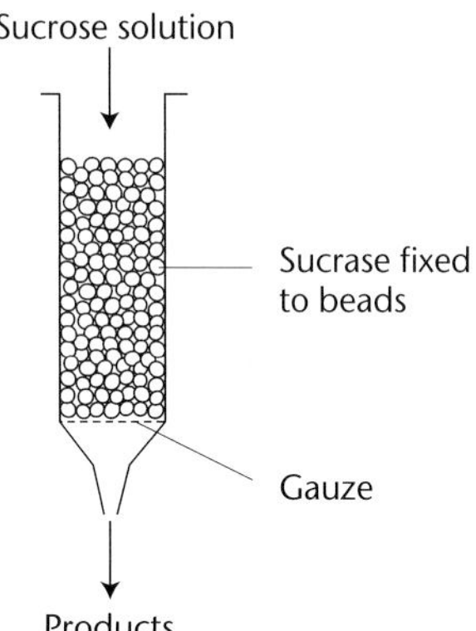

Describe a biochemical test to find out if the solution collected from the apparatus contains

i the products (2)

ii the enzyme. (2)

Total 7

AQA, January 2005, Unit 1, Question 2

7 The diagram represents an enzyme molecule and three other molecules that could combine with it.

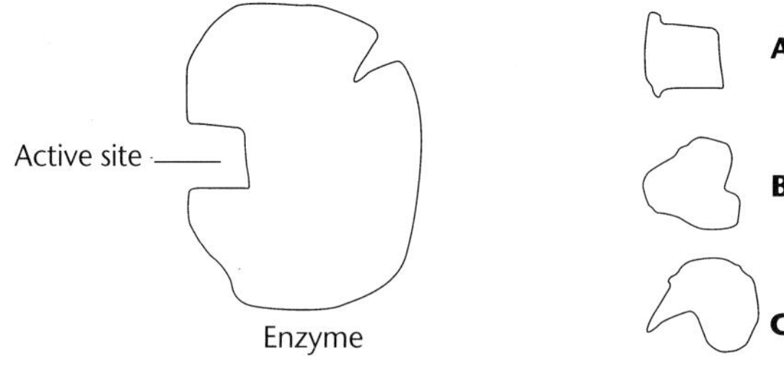

a Which molecule is the substrate for the enzyme? Give a reason for your answer. (1)

b Use the diagram to explain how a **non-competitive** inhibitor would decrease the rate of the reaction catalysed by this enzyme. (3)

c Lysozyme is an enzyme. A molecule of lysozyme is made up of 129 amino acid molecules joined together. In the formation of its active site, the two amino acids that are at positions 35 and 52 in the amino acid sequence need to be close together.

i Name the bonds that join amino acids in the primary structure. (1)

ii Suggest how the amino acids at positions 35 and 52 are held close together to form the active site. (2)

Total 7

AQA, June 2006, Unit 1, Question

8

a Describe and explain how an increase in temperature affects the rate of an enzyme-controlled reaction. (6)

b i Use your knowledge of the tertiary structure of enzymes to explain how a non-competitive inhibitor could reduce the rate of an enzyme controlled reaction. (4)

ii Alcohol dehydrogenase is an enzyme found in the liver. It normally breaks down ethanol (C_2H_5OH) into less harmful products. About 50 deaths each year occur following ingestion of a compound called ethylene glycol ($C_2H_4(OH)_2$) which is found in antifreeze. This compound is not lethal, but it is broken down by alcohol dehydrogenase into highly toxic oxalic acid. Giving a large dose of ethanol as quickly as possible can treat ethylene glycol poisoning.

Use your knowledge of enzyme activity to suggest how this treatment may counteract the effects of ethylene glycol poisoning. (4)

Total 14

AQA, June 2002, Unit 1, Question 6

9

a Complete the structural formula of the amino acid molecule.

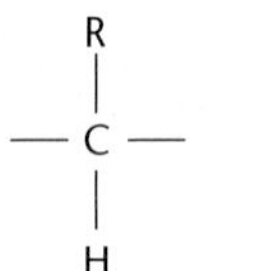

(2)

b In mammals, amino acids are broken down and urea is formed.

Urea from animal waste is often used as a natural fertiliser. Soil bacteria secrete an enzyme called urease that breaks down urea into ammonia and carbon dioxide. Some of this ammonia is released into the atmosphere.

urea + water → carbon dioxide + ammonia

Scientists have studied this reaction because it results in the loss of fertiliser. They have produced a substance called NBPT which is added to urea fertiliser.

NBPT is an enzyme inhibitor which affects the action of the urease produced by soil bacteria. The graph shows the results of an experiment in which a standard amount of urease and of the enzyme inhibitor NBPT were added to different concentrations of urea solution.

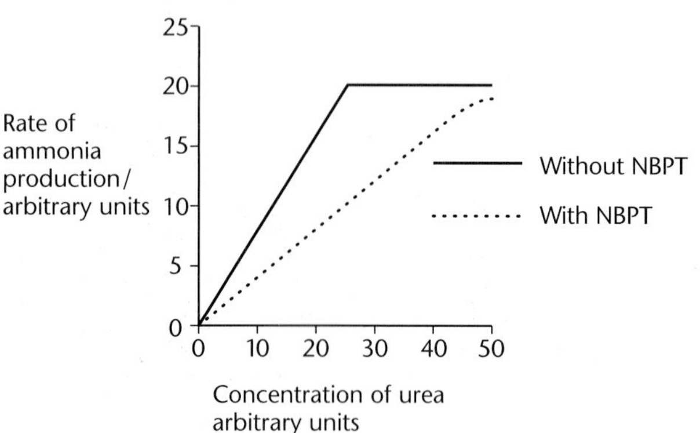

Describe and explain the effect on the rate of the reaction of increasing the urea concentration

i without NBPT present;

ii with NBPT present. (6)

Total 8

AQA, June 2003, Unit 1, Question 6

10 The diagram shows part of the gut wall of an animal.

a i Name the structure labelled **X**. (1)

ii Describe the function of the layer labelled **Y**. (2)

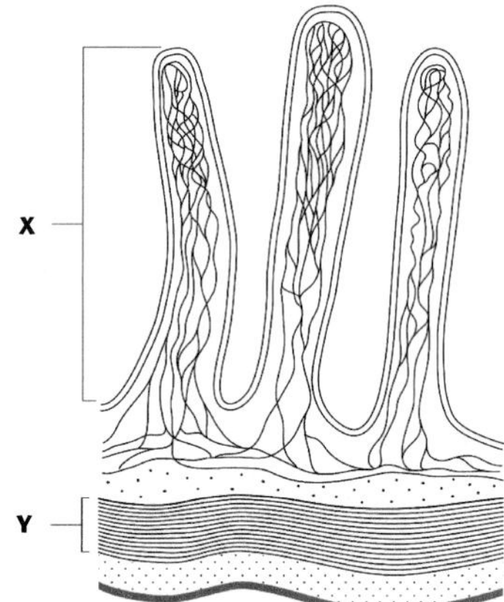

b Describe and explain how **two** features shown in the diagram increase the rate of absorption of digested food.

Feature 1
Description
Explanation

Feature 2
Description
Explanation (4)

Total 7

AQA, January 2005, Unit 1, Question 5

how science works **assignment**

Enzymes for the future

A hydrogen-powered bus built by the Southeastern Technology Center in Augusta, Georgia, USA; it is 'cleaner and greener' than its diesel-powered brothers.

The dream of using hydrogen as an economical pollution-free fuel has moved closer to reality with the discovery of two bacterial enzymes. Working together, the two enzymes use glucose to produce hydrogen gas. During the process, which produces no potentially harmful by-products, the enzymes convert glucose to gluconic acid, a substance used in detergents and some pharmaceuticals.

Thermoplasta acidophilum, a type of bacteria first found in a coal tip, produces the enzyme glucose dehydrogenase. This enzyme can pluck two hydrogen atoms from each glucose molecule it interacts with. One of the hydrogen atoms combines with a carrier molecule while the other goes into solution as a hydrogen ion. The reaction is completed when the second enzyme, hydrogenase, catalyses the reaction between two hydrogen ions to form a molecule of hydrogen gas. Hydrogenase is produced by *Pyrococcus furiosus*, a species of bacteria that flourishes around volcanic vents deep in the Pacific Ocean.

Both the enzymes are unusual in that they can resist quite high temperatures. Glucose dehydrogenase works at up to 60 °C, while hydrogenase can cope with temperatures up to 100 °C. Being able to operate at these temperatures makes the conversion of glucose very rapid. Contamination by other bacteria is also unlikely because hardly any other bacteria can survive at these temperatures.

A1 After removing the hydrogen from glucose, the gluconic acid is made. Which group in gluconic acid makes it acid?

A2 Sketch graphs to show the likely rates of reaction for glucose dehydrogenase and hydrogenase at different temperatures.

A3 If both processes go on at the same time in the same container, what would be the optimum temperature to use? Explain your answer.

A4 One possible source of glucose for this process is the cellulose in waste paper. How could glucose be obtained from the cellulose?

A5 Glucose obtained from cellulose costs about 30 cents per kilogram in the United States. Gluconic acid sells for $4 per kilogram.

a Assuming that each kilogram of glucose yields approximately one kilogram of gluconic acid, what percentage profit would be produced?

b About seven million tonnes of cellulose from waste paper could be available. What could be the value of the gluconic acid produced?

c What other source of profit would there be from the process?

d The demand for gluconic acid would be limited and would be less than the amount produced. Explain how this could affect the economics of the process.

e What other factors would have to be taken into account in assessing profitability?

5 Breathing and lung diseases

South Africa is considering forcibly detaining people who carry a deadly strain of tuberculosis (TB) that has already claimed hundreds of lives. The extreme drug-resistant TB strain (XDR-TB) threatens to cause a global pandemic. But the plan pits public protection against human rights. The country's health department says it has discussed with the World Health Organisation and South Africa's leading medical organisations the possibility of placing carriers of XDR-TB under guard in isolation wards until they die. Pressure to take action has been growing since a woman diagnosed with the disease discharged herself from hospital and probably spread the infection before she was finally coaxed back when she was threatened with a court order.

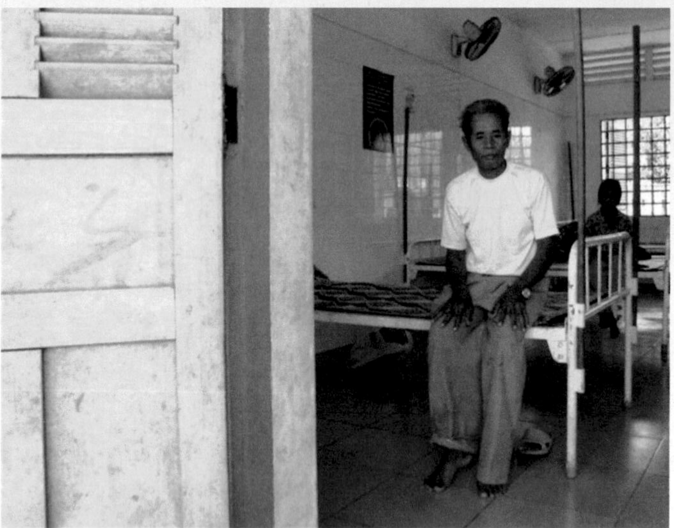

A patient in a TB ward in Cambodia.

XDR-TB is a highly infectious disease, which is spread by airborne droplets; it kills 98% of those infected within about two weeks. More than 300 cases have been identified in South Africa. But doctors believe there have been hundreds more and numbers are growing among the millions of people with HIV, who are particularly vulnerable to the disease. Doctors fear that patients with XDR-TB who are told that there is little that can be done for them will leave the isolation wards and go home to die. But while they are still walking around they risk spreading the infection.

A spokesman from Witwatersrand University's medical school supports enforced quarantine.

'You can look at it from two points of view. From the patient's point of view, you are expected to stay in some awful place, you can't work and you can't see your family. You will probably die there. From the community's point of view such a person is infectious. If they go to the shops or wander around with their friends they can spread it, potentially to a large group of people.'

However, the constitution also guarantees communal rights, including protection from infection and the right to a safe environment.

TB is just one of the diseases that affect the lungs. Most lung diseases reduce the rate at which oxygen enters the blood, which in turn reduces the patient's activity.

In this chapter you will learn how oxygen enters the blood and how this is affected by lung diseases.

5.1 Alveoli

Larger organisms such as humans have evolved systems that increase the surface area available for gas exchange. This area is called the **respiratory surface**. The respiratory surface in the lungs is large enough to collect enough oxygen to supply all of the body's tissues and to remove carbon dioxide before it builds up to toxic levels.

The lungs have a very large surface area because of the thousands of tiny air sacs they contain. These air sacs are the **alveoli** (Fig. 1). When we breathe in, air enters the alveoli and the oxygen in that air diffuses across the thin walls and into the blood. At the same time, carbon dioxide from the blood diffuses into the alveoli to be breathed out.

Alveoli are well adapted to their function as an exchange surface for gases.

- The epithelial cells that make up the walls of the alveoli support a rich network of blood capillaries. This blood supply is essential to collect the oxygen from the air and to deliver the carbon dioxide to the alveoli.
- The alveolar walls are very thin and the cells of the wall are flattened. This ensures a short diffusion pathway, a large diffusion gradient and so a rapid rate of diffusion.
- The walls of the alveoli are fully permeable. This allows oxygen to pass into the cells but also lets water out, so the exchange surface is always moist. Oxygen dissolves in the layer of moisture and then diffuses through the wall into the blood.

Fig. 1 Gas exchange in the lungs

Bronchiole (supported by cartilage)

Bronchus

Terminal bronchiole (made from smooth muscle)

Alveoli

Single alveolus

Blood to pulmonary vein

O_2

Diffusion

CO_2

Lining epithelium of alveolus

Blood from pulmonary artery

Red cells in capillary

Each alveolus is a tiny air sac that has thin, flat walls. Oxygen from the air dissolves in the liquid that lines the alveolus and then diffuses across into the blood capillary. Carbon dioxide leaves the blood by the reverse route.

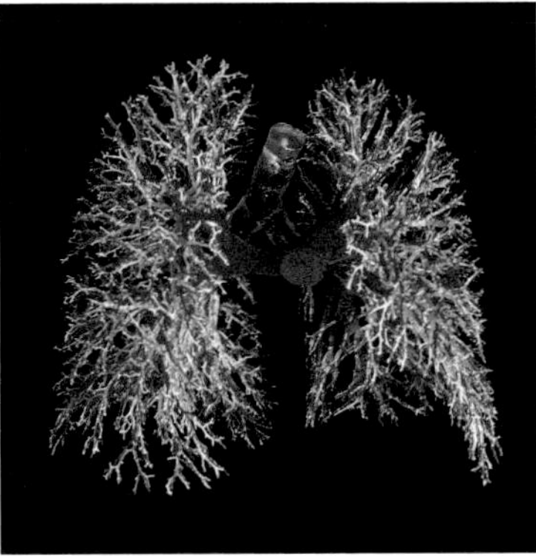

This resin cast reveals the intricate network of blood vessels that surround the alveoli of the lungs.

1 Will the layer of moisture lining the alveoli increase or decrease the rate of diffusion of oxygen into the blood? Explain your answer.

2 How does the blood supply to the alveoli affect the rate of diffusion of oxygen? Explain your answer.

Having a layer of liquid inside the alveoli presents a problem. If the liquid was just water, surface tension would cause the walls of the alveolus to stick to each other and the air sac would tend to collapse in on itself, reducing the surface area for gas exchange. However, in healthy lungs this does not happen because the liquid in the alveoli contains a **surfactant**, a chemical that reduces the surface tension of the liquid. This surfactant is produced in a baby's lungs from about the seventh month of pregnancy onwards. The lungs of babies born before 7 months' gestation do not produce surfactant and many of these babies have breathing difficulties. Treatment with artificial surfactant now allows more premature babies to survive and helps to prevent brain damage from lack of oxygen in the first few vital weeks.

key facts

- The **alveoli** of the lungs form the respiratory surface in humans and other animals.
- The alveoli and the lung capillaries that surround them have a large surface area for gas exchange.
- Alveolar walls are very thin. This makes the diffusion pathway between the air in the lungs and the blood very short.
- The walls of the alveoli that are in contact with air are moist. Oxygen dissolves in this liquid before diffusing across the alveolar wall and into the blood.
- The liquid contains **surfactant,** which reduces surface tension and prevents the air sacs collapsing.

5.2 Breathing

Fig. 2 Ventilating the lungs

Inspiration
Volume of lung increases
During exercise, external intercostal muscles contract (only a small number shown) and pull the rib cage upwards and outwards, increasing the volume of the thorax
Internal intercostal muscles relax (only a small number shown)
Diaphragm muscles contract and pull the diaphragm down, increasing the effective volume of the thorax
Expiration
Volume of lung decreases
External intercostal muscles relax and the rib cage moves downwards
During exercise, internal intercostal muscles contract
Diaphragm muscles relax and the diaphragm moves upwards

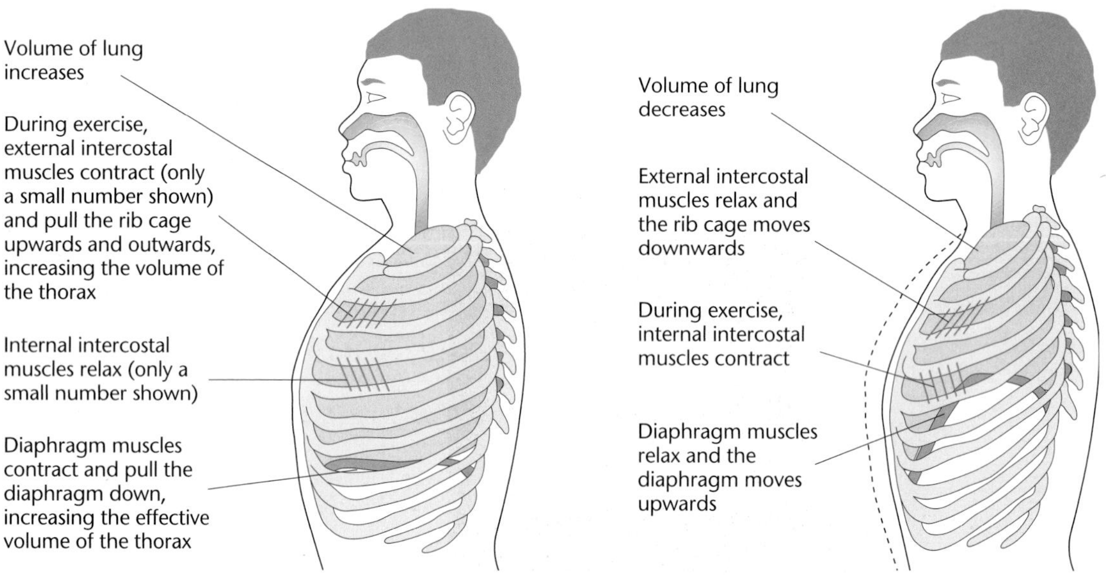

Breathing draws fresh air into the lungs and forces stale air out. This process is also called **ventilation**. Ventilation ensures there is always a good supply of 'fresh' air inside the lungs. This maintains large diffusion gradients for oxygen and carbon dioxide between air and blood. Large diffusion gradients mean efficient gas exchange. Notice the difference between *ventilation* and *respiration.* Respiration is a series of oxidation reactions that occur in cells.

Breathing in is known as **inspiration**, breathing out is called **expiration** (Fig. 2). Because the flow of air through the lungs occurs in both directions – in and out – we say that the ventilation is **tidal**. Air passes into and out of the alveoli via the **trachea**, **bronchi** and **bronchioles**. Except for the narrowest bronchioles, the walls of these three air tubes contain rings of **cartilage**. Cartilage is a strong, slightly flexible tissue. It helps to keep the tubes open during the pressure changes that take place as we breathe in and out.

Inspiration is always an active process. When the body is at rest, expiration is a passive process. During gentle expiration, the **diaphragm** muscles relax and the elastic recoil of the lungs and chest wall returns the **thorax** to its original shape. When the body is exercising, expiration is boosted as the **internal intercostal muscles** contract, pulling the ribs

Fig. 3 Ventilating the lungs

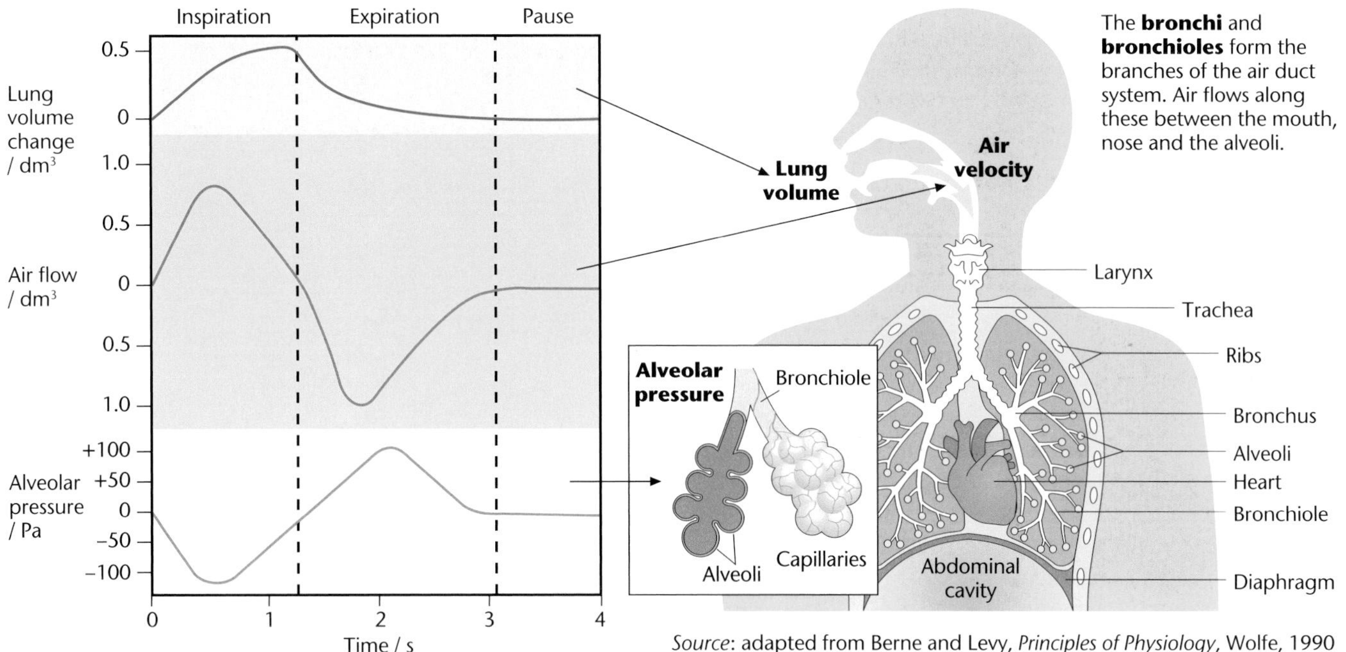

Source: adapted from Berne and Levy, *Principles of Physiology*, Wolfe, 1990

downwards, and the muscles in the abdomen wall contract, forcing the diaphragm upwards. These contracting muscles decrease the volume of the thorax and so increase the volume of air exhaled with each breath.

Increasing the volume of the thorax decreases air pressure in the alveoli to below atmospheric pressure (Fig. 3). Air is forced into the lungs until the pressure in the alveoli equals that of the atmosphere. The inflow of air inflates the lungs. Decreasing the volume of the thorax has the reverse effect. As air pressure in the alveoli rises above atmospheric air pressure, air is squeezed out until the alveolar pressure equals atmospheric pressure.

5.3 Breathing rate and exercise

It is very important that your breathing rate is matched to your activity. Moving muscles use up energy quickly; and the faster you move, the quicker your cells need to burn fuel. This means they need extra oxygen, so the body makes the lungs take in more air to supply that need. Air is only 20 per cent oxygen, and so only a fifth of the air that you breathe in can be used. The air in the alveoli contains even less oxygen – approximately 14.5 per cent. This is because breathing replenishes only 30 per cent of the air in the lungs when we are not active. Even when we are exercising and breathing deeply, only 50 per cent of the air in the lungs is changed with each breath. This low concentration of oxygen in the alveoli gives quite a shallow diffusion gradient for oxygen between air in the alveoli and blood in the lung capillaries. Consequently, we only extract about 25 per cent of the oxygen from the air we breathe in. The air we breathe out still contains about 15 per cent oxygen.

These limitations of the breathing system mean that the body often passes the point at which its need for oxygen is satisfied. At this point, the muscles must respire without oxygen. You get muscle fatigue when a muscle cannot get enough oxygen to produce all the ATP it needs, despite all the puffing and panting. A substance called lactic acid can build up in the muscles. Hydrogen ions from the dissociation of lactic acid may damage the muscle tissue.

Training can improve the ability to exercise for long periods without getting muscle fatigue. One of the main aims of training for long-distance running is to improve the oxygen supply to the muscles. Training increases muscle mass, it increases the blood supply to the muscles and it improves the rate at which oxygen gets into the blood in the lungs.

5.4 The effect of training

The amount of oxygen that a person can take in per minute depends to some extent on their size; a large person has larger lungs and can breathe in more air than a smaller person. The volume of air inhaled and exhaled with each normal breath is called the **tidal volume**. In an average sized adult this is about 500 cm^3. At rest we breathe in and out 14 times per minute (the **ventilation rate**). This gives a **pulmonary ventilation** of 7 litres per minute. But a reasonably fit and active adult can take in 12 litres of air in a minute, which delivers 3 litres of oxygen per minute to the blood. With the right training, long distance runners can deliver over six litres of oxygen to the blood per minute. This increase in rate is due to an increase in the capacity of the lungs and an increased rate of blood flow through the lungs. The increase in capacity of the lungs depends on three factors:

- the **vital capacity** of the lungs – this is the maximum volume of air that can be exhaled from the lungs in each expiration
- the **ventilation rate** – this is the number of inspirations and expirations in a given time
- the **maximum ventilation capacity** – this is the maximum amount of air that can move in and out of the lungs in a given time.

All three are determined by the volume of the lungs and also by the strength of the muscles that move air in and out of the lungs.

Three sets of muscles are actively concerned with breathing (Fig. 4):

- the muscles of the diaphragm
- the external intercostal muscles
- the internal intercostal muscles.

Training can increase the strength of these muscles, raising vital capacity and improving maximum breathing capacity.

Fig. 4 Muscles involved in breathing

Intercostal muscles are connected to the ribs.

The **diaphragm** is the main muscle for inspiration and is solely responsible for inspiration when the body is at rest.

The lungs and chest wall are elastic. When the muscles relax, the lungs and chest fall, or 'recoil', back to shape.

The **thorax** is the chest area.

The flow of air in and out of the lungs is **tidal**.

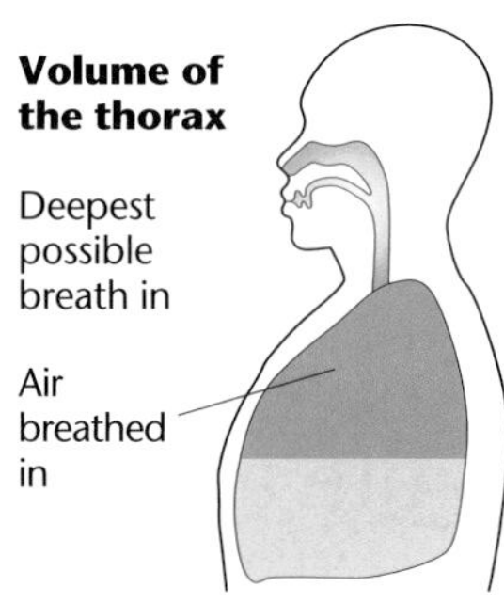

3 The rate of oxygen uptake into an athlete's blood during a training programme is measured in litres of oxygen per minute. Can you think of fairer units in which to measure improvements due to training?

4 In 3 months of training, an athlete increased her maximum ventilation capacity from 130 litres min^{-1} to 180 litres min^{-1}. Calculate the percentage improvement in her maximum ventilation capacity.

key facts

- Breathing maintains large diffusion gradients for oxygen and carbon dioxide between air and blood.
- The **diaphragm muscles**, the **intercostal muscles** and the **abdominal muscles** work together to ventilate the lungs.

how science works

Are women catching up?

The table shows the winning times for the 200 m event in the Olympic games.

	Men		Women	
Year	**Athlete**	**Time/s**	**Athlete**	**Time/s**
1948	Mel Patton, USA	21.1	F. Blankers-Koen, Netherlands	24.4
1952	Andrew Stanfield, USA	20.7	Marjorie Jackson, Australia	23.7
1956	Bobby Morrow, USA	20.6	Betty Cuthbert, Australia	23.4
1960	Livio Berruti, Italy	20.5	Wilma Rudolph, USA	24.0
1964	Harry Car, USA	20.3	Edith McGuire, USA	23.0
1968	Tommie Smith, USA	19.83	Irena Szewinska, Poland	22.5
1972	Valeri Borzov, USSR	20.00	Renate Stecher, E. Germany	22.40
1976	Donald Quarrie, Jamaica	20.23	Barbel Eckert, E. Germany	22.37
1980	Pletro Mennes, Italy	20.19	Barbel Wockel, E. Germany	22.03
1984	Carl Lewis, USA	19.80	Valerie Brisco-Hooks, USA	21.81
1988	Joe Deloach, USA	19.75	Florence Griffith-Joyner, USA	21.34
1992	Mike Marsh, USA	20.01	Gwen Torrence, USA	21.81
1996	Michael Johnson, USA	19.32	Marie-Jose Perec, France	22.12
2000	Konstantinos Kenteris, Greece	20.09	Marion Jones, USA	21.84
2004	Shawn Crawford, USA	19.79	Veronica Campbell, Jamaica	22.05

5 Which type of graph should be used to present these data in the clearest way?

6

a Describe the patterns shown by these data.

b Explain these patterns in terms of body structure.

5.5 Tuberculosis

Tuberculosis (TB) is an infectious disease caused by bacteria called *Mycobacterium tuberculosis*. TB most commonly affects the lungs. Many years ago, this disease used to be called 'consumption' because without effective treatment affected patients would often waste away. Today, most strains of tuberculosis can usually be treated successfully with antibiotics.

TB bacteria get into the air when someone who has a TB lung infection coughs, sneezes, shouts or spits. People who are nearby can then possibly breathe the bacteria into their lungs. Drinking unpasteurised milk transmits another strain of TB. Previously, this strain was a major cause of TB in children, but it rarely causes TB in the UK now because most milk is pasteurised (a heating process that kills the bacteria).

Course of infection

When the TB bacteria enter the lungs, they can multiply and cause a local lung infection. At this stage the person often develops pneumonia. In addition, TB can spread to other parts of the body. The body's immune system usually stops the bacteria from spreading. It does so ultimately by forming scar tissue around the bacterial colony and isolating it from the rest of the body. The scar tissue can be detected by radiography (shown in the X-ray photograph below).

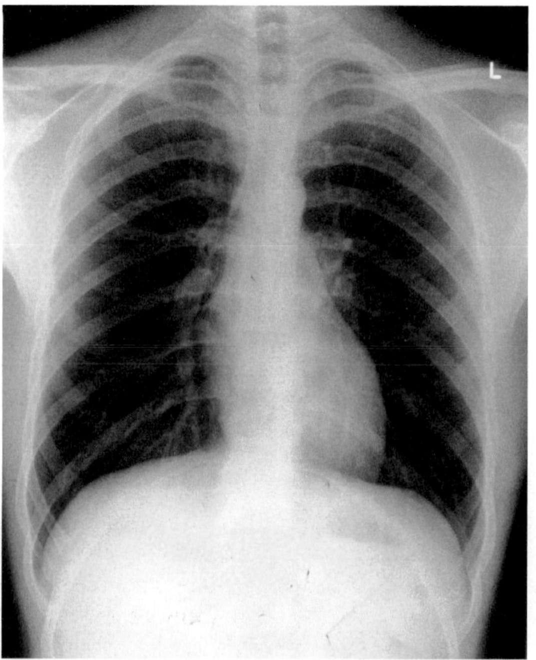

X-ray photograph of a lung showing massive scarring caused by TB.

If the body is able to form scar tissue around the TB bacteria, then the infection is contained in an inactive state. Such a person typically has no symptoms and cannot spread TB to other people. The scar tissue and lymph nodes may eventually harden. This is due to the process of calcification of the scar tissue – deposition of calcium from the bloodstream in the scar tissue.

If the body's immune system becomes weakened, TB bacteria may break through the scar tissue. For example, the immune system can be weakened by old age or the development of another infection such as HIV. The breakthrough of bacteria can result in recurrence of pneumonia and the spread of TB to other locations in the body. The kidneys, bone and lining of the brain and spinal cord are the most common sites affected by the spread of TB beyond the lungs.

The risk factors for acquiring TB include close-contact situations, drug abuse, certain diseases (e.g. diabetes, cancer and HIV) and occupations (e.g. health care workers).

Inactive TB may be treated with an antibiotic, to prevent the TB infection from becoming active. Active TB is treated, usually with several antibiotics.

Symptoms

It usually takes several months from the time of the infection until symptoms develop. The usual symptoms that occur with an active TB infection are tiredness or weakness, weight loss, fever and night sweats. If the infection in the lung worsens, then further symptoms can include coughing, chest pain, coughing up of sputum (material from the lungs) and shortness of breath.

Shortness of breath is caused by the restriction of gaseous exchange in the affected parts of the lungs caused by the scar tissue.

7 Explain why one of the symptoms of TB is a fever.

8 Explain why someone with TB may be short of breath.

Fig. 5 Trends in TB notifications in London and the rest of England 1981–2001 (the date of the last census)

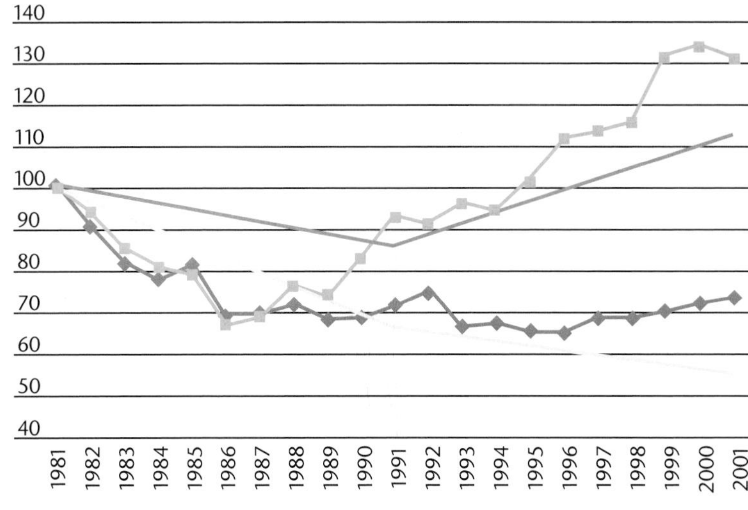

Source: TB data from Health Protection Agency. Overcrowding data from Census 1981, 1991, 2001, based on the person per room rating. Census data is Crown Copyright.

Note: "Recent trends in tuberculosis notification should be interpreted with caution since the decline in notifications is most likely attributable to changes in surveillance practice at local level". *Health Protection Agency*

9 Look at the graph (Fig. 5).

a Suggest an explanation for the rise in TB notifications in London during the 1990s.

b Suggest an explanation for the rise in the number of TB notifications in the rest of England during the 1990s.

key facts

- Tuberculosis (TB) is caused by a bacterial infection.
- It is contracted by breathing in infected air.
- The bacteria multiply in the lungs, causing the development of scar tissue, which isolates these bacteria.
- Symptoms include tiredness or weakness, weight loss, fever and night sweats and eventually shortness of breath.
- If the immune system is weak, TB bacteria may spread to other parts of the body.

5.6 Pulmonary fibrosis

Pulmonary fibrosis describes a group of diseases that produce lung damage. The disease may lead to fibrosis in which an inelastic matrix develops between the cells. This eventually leads to loss of the elasticity of the lungs. The peak age of people with the disease is 50–70 years, and men and women are affected equally.

There are many known occupational causes of pulmonary fibrosis. People working with substances such as asbestos, coal dust and metal dust are particularly affected. Farmers may be affected by mould spores from hay or straw. Smokers are more susceptible to fibrosis than non-smokers.

It is thought that the changes to the lung tissue in fibrosis are associated with irritation caused by small particles or certain chemicals.

The most common symptom of pulmonary fibrosis is shortness of breath or a cough.

Investigation of a patient's breathing will usually show:

- reduced vital capacity
- reduced maximum ventilation capacity
- reduced rate of oxygen transfer into the blood.

At present, there is no treatment that has been shown to improve the condition. A lung transplant is the only option that improves long-term survival.

10 Explain why each of vital capacity, maximum ventilation capacity and rate of oxygen transfer into the blood is reduced by fibrosis.

key facts

- Pulmonary fibrosis is a condition that results in lung tissue becoming inelastic.
- The most common cause is irritation caused by small particles or irritant chemicals.
- The condition results in:
  - reduced vital capacity,
  - reduced maximum ventilation capacity,
  - reduced rate of oxygen transfer into the blood.

5.7 Asthma

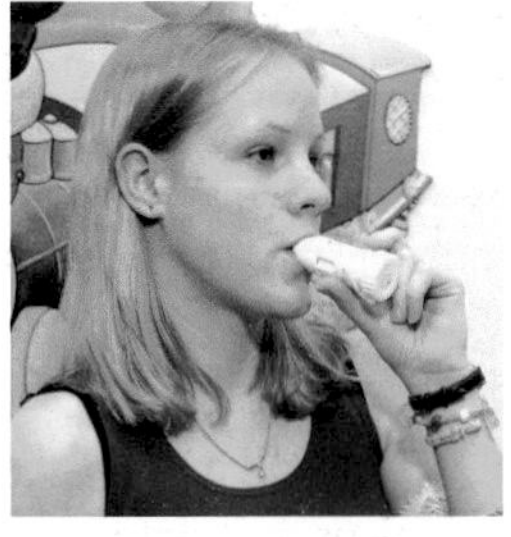

An inhaler contains a drug to help open up the airways.

Asthma is a condition that affects the bronchioles – the small tubes that carry air in and out of the lungs. About 5.2 million people in the UK are currently receiving treatment for asthma. This often involves using an inhaler.

When a person with asthma comes into contact with something that irritates their bronchioles (an asthma trigger), the muscles around the walls of the bronchioles contract so that the bronchioles become narrower. The lining of the bronchioles also becomes inflamed and starts to swell. Sometimes sticky mucus or phlegm builds up, which can further narrow the bronchioles. This makes it more difficult for air to flow to and from the alveoli and causes the symptoms of asthma.

Asthma symptoms include:

- coughing
- shortness of breath
- tightness in the chest.

The exact cause of asthma is not fully understood as yet. Sometimes, the symptoms flare up for no obvious reason, but attacks are most frequently triggered by things that irritate the bronchioles. These can include:

- irritants such as dust and cigarette smoke
- chemicals found in the workplace
- pollen, animals, house dust mites.

Fig. 6 The effect of asthma on bronchioles

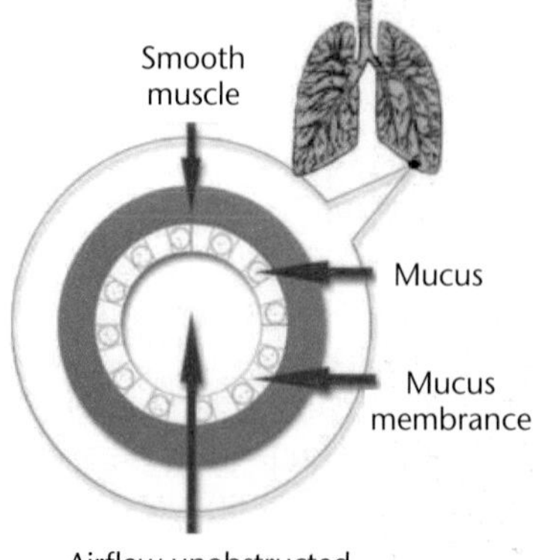

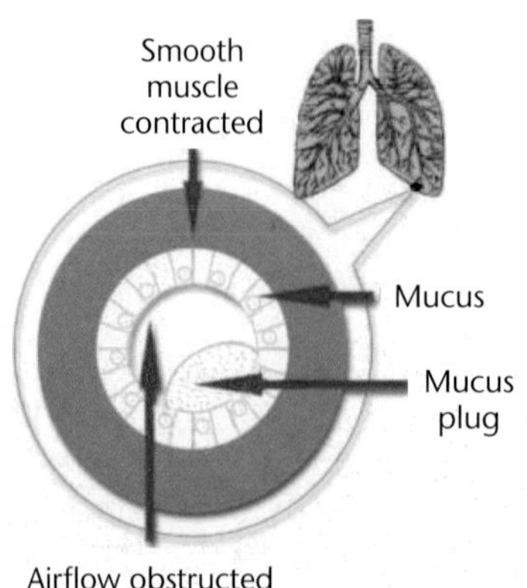

11 The drugs in inhalers used by people with asthma relieve the symptoms. Suggest which part of the bronchioles these drugs affect. Explain your answer.

Is asthma caused by air pollution?

Traffic pollution may boost the risk of children getting asthma – if they have genes that make them vulnerable.

A team of scientists at the University of Southern California studied the health records and genetic profiles of 3000 children. Those with a gene variation were slightly more at risk, but if they lived near a main road, the risk was increased even further. But UK asthma experts said the link remained unclear.

Scientists exploring how asthma develops have highlighted the importance of genes that control key body chemicals linked to 'clean-up' functions in the body. Enzymes called EPHX1 and a gene called *GSTP1* appear to have some responsibility for getting rid of harmful chemicals that we breathe in. The researchers found that those who had high levels of EPHX1 were one and a half times more likely to have been diagnosed with asthma, while those who also had variations in *GSTP1* were four times as likely to have asthma.

However, living close to a main road appeared to make this effect even greater. Children with very active EPHX1 who lived within 75 metres of a road had a doubled risk of asthma compared with those who had low EPHX1 levels. Having active EPHX1, variations in the gene, and a home near a road meant a risk nine times greater.

Their conclusion was that while children with the 'wrong' genes and enzyme activity were more prone to having asthma, living near a road seemed to compound that risk.

There has been a long-running dispute about a link between asthma and exhaust fumes.

A British scientist said that more work was needed. 'This study is very promising as it is one of the first to look specifically at how genetic susceptibility to respiratory disease and environmental traffic fumes can cause childhood asthma. People with asthma tell us that traffic fumes make their asthma worse and although this research only looks at individuals with a certain genetic make-up, we await further robust research in this new and exciting area to help us find better ways to treat asthma.'

The map in Fig. 7 shows the global distribution of asthma.

12 Does the map show evidence of a link between asthma and air pollution? Explain your answer.

13 There is a theory (called the hygiene hypothesis) that people in developed countries are no longer exposed to the kinds of infections they would have had to deal with in the past, so the immune system over-reacts to harmless substances. It is difficult to test this theory. Explain why.

14 Suggest why the two groups of scientists disagree.

15 Is there an association between road traffic fumes and asthma, or do road traffic fumes cause asthma? Explain the reasons for your answer.

16 Suggest what is meant by 'more robust research'.

Fig. 7 Global asthma rates

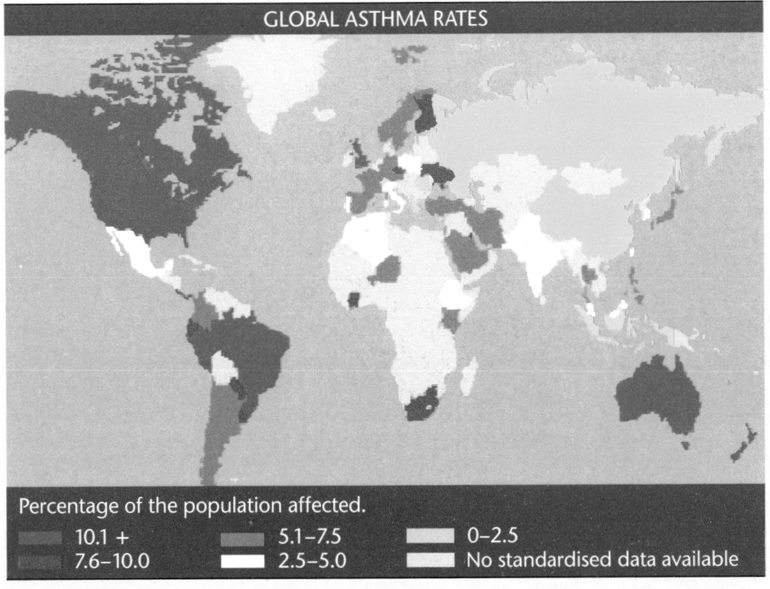

key facts

- In an asthma attack, the muscles in the bronchioles contract causing constriction of the bronchioles.
- This constriction reduces the flow of air to the alveoli.
- Irritants such as dust, pollutants and workplace chemicals often cause asthma attacks.

5.8 Emphysema

Air pollution and smoking are responsible for irritation of the lungs, which can lead to chronic bronchitis and eventually to emphysema (Fig. 8).

Fig. 8 Development of lung disease

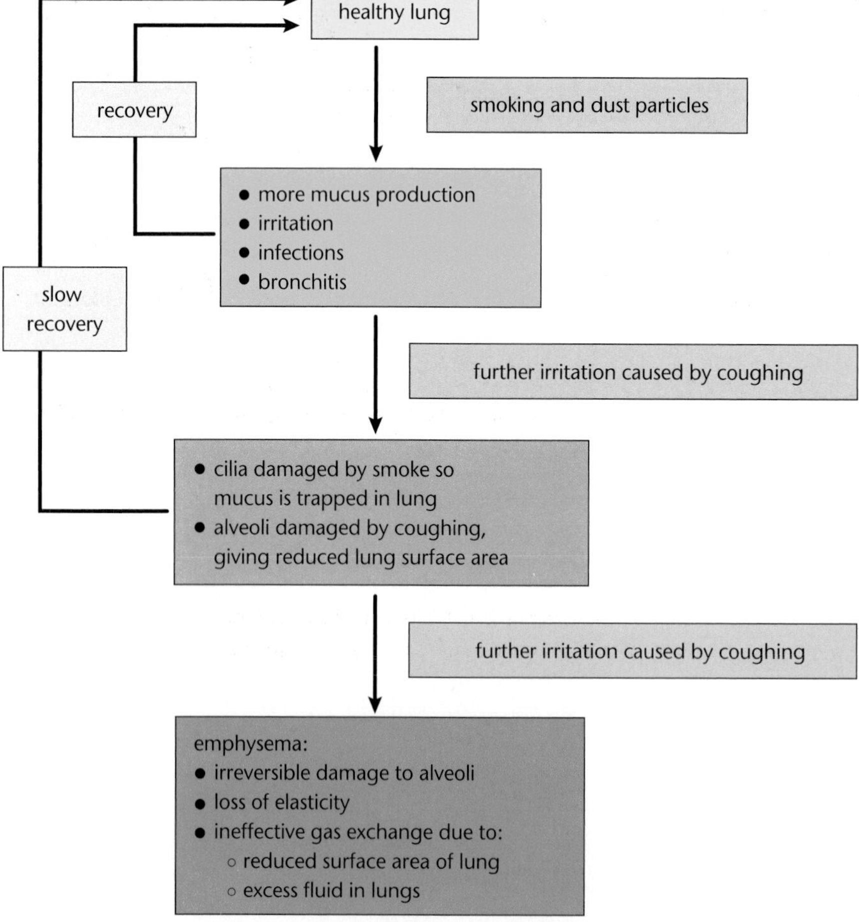

Cilia in the bronchial tubes beat rhythmically to produce a constant upward flow of mucus from the bronchi to the back of the throat, where it is swallowed. This healthy mucus flow removes trapped dust particles, microbes and other irritants from the bronchi and bronchioles, so they do not reach the alveoli.

Smoking and air pollution paralyse the cilia so that the mucus builds up into clumps that are coughed up. The lining of the bronchial tubes becomes irritated and inflamed, with excess production of thick mucus. Infections of the breathing system such as pneumonia are then much more likely. Smoking also damages the walls of small bronchioles and alveoli and causes the growth of fibrous tissue around the bronchioles, narrowing the air passageway. This makes it harder to breathe. Bronchitis will get better if a person stops smoking.

17

a What damage does smoking do to the bronchi and alveoli?

b What is the effect of this damage?

Smoke and air pollutants also irritate the delicate moist surfaces of the lungs. Physical damage by repeated coughing, together with loss of elastin from the walls of the alveoli, lead to emphysema. In this condition, less surface area is available for the exchange of gases, which causes breathlessness. Emphysema is common in people who have smoked for many years, although it is also linked to a number of occupations, such as coal mining. Emphysema is irreversible. Patients with emphysema are advised to stop smoking and to avoid environments with high levels of atmospheric pollutants.

18 Describe how emphysema develops.

19 Why does someone with emphysema become breathless easily?

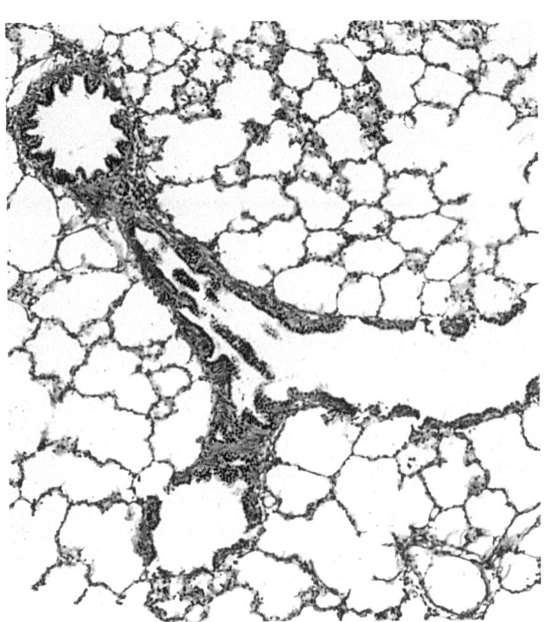

In this healthy lung tissue, the alveoli are folded because of the presence of elastin. They expand during inspiration and contract during expiration to push air out of the lungs.

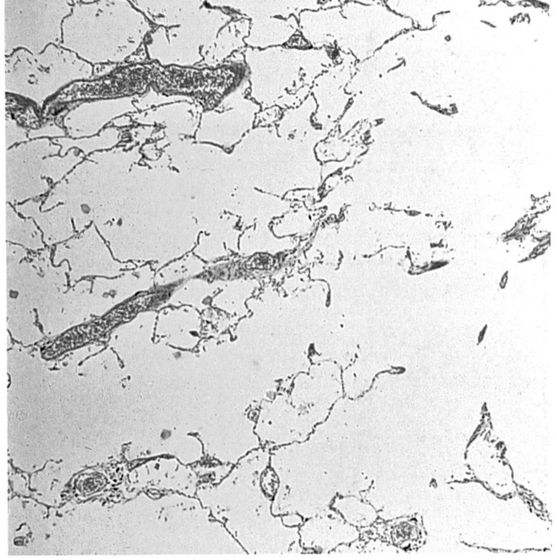

Loss of elastin in emphysema means that the alveoli cannot contract during expiration. Poor ventilation and trapped air cause further damage to airways and alveoli.

key facts

- Emphysema is a disease caused by air pollutants and smoking.
- It is characterised by irreversible damage to the alveoli.
- This is caused by loss of elasticity.
- Gaseous exchange is reduced due to reduced total surface of the alveoli and excess fluid in the lungs.

examination questions

1

a Describe how tuberculosis is transmitted. (2)

b Describe the course of infection of tuberculosis. (3)

c The graph shows the number of new reported cases of tuberculosis in England and Wales between 1988 and 2005.

i Describe the trend shown by the data. (2)

ii Suggest one explanation for this trend. (2)

Total 9

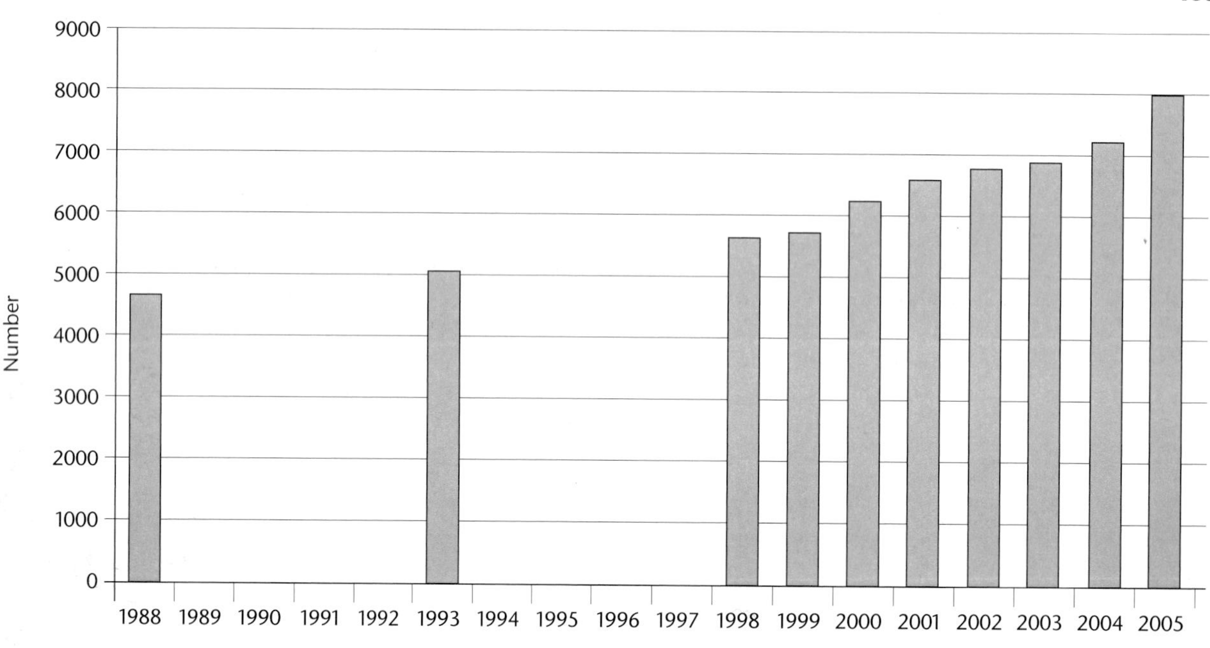

2

a **i** Describe the symptoms of asthma (2)

ii Explain how these symptoms are produced. (3)

b Read the passages from a national newspaper about swiming and asthma, and the reply from a swimming pool official.

Newspaper article

The chlorine in indoor swimming pools may be contributing to the rising number of children with asthma, new research suggests.

A study spanning 21 European countries found a close association between access to indoor swimming pools and respiratory problems in children.

Prevalence of childhood asthma and wheeze rose by around 2 to 3 per cent for every indoor swimming pool provided per 100,000 people.

The findings strongly support the theory that chlorine in the water and air at indoor swimming pools can trigger asthma. Childhood asthma in the UK soared by 400 per cent in the 1980s and 1990s, although there are signs that rates are now slowing.

Swimming pool official

The report says that the rise in childhood asthma associated with childrens' use of swimming pools could be a significant factor in the general increase in the prevalence of asthma in all children. The rise in asthma is across all sectors and age groups of society, hence there must be some other factor at work. As children have been swimming in pools in school classes for over a hundred years, ISRM would question why only now has this resulted in an increase in asthma in children, when the quality of water in pools in the UK has improved dramatically.

Using information from the passages, evaluate the evidence for a link between swimming pools and asthma. (5)

Total 10

3

a Describe the change in lung structure caused by emphysema. (2)

b Explain how this change in structure affects the functioning of the lungs. (2)

c Read the passage about research into emphysema.

> Rats fed a vitamin A-deficient diet develop emphysema. Benzopyrene is a common carcinogen found in cigarette smoke.
>
> A scientist called Baybutt exposed a group of rats to cigarette smoke and found that those rats became vitamin A deficient.
>
> Baybutt observed that when the lung content of vitamin A was low, the area of emphysema in the lungs was high. His hypothesis was that smokers develop emphysema because of a vitamin A deficiency.
>
> Baybutt then fed the rats exposed to cigarette smoke a diet with higher levels of vitamin A.
>
> "We saw that the areas of emphysema were effectively reduced," he said.
>
> Baybutt said he believes this might help explain the occurrence of emphysema.
>
> "There are a lot of people who live to be 90 years old and are smokers," he said.
>
> "Why? Probably because of their diet."

Do the results of the experiment prove that emphysema is caused by vitamin A deficiency? Explain the reasons for your answer. (3)

Total 7

4 The drawing shows some of the structures involved in ventilating human lungs.

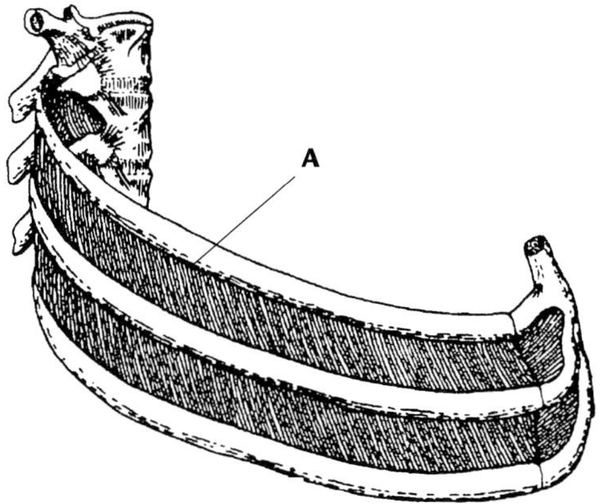

a Name structure **A** (1)

b **i** Describe the role of structure **A** in inspiration. (3)

ii Explain how ventilation increases the rate of gas exchange in the alveoli. (2)

Total 6

AQA, June 2003, Unit 1, Question 4

5 The diagram shows a section through an alveolus and a blood capillary.

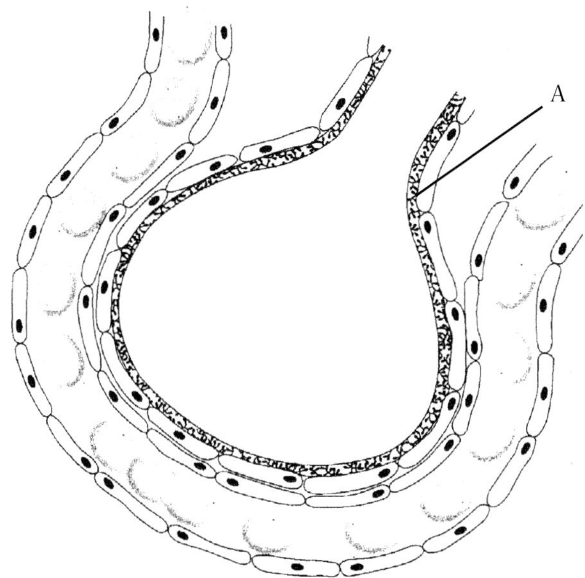

a What structural features, visible on the diagram, increase the rate of diffusion of oxygen into the blood? (3)

b Explain the function of the layer labelled A. (2)

c Explain how a diffusion gradient for oxygen is maintained between the air in the alveolus and the blood. (3)

Total 8

6

a Describe how air is taken into the lungs. (3)

The volume of air breathed in and out of the lungs during each breath is called the tidal volume. The breathing rate and tidal volume were measured for a cyclist pedalling at different speeds. The graph on the following page shows the results.

Tidal volume

Breathing rate

Tidal volume/dm³ (0, 1, 2, 3)

Breathing rate/breaths per minute (0, 10, 20, 30)

Cycling speed/km h⁻¹ (0, 5, 10, 15, 20, 25)

b Describe the **two** curves.

i Tidal volume

ii Breathing rate (2)

c Calculate the total volume of air breathed in and out per minute when the cyclist is cycling at 20 km h^{-1}. Show your working.

_________ dm^3 (2)

Total 7

AQA, June 2005, Unit 1, Question 6

7 **Graph 1** shows how lung volume of a human changes during inspiration and expiration.

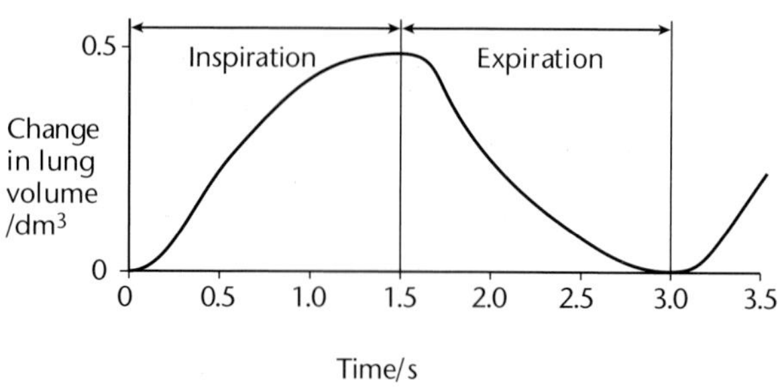

Graph 1

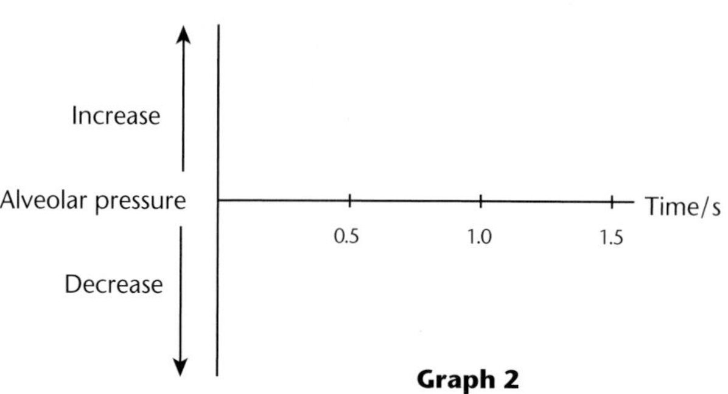

Graph 2

i Sketch, on Graph 2, a curve to show the changes in alveolar pressure during inspiration. (2)

ii Use **Graph 1** to calculate the rate of breathing in breaths per minute.

_________ breaths per minute (1)

Total 3

AQA, June 2002, Unit 1, Question 4

hsw

how science works **assignment**

Altitude sickness

Climbing a mountain is hard work. As you climb, the concentration of oxygen in the air remains the same (about 21 per cent) but the number of oxygen molecules that you can take in with each breath decreases as the pressure of air decreases. At sea level, air pressure is about 100 kPa, but at 3500 metres (about 12 000 ft) this has fallen to about 63 kPa. Your body compensates by increasing your breathing rate, but if you ascend too quickly, you can start to develop altitude sickness. Climbing above 3000 metres causes the symptoms of mild altitude sickness in many people. These symptoms include headache, dizziness, shortage of breath, nausea and disturbed sleep. They arise because when the air pressure falls to a level below the pressure in the lung capillaries, fluid leaks from the capillaries into the alveoli. Within a couple of days the body begins to compensate, and the symptoms start to disappear after about 3 days. This process is known as acclimatisation.

If you ascend very quickly to a very high altitude (4000–5000 metres), the symptoms are much worse – shortness of breath even when resting, mental confusion and the inability to walk. This is acute altitude sickness and is caused by fluid leaking from the capillaries into the cavities in the brain. Someone with these symptoms should be returned to a lower altitude immediately. If this is not possible then they should be placed in a Gamow bag. The bag is inflated with a foot pump. Conditions inside the bag simulate lower altitudes and the patient quickly recovers. Most high-altitude climbing expeditions now carry at least one of these bags.

Altitude sickness can be prevented by acclimatising to the change in altitude in stages. This involves spending 2–3 days at a particular altitude before trying anything strenuous like rock climbing or skiing and before moving to higher levels. Acclimatisation occurs because when the body is at high altitudes it produces a hormone called erythropoietin. This hormone stimulates the body to produce more red blood cells to compensate for the reduced capacity to take in oxygen from the air. Many athletes train at high altitude. The increased number of red blood cells lasts for about 2 weeks and enhances performance, particularly in middle- and long-distance running events.

It is possible to inject erythropoietin to obtain the same effect, but athletics authorities forbid this. However, athletes who want to simulate the effects of going to high altitudes without having to climb a mountain can use a high altitude-sleeping chamber quite legitimately. An athlete training at sea level can gain the advantage of training at high altitude by sleeping in a Gamow bag set up to simulate high altitude.

An inflated Gamow bag

A1 Explain why the body obtains less oxygen per breath at high altitude than at sea level.

A2 Explain how fluid leaking from capillaries into the alveoli affects gas exchange.

A3 Explain what is meant by acclimatisation.

A4 Explain how you would recognise if a companion has acute altitude sickness.

A5 Explain how the Gamow bag helps a person who has acute altitude sickness.

6 The heart and heart disease

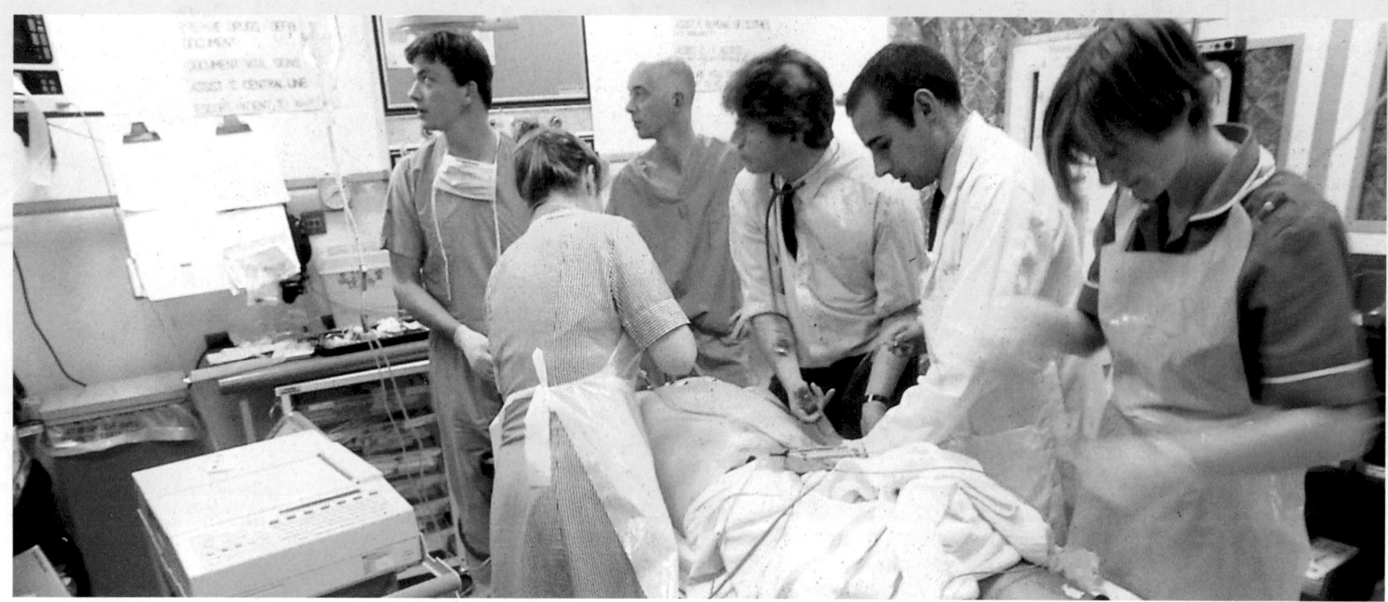

Most hospitals have a team of doctors and nurses who can respond rapidly to a patient having a heart attack. They are known as the crash team.

Doctors in hospital emergency departments cope skillfully with emergencies such as heart attacks. Nevertheless, a heart attack can cause a great deal of damage to the heart and other parts of the body, and about 40 per cent of patients who have a heart attack die as a result. So the best thing to do is to prevent heart attacks from happening in the first place.

Like all muscles, heart muscle needs oxygen. Oxygenated blood is carried to heart muscle by the coronary arteries. In coronary heart disease, layers of fatty material build up in the coronary arteries, causing them to narrow. This fatty material is known as atheroma. If it narrows the arteries enough, the oxygen supply to the heart muscle can be cut off. If a muscle is deprived of oxygen we experience pain in that muscle; that is why one symptom of a reduced blood supply to the heart is chest pain. If the blood supply becomes very poor, sections of heart muscle can die; when this happens, the person has a heart attack.

But if only the heart muscle dies, why does a heart attack cause damage to other parts of the body? The heart is essentially a muscular pump that pumps blood around the body. This blood carries the oxygen and other substances that organs and tissues need, and takes away waste. When the heart is damaged, the efficiency of the pump is reduced and other organs can be damaged because their supply of oxygen fails. The brain is the most vulnerable organ; it is badly damaged if deprived of oxygen for as little as 4 minutes.

In this chapter you will study the structure and function of the heart and consider coronary heart disease.

6.1 Structure and function of the heart

The heart contains two muscular pumps (Fig. 1). Each pump has two chambers, an upper **atrium** and a lower **ventricle**, and two valves, an **atrioventricular** valve and a **semilunar** valve. The valves prevent back-flow of blood into the atria and ventricles. The right side of the heart pumps deoxygenated blood from the body to the lungs along the **pulmonary artery**.

Deoxygenated blood has a low concentration of oxygen and a high concentration of carbon dioxide. The left side of the heart pumps oxygenated blood from the lungs to the rest of the body along the **aorta**. Oxygenated blood has a high concentration of oxygen and a low concentration of carbon dioxide.

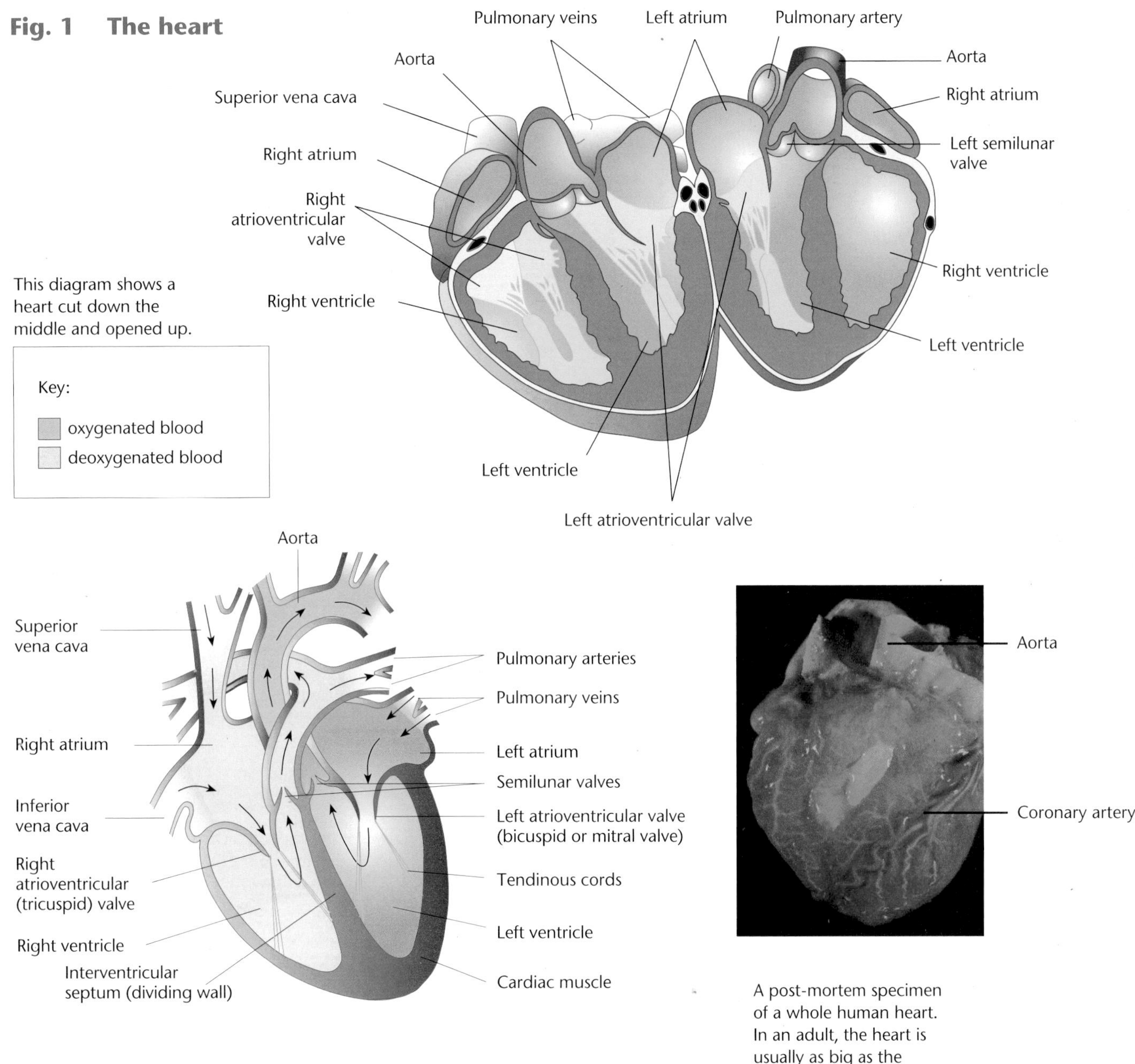

A post-mortem specimen of a whole human heart. In an adult, the heart is usually as big as the person's clenched fist.

6.2 The cardiac cycle

Blood enters the atria from the veins. Contraction of the muscles of the atrial wall forces blood into the ventricles. Blood is pumped out of the heart by contraction of the ventricle muscles. Contraction of atrial muscle is called atrial **systole**, and contraction of ventricle muscle is known as ventricular systole. Relaxation of heart muscle is called **diastole** (Fig. 2).

1 Use Fig. 2 overleaf to calculate how many complete cardiac cycles (heartbeats) there are per minute.

Fig. 2 The cardiac cycle

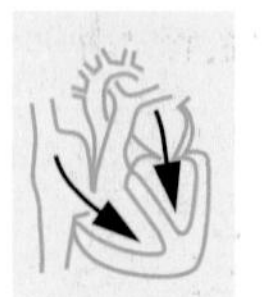
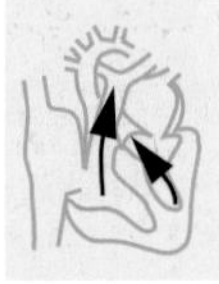

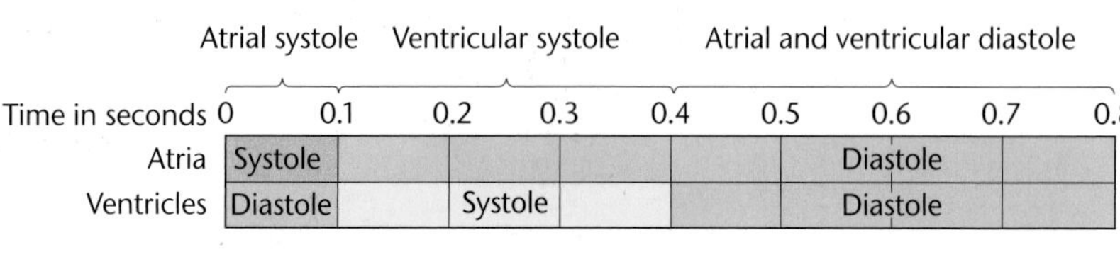

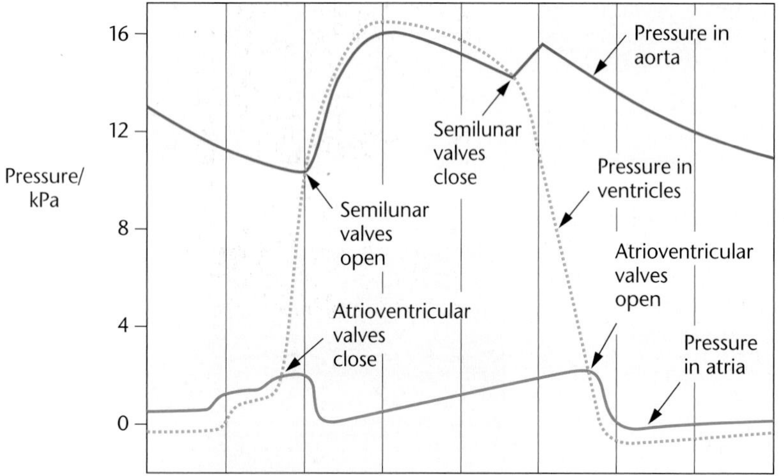

The cardiac cycle

- Pacemaker cells initiate systole (contraction).
- The squeezing of the muscle walls reduces the volume and so increases the pressure in the chamber(s), forcing blood in a particular direction.
- The direction of blood flow causes the valves to open or close.

As pressure builds up in the ventricles the atrioventricular valves are forced shut; this prevents backflow of blood from the ventricles to the atria. The semilunar valves prevent blood flowing back into the heart from the arteries. Fig. 3 shows how the valves work.

2 Suggest why tendinous cords ('heart strings') are attached to the edges of the valve flaps.

Fig. 3 Valves

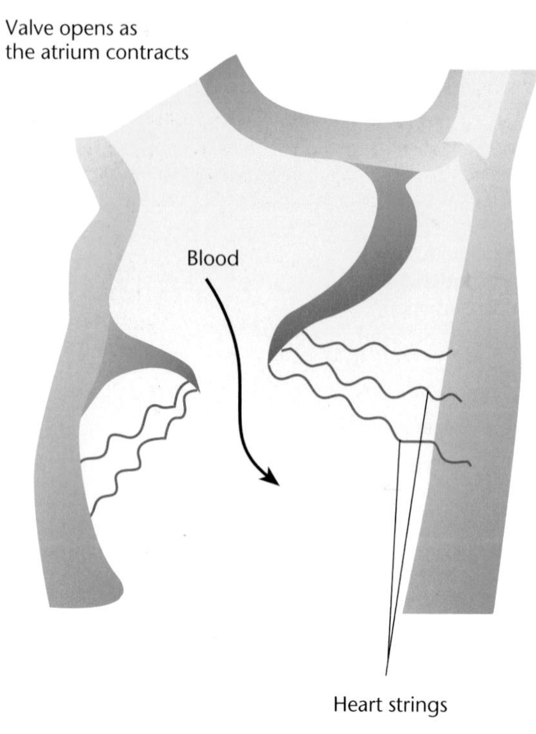

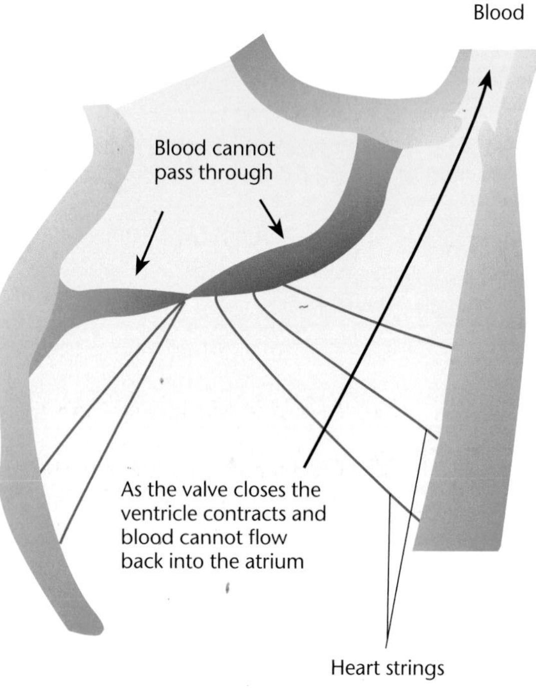

how science works

The active heart

A fetus does not breathe air. It obtains all its oxygen from its mother's blood via the placenta. The heart of the fetus and the main blood vessels are modified as shown in Fig. 4. Before birth there is a 'hole', the foramen ovale, between the right atrium and the left atrium, and a 'shunt', the ductus arteriosus, between the left pulmonary artery and the aorta. Usually, when the baby takes its first breath the 'hole' and the 'shunt' close but sometimes this does not happen. Such babies are said to have a 'hole in the heart'. These defects can now be detected shortly after birth. Many heal naturally; others are corrected by surgery.

Fig. 4 Circulatory system of a fetus

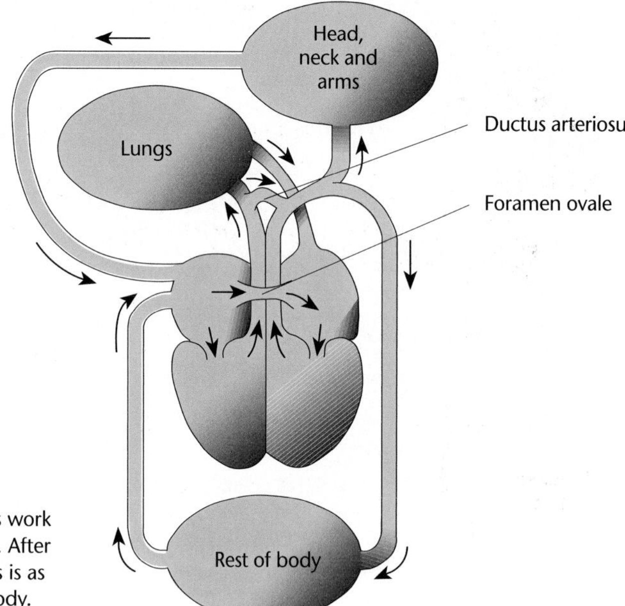

Before birth, the foramen ovale and the ductus arteriosus work together to greatly reduce the blood supply to the lungs. After birth, the amount of blood that passes through the lungs is as large as the amount that travels around the rest of the body.

3

a Suggest the advantage to the fetus of the 'hole' and the 'shunt'.

b Suggest how the circulation of blood is affected if the hole and the shunt do not close after birth.

4 Elderly people sometimes develop problems with heart valves. Suggest how each of the following problems can affect the circulation of the blood.

a The left atrioventricular valve does not close fully.

b The left semilunar valve does not close fully.

5 The healthy heart is a remarkably efficient pump. Suggest why the wall of the left ventricle is thicker than the wall of the right ventricle.

6

a At rest the heart pumps out about 75 cm^3 of blood at each beat. Measure your own resting heart rate by taking your pulse whilst sitting down. Count how many pulse beats you can feel during a minute. Do three readings and work out the average. Calculate how much blood your heart would pump out if you remained sitting or lying in bed for 24 hours.

b Your heart pumps out far more blood than this during a day. The more active you are, the faster the heart rate and the greater the volume of blood pumped out at each beat. What is the advantage to the body of an increased heart rate and beat volume?

7 Suggest what causes:

a the atrioventricular valves to close

b the semilunar valves to open

c the semilunar valves to close

d the atrioventricular valves to open.

6.3 Controlling the heart rate

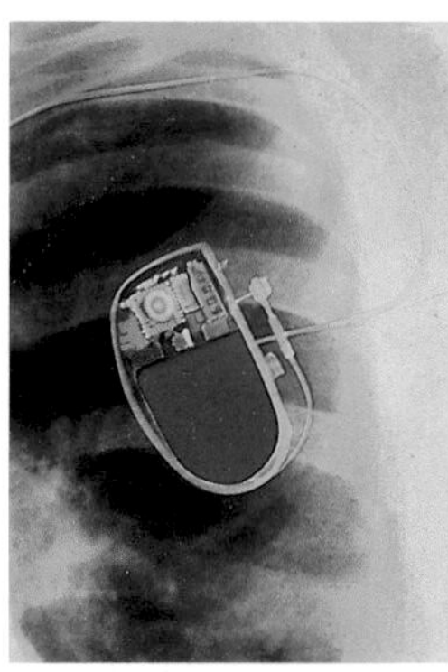

A coloured X-ray of the chest of a patient who has had a heart pacemaker fitted. The electronic battery-run device can be seen above the ribcage, with a yellow lead connecting it to the heart on the lower right of the picture. Most pacemakers are internal, implanted into the chest like this one, but they can also be external and worn on a belt fastened around the body. Pacemakers can supply a fixed rate of impulses or can discharge only when a heart beat is missed.

The beating of the heart is automatic; it happens without us thinking about it. However, unlike the muscles used in breathing movements, heart muscle does not need nerve impulses from the brain to keep up steady contractions. The heart has its own internal pacemaker, the **sinoatrial node** (SAN) (Fig. 5). The SAN is a group of cells in the wall of the right atrium that produces electrical impulses at regular intervals. These spread out across the muscles of the heart, causing the atria to contract first, followed by the ventricles. This is known as myogenic stimulation.

The electrical impulses are carried by specialised muscle cells that behave like nerve cells. As impulses travel along these muscle cells, the surrounding atrial muscles cells contract. This has a sort of 'domino' effect, causing neighbouring muscle cells to contract too. When the atria contract, the heart is said to be in **atrial systole**.

The impulses are prevented from spreading directly to the ventricle muscles by the layer of connective tissue that separates the atria and the ventricles. A group of receptor cells called the **atrioventricular node** (AVN) is found near to this junction. When impulses from the contracting atrial muscle cells reach the AVN, this starts impulses in a group of specialised conductive muscle fibres called the **Purkyne fibres**. These fibres group together to form the **bundle of His**, which passes down the wall between the two ventricles.

At the base of the ventricles the bundle of His divides into two branches, one passing to each ventricle. Fibres fan out from each of these and as impulses reach them from the bundle of His, the electrical activity causes the surrounding ventricle muscle to contract. Because the fibres in the bundle of His do not affect the muscle in the wall between the two ventricles, muscle contraction starts at the base of the ventricles and spreads upwards. Contraction of the ventricles is called **ventricular systole**.

The heart's natural pacemaker can develop problems – one of the most common is heart block. Damage to the electrically conductive tissue of the heart blocks the conduction of impulses from the atria to the ventricles. Hundreds of thousands of people in the UK have heart block. It can be treated by implantation of an artificial pacemaker. This device overcomes the heart block by sensing the level of electrical activity in the atrium and delivering an electrical impulse at the correct rate to the ventricle. The pacemaker mimics the natural activity of the SAN, making the heart rate speed up or slow down to suit the activity of the body.

Fig. 5 The internal pacemaker

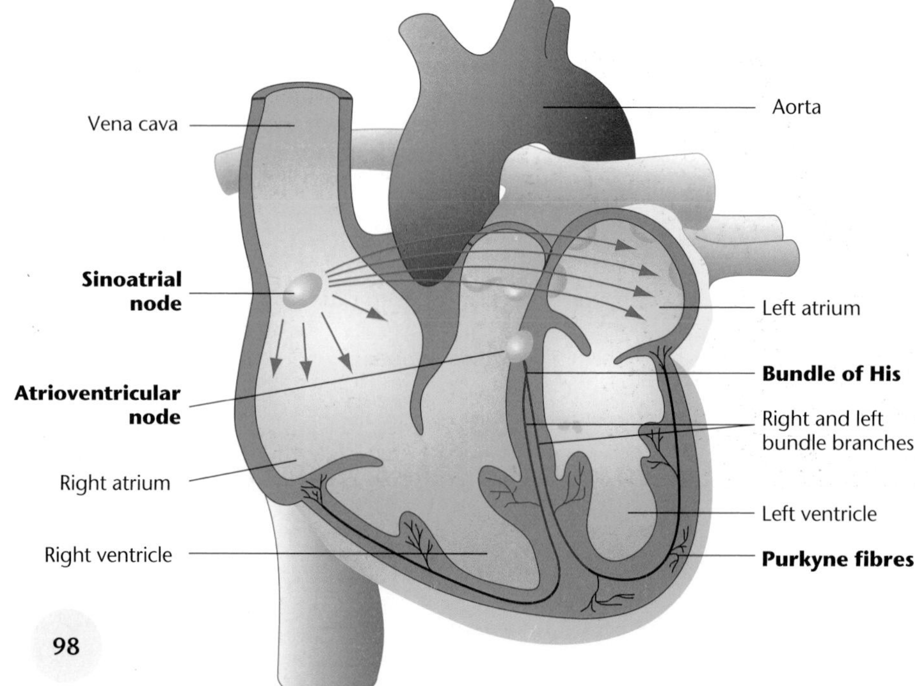

Impulses produced by the sinoatrial node are conducted through the atria by special muscle cells. The impulses stimulate the atrial muscles to contract, causing atrial systo

⇩

The impulses stimulate the atrioventricular node at the base of the atria.

⇩

Impulses from the atrioventricular node pass down fibres, collectively called the bundle of His, in the muscle separating the right and left ventricles.

⇩

The bundle of His divides into right and left branches, whic spread through the walls at the base of the ventricles.

⇩

Purkyne fibres carry impulses up through the walls of the ventricles.

⇩

Impulses from the Purkyne fibres cause the ventricle muscles to contract from the base upwards, causing ventricular systole.

key facts

- The **sinoatrial node** (SAN) is the 'natural' pacemaker of the heart.
- The SAN does not require impulses from the nervous system to initiate electric impulses.
- Impulses are conducted from the SAN to the atria, and then on to the ventricles.
- These impulses cause the atria to contract, followed by the ventricles.

A practice ISA on the effect of a specific variable on human heart rate or pulse rate can be found at www.collinseducation.co.uk/advancedscienceaqa

how science works

The electrocardiogram (ECG)

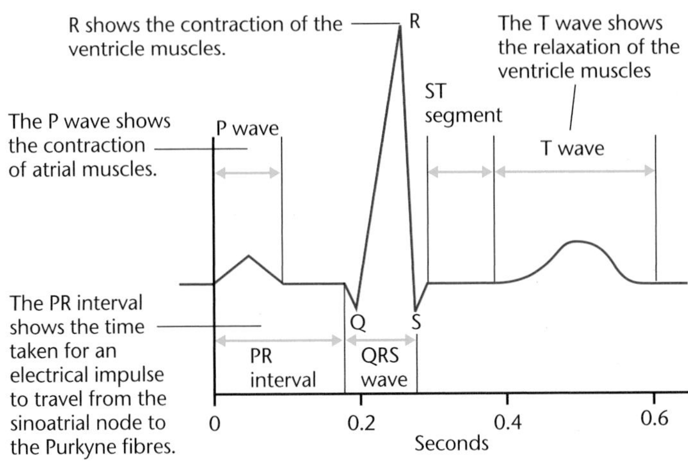
R shows the contraction of the ventricle muscles.
R
The T wave shows the relaxation of the ventricle muscles
ST segment
The P wave shows the contraction of atrial muscles.
P wave
T wave
The PR interval shows the time taken for an electrical impulse to travel from the sinoatrial node to the Purkyne fibres.
Q
S
PR interval
QRS wave
0
0.2
0.4
0.6
Seconds

A machine called an electrocardiograph can record electrical impulses as they pass from the SAN through the heart. The diagram shows how an ECG is related to pressure changes in the cardiac cycle.

If a thrombosis cuts off the blood supply to Purkyne fibres, they die and prevent electrical impulses from passing to healthy muscle cells. This is the main cause of death from heart attack. Damage to any part of the heart's muscle can be recognised by changes to the ECG of the patient. If an ECG shows a large PR interval, it probably means that there has been damage to the conducting fibres in the region of the atrioventricular node. On seeing such a trace a doctor would consider fitting the patient with a pacemaker.

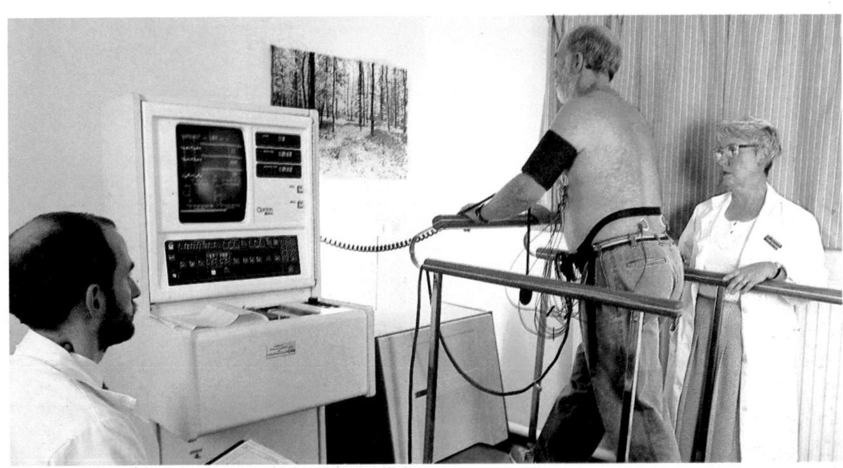

This patient is exercising whilst an electrocardiograph records the electrical activity of his heart via the various electrodes taped over his chest and back.

6.4 The biological basis of heart disease

You have already learnt about the causes of heart disease in Chapter 1. Coronary arteries are the blood vessels that supply blood to the muscles in the walls of the heart. Like other vessels in the body, these arteries may become partly blocked by a build up of fatty tissue, called **atheroma**. This restricts the blood supply to the heart muscle. Any muscle in the body that is short of oxygen gives rise to pain and the heart muscle is no exception. In more severe cases the pain may be experienced even at rest. This is a condition known as **angina**.

A blood clot, called a thrombosis, may develop on the surface of the plaques of atheroma. When this happens, the blood vessel can become completely blocked and oxygenated blood is no longer able to reach the heart muscle. Without oxygen, the heart muscle dies and the result is a **myocardial infarction**, or heart attack. If the area of muscle affected is small, the person may recover; however, heart attacks that are caused by the death of large parts of the cardiac muscle are usually fatal.

High blood cholesterol is a risk factor for coronary heart disease (CHD). However, a much more reliable indication comes from the levels of two lipoproteins found in the plasma; low-density lipoprotein (LDL) cholesterol and high-density lipoprotein (HDL) cholesterol.

LDL cholesterol is 'bad' because it causes cholesterol to build up inside blood vessels. HDL cholesterol is 'good' because it actually removes cholesterol from the walls of blood vessels and brings it back to the liver to be safely excreted. HDL helps to 'clean' your arteries from cholesterol build up. The ratio between your total cholesterol and HDL cholesterol is an important indicator of heart disease. The lower this ratio the better.

Fig. 7 shows the effect of some of the factors on the risk of CHD.

8 List the risk factors for coronary heart disease.

Look at Fig. 7.

9 What is the risk of developing CHD for a male non-smoker aged 68 years with a systolic blood pressure of 160 mmHg and a total cholesterol/HDL ratio of 4?

10 Describe, in as much detail as you can, the effect of each of these on the risk of CHD:

a gender

b smoking

c age

d total cholesterol/HDL cholesterol ratio.

Fig. 6 Healthy and unhealthy arteries

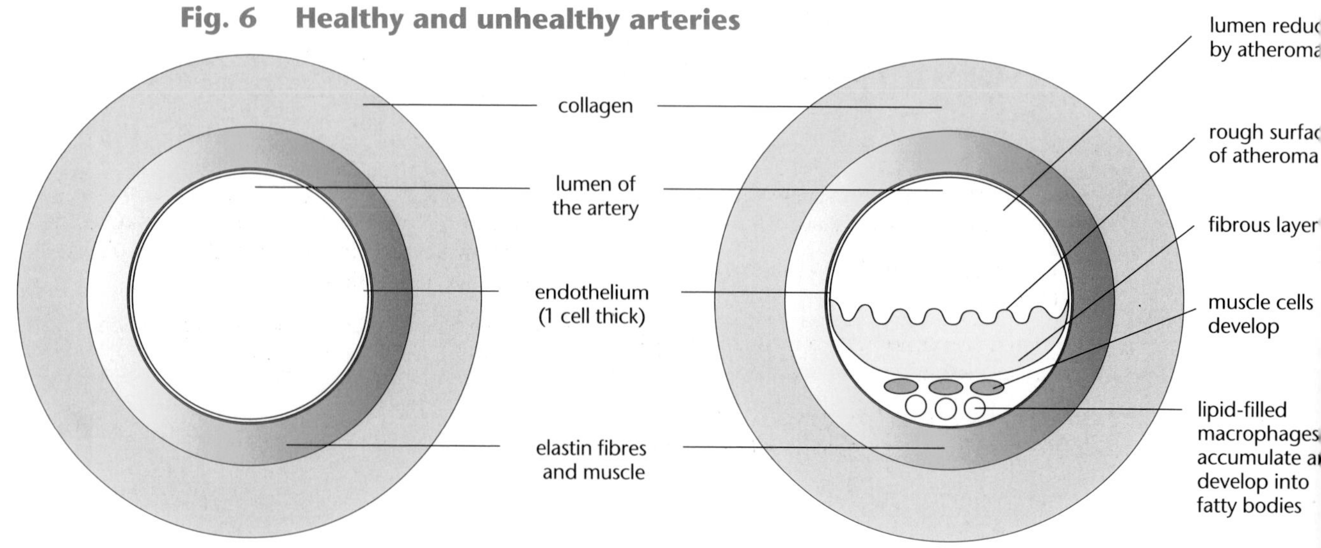

Fig. 7 Coronary risk prediction chart

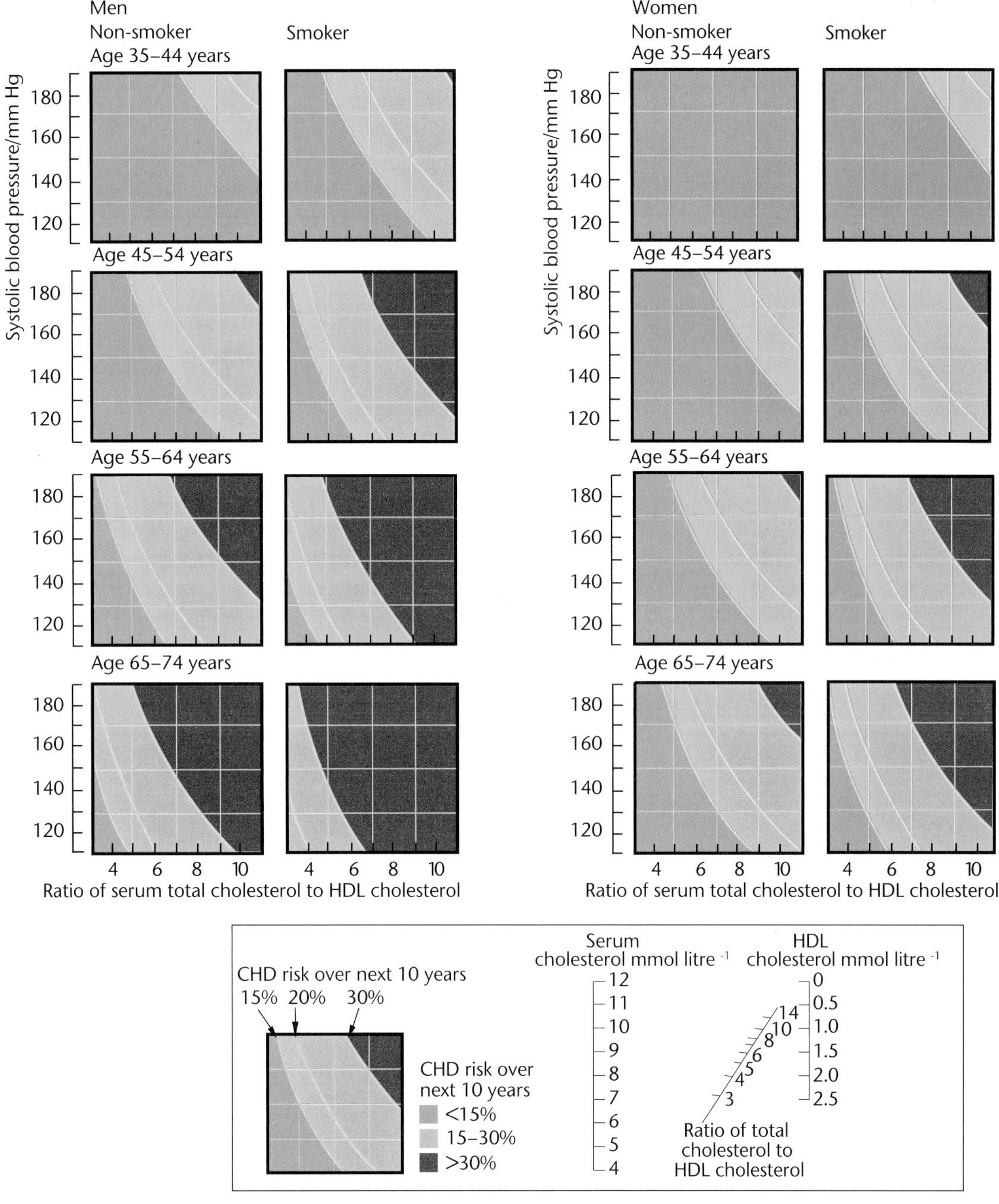

key facts

- Diseases of the heart and circulation system are called cardiodiovascular diseases.
- Atheriosclerosis is a build-up of atheroma in blood vessels.
- A weakened and ballooning area of an artery is called an aneurysm.
- Blood clots may block blood vessels. Clots may be mobile (embolus) or stationary (thrombus).
- Coronary heart disease (CHD) is caused by a blockage in the coronary arteries. This may cause a heart attack due to the death of muscle tissue (myocardial infarction).
- Risk factors for CHD include diet, high blood cholesterol, smoking and high blood pressure.

how science works

Keep those arteries open

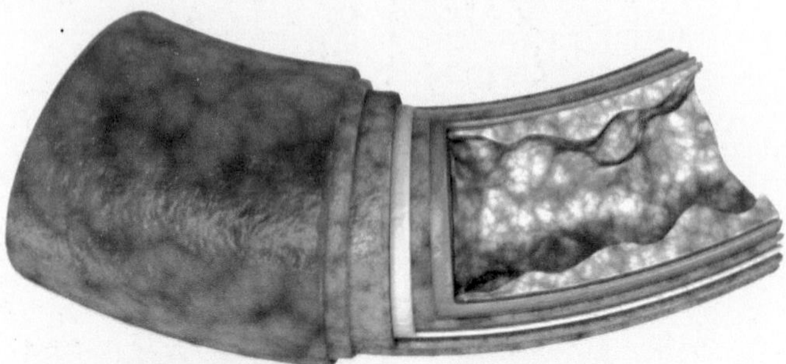

An artery narrowed by atheroma

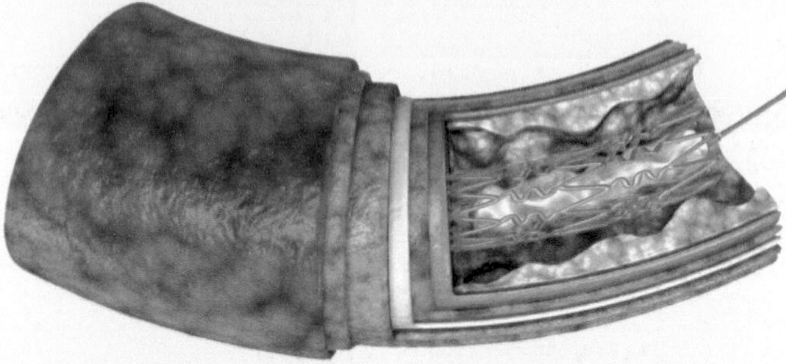

A narrowed artery with catheta in place and balloon inflated

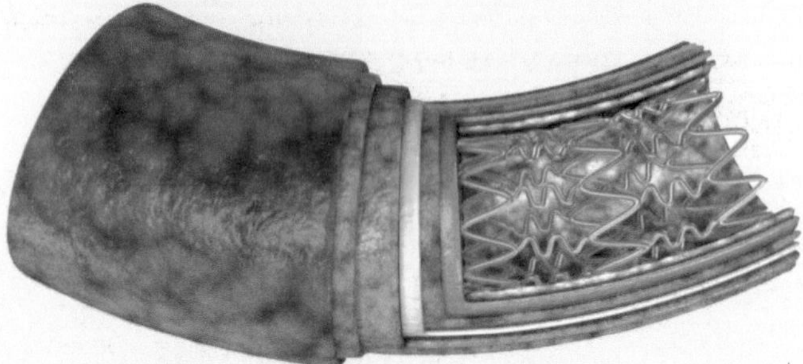

An artery with a stent in place

Two surgical methods are used to keep diseased coronary arteries open and so prevent heart attacks.

- **Balloon angioplasty**. Cardiologists insert a balloon catheter into the blood vessel where blood flow is obstructed. When inflated, the balloon expands the inside walls of the vessel, compressing the fatty material blocking the vessel, and clearing some of it away. When the balloon is removed, the vessel has a much wider diameter. However, the technique involves some slight damage to the inside of the vessel and scarring can occur. This encourages more fatty material to be laid down and the narrowing recurs, sometimes within weeks. This condition is known as **restenosis**.
- **Stenting**. This is a modification of the angioplasty technique in which surgeons use the balloon but they insert a mesh tube into the blood vessel at the same time. This rigid tube is left permanently inside the vessel to keep it open.

People who have had stents placed in their blood vessels have fewer heart attacks and require fewer treatments to restore blood flow to the heart.

One Dutch research project studied 227 heart attack patients: 112 were randomly selected to receive stents while the remaining 115 were treated with conventional balloon angioplasty. One patient with a stent had another heart attack, compared with eight of the people who had had angioplasty. Only four patients with stents needed further treatment for blocked arteries, compared with 19 patients who had had angioplasy.

11 What are the ethical issues involved in using different treatments on two groups of patients?

examination questions

1 The diagram shows the human heart.

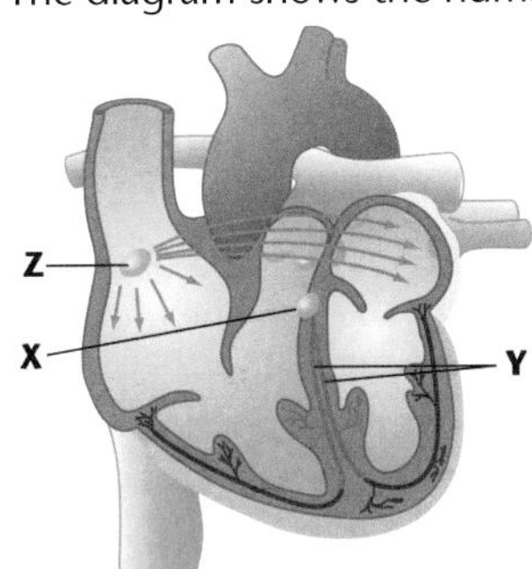

a Name the structures labelled **X, Y** and **Z**. (3)

b Explain the role of structures X, Y and Z in initiating and controlling heartbeat. (5)

c The table shows the maximum heart rate, maximum stroke volume and maximum cardiac output for a sedentary student and an athletic student.

Student	Maximum heart rate / beats min^{-1}	Maximum stroke volume / cm^3	Maximum cardiac output / dm^3 min^{-1}
Sedentary	200	100	20
Athlete	190		30.4

i Calculate the maximum stroke volume for the athletic student. (2)

ii Explain the advantage to the athletic student of the higher maximum cardiac output. (4)

Total 14

2

a Explain how each of the following increase the risk of heart disease:

i high blood pressure; (2)

ii high blood cholesterol; (2)

iii cigarette smoking. (2)

b The diagram shows the effect of different factors on the risk of heart disease.

High Blood Pressure 1.5 × greater
3.5 × greater
High Cholesterol 2.3 × greater
6.2 × greater
2.8 × greater
4.0 × greater
Diabetes 1.8 × greater

Describe in detail the effect on the risk of heart disease of

i having two risk factors; (2)

ii having three risk factors. (2)

c The graph shows trends in risk factors for heart disease in the USA between 1960 and 2000

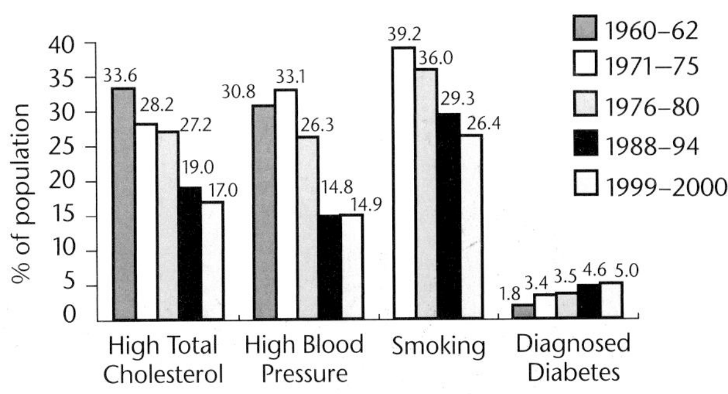

i Describe the trend for high blood pressure between 1960 and 2000. (2)

ii Use information from the graph to suggest and explanation for the trend in blood pressure. (2)

Total 14

3 The diagram shows a mammalian heart. Structures **A** and **B** are involved in coordinating the heart beat.

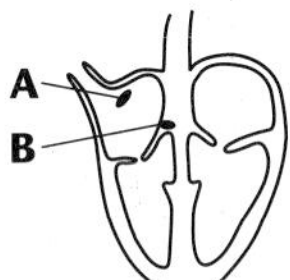

a Name structures **A** and **B**. (2)

b Describe the role of structure **A** in coordinating the heart beat. (2)

c The graph shows the changes in pressure which take place in the left side of the heart.

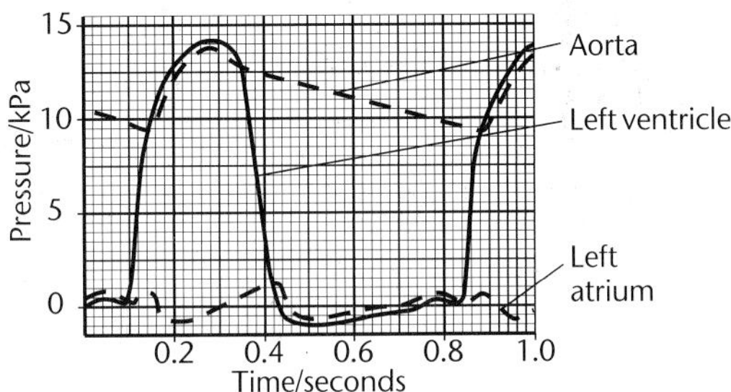

i Use the graph to calculate the heart rate in beats per minute. Show your working. (2)

ii The atrioventricular valve closes at 0.1 seconds. Explain the evidence from the graph which supports this statement. (1)

d The blood pressure in the aorta is higher than in the pulmonary artery.
Explain what causes the blood pressure in the aorta to be higher. (1)

Total 8

AQA, January 2002, Unit 3, Question 5

4

a The times taken in the various stages of a complete cardiac cycle are shown in the table on the next page.

examination questions

Stage of cardiac cycle	Time taken/s
Contraction of the atria	0.1
Contraction of the ventricles	0.3
Relaxation of both atria and ventricles	0.4

i Use the information in the table to calculate the heart rate in beats per minute. (1)
ii If the same rate of heartbeat were maintained throughout a twelve-hour period, for how many hours would the ventricular muscle be contracting? Show your working. (2)

b Although the heart does have a nerve supply, the role of the nervous system is not to initiate the heartbeat but rather to modify the rate of contraction. The heart determines its own regular contraction.
i Describe how the regular contraction of the atria and ventricles is initiated and coordinated by the heart itself. (5)
ii Describe the role of the nervous system in modifying the heart rate in response to an increase in blood pressure. (5)

c An interventricular septal defect is an opening in the wall (septum) that separates the left and right ventricles. Suggest and explain the effect of this defect on blood flow through the heart. (2)

Total 15

AQA, June 2002, Unit 3, Question 7

5

a The diagram shows a human heart.

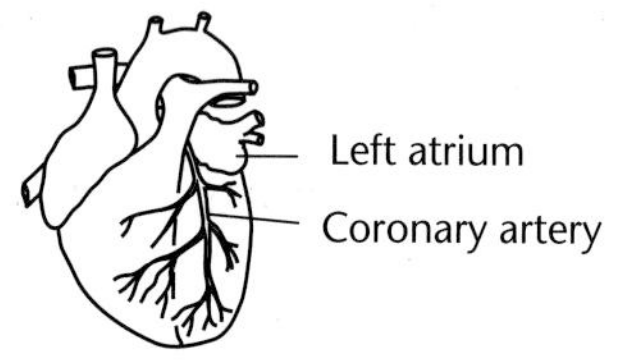

i Use a guideline and label to show the position of the sinoatrial node (SAN). (1)
ii What is the function of the coronary artery? (1)

b Impulses spread through the walls of the heart from the SAN. The table shows the rate of conduction of impulses through various parts of the conducting tissues.

Part of pathway	Rate of conduction/ms^{-1}	Mean distance /mm
From SAN to atrioventricular node (AVN) across atrium	1.0	40
Through AVN	0.05	5
From AVN to lower end of bundle of His	1.0	10
Along Purkyne fibres in ventricle walls	4.0	–

i Calculate the mean time taken for an impulse to pass from the SAN to the lower end of the bundle of His. Show your working. (2)
ii Explain the advantage of the slow rate of conduction through the AVN. (2)
iii Suggest **one** advantage of the high rate of conduction in the Purkyne fibres which carry impulses through the walls of the ventricles. (1)

c How would cutting the nerve connections from the brain to the SAN affect the beating of the heart? (1)

Total 8

AQA, June 2003, Unit 3, Question 4

6

The graph shows changes in pressure in different parts of the heart during a period of one second.

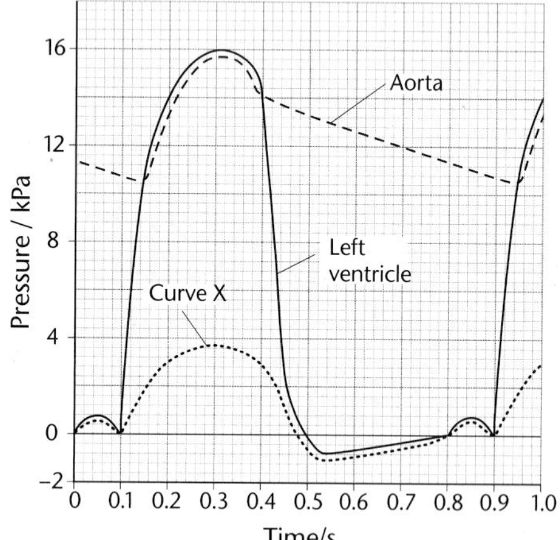

a **i** At what time do the semilunar valves close? (1)
ii Use the graph to calculate the heart rate in beats per minute. Show your working. (1)
iii Use the graph to calculate the total time that blood flows out of the left side of the heart during one minute when beating at this rate. Show your working. (1)

b What does curve **X** represent? Explain your answer. (2)

c The volume of blood pumped out of the left ventricle during one cardiac cycle is called the stroke volume. The volume of blood pumped out of the left ventricle in one minute is called the cardiac output. It is calculated using the equation

Cardiac output = stroke volume × heart rate

After several months of training, an athlete had the same cardiac output but a lower resting heart rate than before. Explain this change. (2)

Total 7

AQA, June 2006, Unit 3, Question 2

Give statins to children from the age of eight

(Daily Express, 7 August 2007)

Statins should be given to at-risk children

(The Daily Telegraph, 7 August 2007)

Heart kids' drug hope

(Daily Mirror, 7 August 2007)

Familial hypercholesterolaemia is a hereditary condition resulting in high blood cholesterol levels, even in children.

Statins are drugs used to help reduce the amount of cholesterol in the blood. High levels of cholesterol are a risk factor for heart disease. By lowering unhealthy levels of cholesterol, the risk of coronary heart disease (CHD) and heart attacks is reduced.

The headlines refer to an investigation conducted by Dr Rodenburg and colleagues from the University of Amsterdam in the Netherlands. The doctors objective was to evaluate the safety and efficacy of statin treatment in children.

The investigation involved a double-blind randomised controlled trial of 186 children with familial hypercholesterolaemia (age range 8–18.5 years) comparing a drug called pravastatin with a placebo. The mean age of the 186 children was 13.7 years, and 49% were boys. Children under the age of 14 years in the treatment group received 20 mg pravastatin; children aged 14 years and over received 40 mg. The trial lasted 2 years, after which children in the treatment group continued on pravastatin, while children in the placebo group were also prescribed pravastatin. Children were followed up for at least 2 years after the completion of the original trial. The doctors measured changes in the thickness of the carotid intima media (part of the wall of the carotid arteries – an indicator of the risk of cardiovascular disease), cholesterol and triglyceride levels.

On average, statin treatment reduced total cholesterol by 22.5%, low-density lipoprotein (LDL) cholesterol by 29.2% and triglycerides by 1.9%. High-density lipoprotein (HDL) cholesterol increased by 3.1%.

Statistical analysis showed that combined carotid intima media thickness and age at start of statin treatment, gender, and the duration of statin use, were all independent predictors of carotid intima media thickness. This implies that the younger that statin therapy is started, the smaller the increase in carotid intima-media thickness at follow-up.

No serious adverse events were reported; none of the children discontinued treatment because of an adverse event.

The doctors concluded that starting statin treatment early delays the progression of carotid intima media thickening in adolescents and young adults with familial hypercholesterolaemia. They also acknowledged that the optimal age for starting statin therapy is still unknown, and that further longer term follow-up of children who receive early treatment is needed.

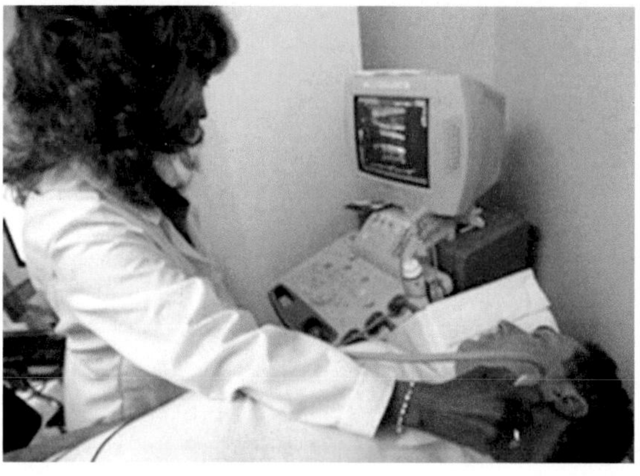

Ultrasound is used to measure the thickness of the carotid intima media.

A1

a Explain what is meant by a 'double-blind randomised controlled trial'.

b What would be the placebo in this investigation?

A2 What was the independent variable in this investigation?

A3 What were the dependent variables in this investigation?

A4 Suggest what is meant by 'independent predictors'.

A5 Comment on the safety of giving statins to children.

A6 Comment on the effectiveness of the statins.

A7 Comment on the reliabilty of the results of this investigation.

A8 Comment on the accuracy of the three headlines that reported the results of this investigation.

7 Protection against disease

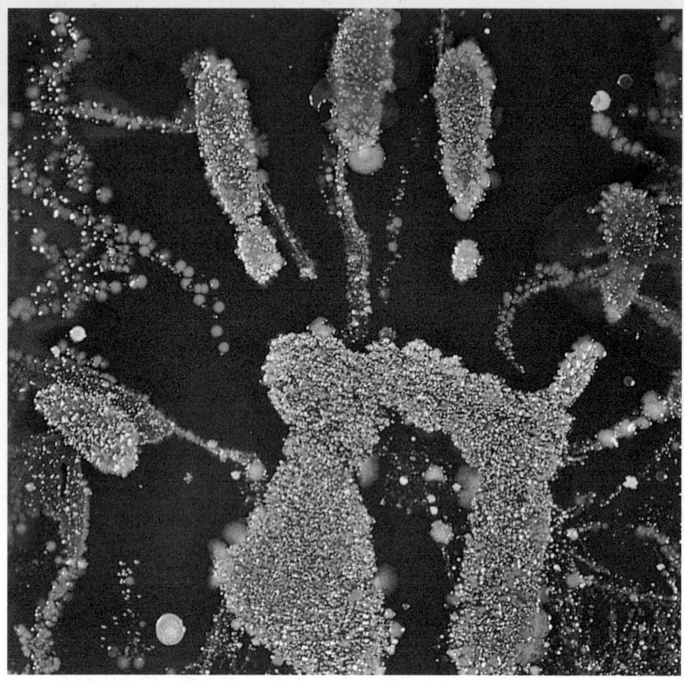

The hills are alive – with the sound of microbes. And it's not only the hills. The environment is literally alive with countless billions of bacteria, viruses, fungi, and single-celled animals and plants. Every surface, every breath of air and everything we eat is covered in microorganisms of one type or another. However, we are not continually suffering from infections. One reason is that most of the microorganisms we meet are harmless to humans. Another reason is that the body has a range of ways to fight the ones that could harm us. As well as mechanisms to stop bacteria getting into our body, such as skin, we also have ways to fight them off if they do get inside. Our immune system is the main player in that battle. It doesn't always succeed of course, and certain infections can cause serious illness or even death. But most of the time it does its job of keeping us free from infection.

Bacterial colonies growing in the print of a human hand on agar gel. Normally the skin is populated by its own colonies of beneficial bacteria. They help to defend the skin against harmful bacteria. Bacteria dislodged from the hand's surface have fed on the nutrients in the agar gel and multiplied quickly.

7.1 Natural defence mechanisms

Fig. 1 Organs of the immune system

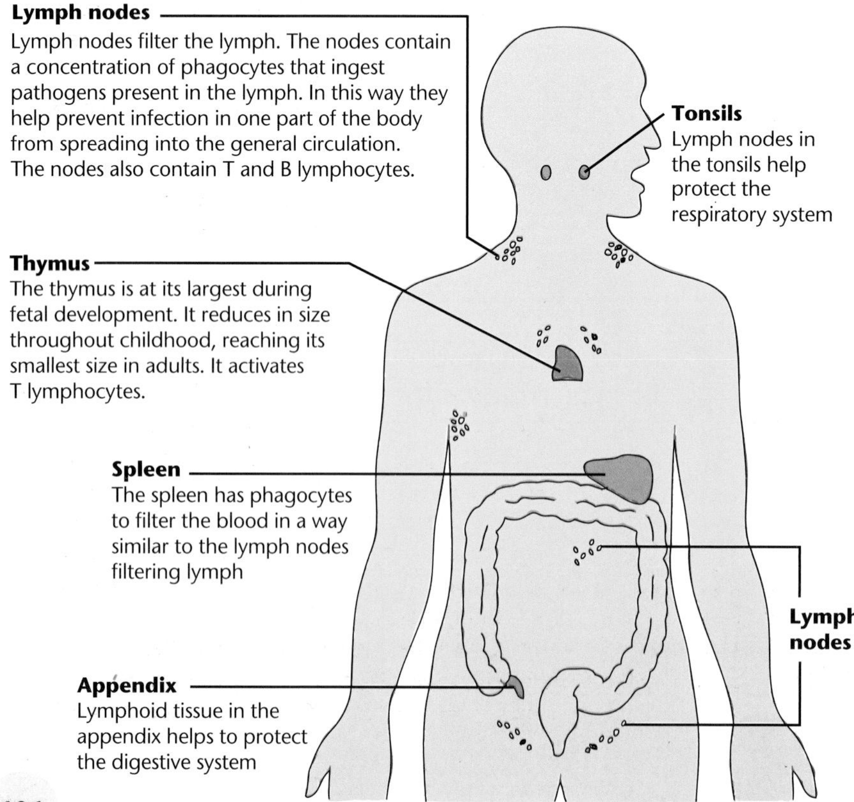

Lymph nodes
Lymph nodes filter the lymph. The nodes contain a concentration of phagocytes that ingest pathogens present in the lymph. In this way they help prevent infection in one part of the body from spreading into the general circulation. The nodes also contain T and B lymphocytes.
Tonsils
Lymph nodes in the tonsils help protect the respiratory system
Thymus
The thymus is at its largest during fetal development. It reduces in size throughout childhood, reaching its smallest size in adults. It activates T lymphocytes.
Spleen
The spleen has phagocytes to filter the blood in a way similar to the lymph nodes filtering lymph
Lymph nodes
Appendix
Lymphoid tissue in the appendix helps to protect the digestive system

Our immune system protects us from being infected by the bacteria that we are continually exposed to. The white blood cells play a key role in the functions of the immune system. The main organs in the immune system are the **thymus**, **spleen** and **lymph nodes** (Fig. 1).

Phagocytosis

Several types of white blood cells, principally **neutrophils** and **macrophages**, can take in bacteria and viruses by protruding **pseudopodia** (cytoplasmic 'arms') to flow around the pathogen (Fig. 2).

The pseudopodia then fuse with a membrane enclosing the pathogen to form a structure called a **phagosome**. The phagosome moves deeper into the cell, and fuses with a **lysosome**, forming a **phagolysosome**. The lysosome contains enzymes, principally **lysozyme** and hydrolytic enzymes. Lysozyme destroys the bacterial cell walls, allowing hydrolytic enzymes to digest the rest of the pathogen.

1 What is meant by hydrolytic enzymes?

Fig. 2 Phagocytosis

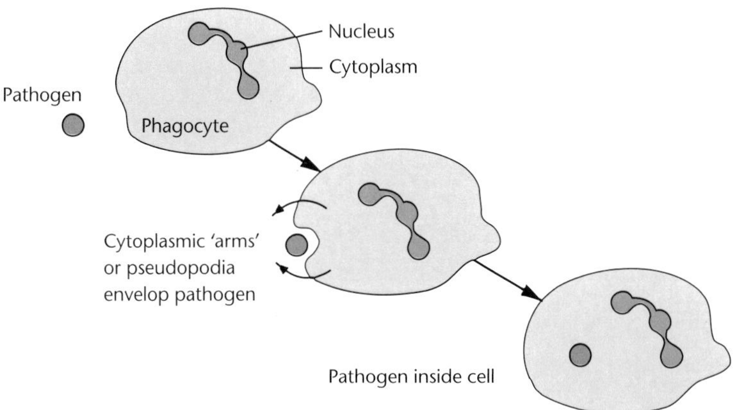

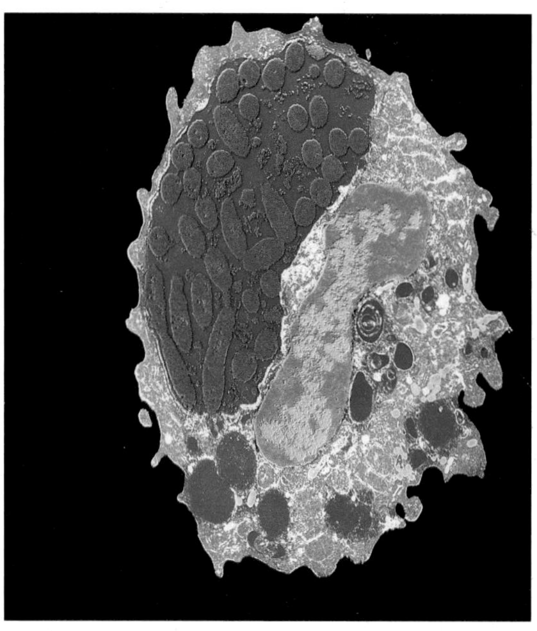

Coloured transmission electron micrograph of a macrophage in a human lung. The macrophage has a dumb-bell-shaped nucleus (orange and green) and has engulfed many cells of *Legionella pneumophila*, the bacteria that cause legionnaires' disease. The bacteria (red) are all in a cell vacuole (dark blue).

Antigens and antibodies

An immune response is triggered when the body detects the presence of foreign material. A substance that causes an immune response is termed an **antigen**. Antigens are large molecules, usually located in the outer cell membrane or cell wall of an invading organism, or on the surface of a virus. Antigens are usually proteins, polysaccharides or **glycoproteins** (combinations of proteins and polysaccharides). Specific immune responses are aimed at particular types of antigen. Some toxins produced by bacteria are also antigens.

The immune response is brought about by white blood cells. The main types are macrophages and **lymphocytes**. Macrophages are phagocytic, but are also involved in the immune response. Some lymphocytes attack pathogens directly; these cells are called **T cells**. Some cells produce **antibodies**, special proteins that can kill invading pathogens; these cells are called **B cells**. There are two types of immune response:

- **cell-mediated response** (when T cells attack pathogens directly)
- **humoral response** (when B cells produce antibodies to kill invading pathogens).

Macrophages

Macrophages ingest foreign cells and viruses. This results in some of the antigen molecules becoming embedded in the macrophage cell membrane and this alerts both T cells and B cells to the fact that the body has been invaded by a particular pathogen.

T cells

T cells develop from **stem cells** in the bone marrow, and then pass to the thymus gland, where they are changed to make them active in the immune response. This activation process is said to make the cells **competent**. T cells are mainly found in the **lymphatic system**. Embedded in their cell membranes are proteins called T-cell receptors. There are millions of types of T cell, each with a different type of T-cell receptor; each T cell can only recognise and respond to one type of antigen.

B cells

B cells, like T cells, are derived from stem cells in the bone marrow. However, whereas T cells are activated in the thymus gland, we do not know what activates B cells in mammals. In birds, they are activated by the organ called the **bursa** (hence the name B cells). The plasma membrane of each B cell has protein molecules that are specific for a particular antigen. These proteins are called antibodies and are released into the lymph. Each B cell has a single antibody. When the antibody matches a particular antigen, the B cell is said to be competent. It is now known as a **plasma cell**.

Cellular response

T cells can detect bacteria but not isolated viruses. When a virus infects a cell, it forces the cell to make viral protein. Some viral protein gets into the cell membrane of the host cell, and is recognised as foreign by certain T cells (Fig. 3).

Humoral response

After infection, a macrophage with antigen embedded in its membrane causes a competent helper T cell to interact with the appropriate competent B cell (Fig. 4).

The increased activity in lymph nodes during times of infection results in a characteristic swelling of lymphatic tissue near the area of infection. A viral or bacterial infection of the upper pharynx (a sore throat) often results in swollen tonsils because the lymph nodes in the tonsils are where the B cells are producing antibodies.

2 What are the main differences between the cellular response and the humoral response?

Interferon

Cells respond to viral infections by producing glycoproteins called **interferons**. They stimulate other blood cells to produce a range of antiviral proteins that stop cells invaded by viruses from manufacturing substances necessary for viral reproduction. Interferons also stimulate certain types of cell to identify and destroy cells that have been infected by viruses. Through **genetic engineering** it has become possible to incorporate the human gene for interferon production into the **DNA** of the bacterium *Escherichia coli* and yeast cells. This has allowed mass production of a potentially life-saving drug, which may have a role in the treatment of cancer, where it seems to help in suppressing the growth of tumours. It has also been studied in terms of its effectiveness in treating the common cold, **herpes zoster**, viral hepatitis and **Kaposi's sarcoma**.

Interferons tend to be host specific. Human interferons only function in humans. Unlike **insulin**, which can be produced using non-human sources (e.g. pigs) for human use, interferon for human use has to come from human DNA.

Fig. 3 Cellular response

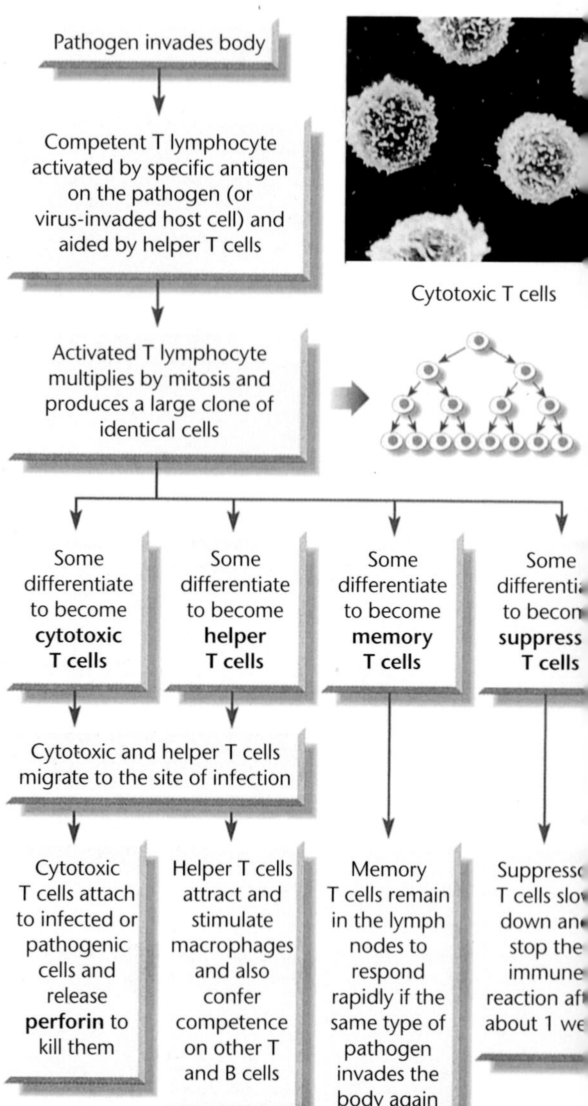

Pathogen invades body
Competent T lymphocyte activated by specific antigen on the pathogen (or virus-invaded host cell) and aided by helper T cells
Cytotoxic T cells
Activated T lymphocyte multiplies by mitosis and produces a large clone of identical cells
Some differentiate to become cytotoxic T cells
Some differentiate to become helper T cells
Some differentiate to become memory T cells
Some differenti to becom suppress T cells
Cytotoxic and helper T cells migrate to the site of infection
Cytotoxic T cells attach to infected or pathogenic cells and release perforin to kill them
Helper T cells attract and stimulate macrophages and also confer competence on other T and B cells
Memory T cells remain in the lymph nodes to respond rapidly if the same type of pathogen invades the body again
Suppress T cells slo down an stop the immune reaction af about 1 we

3 What is a glycoprotein?

4 Why would restricting viral replication help to prevent infection?

5 Suggest why insulin from pigs can be used to treat humans, whereas pig interferon is of no use in treating humans.

6 What is Kaposi's sarcoma?

Fig. 4 Humoral response

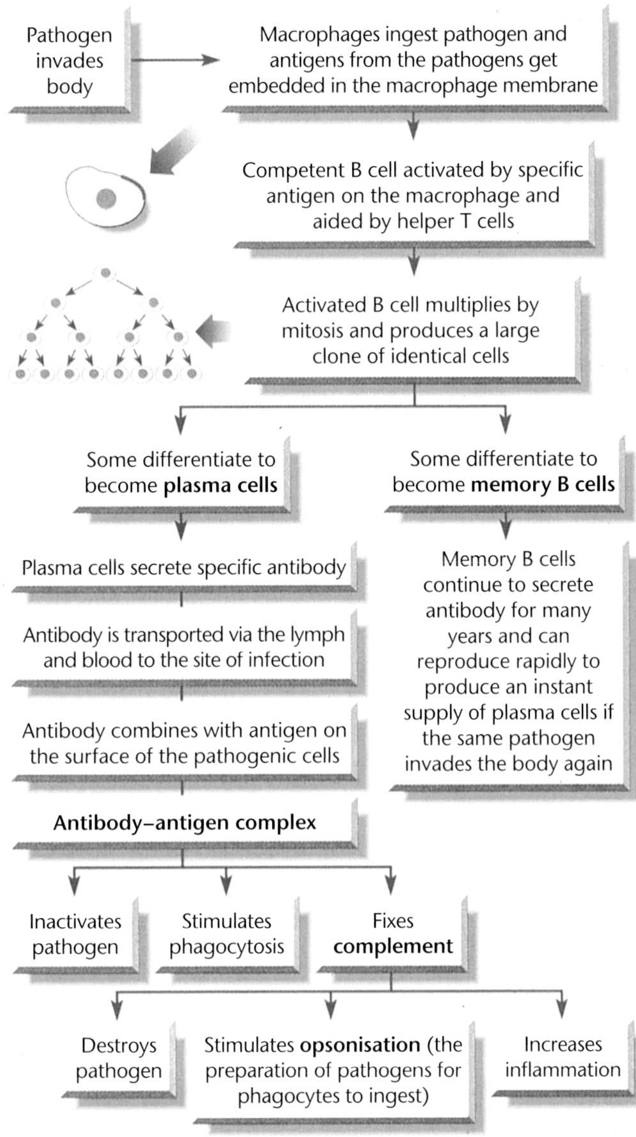

key facts

- The immune system is made up of the thymus, spleen, white blood cells, and the blood and lymph vessels.
- Lymph nodes contain macrophages waiting to ingest foreign material in the lymph – they filter the lymph.
- A substance that causes the immune reaction is called an antigen.
- There are two types of immune response: cellular and humoral.
- The cellular response depends on **cytotoxic** T cells directly attacking pathogens.
- The humoral response depends on B cells producing specific antibodies to destroy pathogens.
- Helper T cells alert both T and B cells to the presence of a pathogen. The alerted cells reproduce rapidly to produce clones of cells. Some of the cells are **memory cells,** which enable the body to recognise the pathogen very quickly if it attacks again.
- Antigens are usually large molecules – proteins, polysaccharides or glycoproteins.

Fig. 5 Antibody structure and function

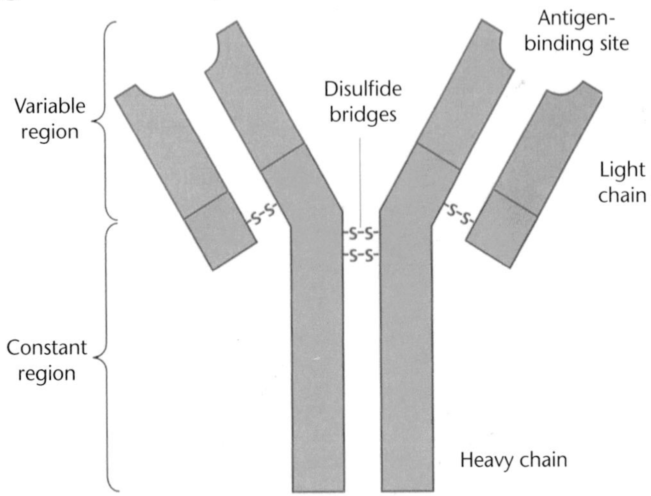

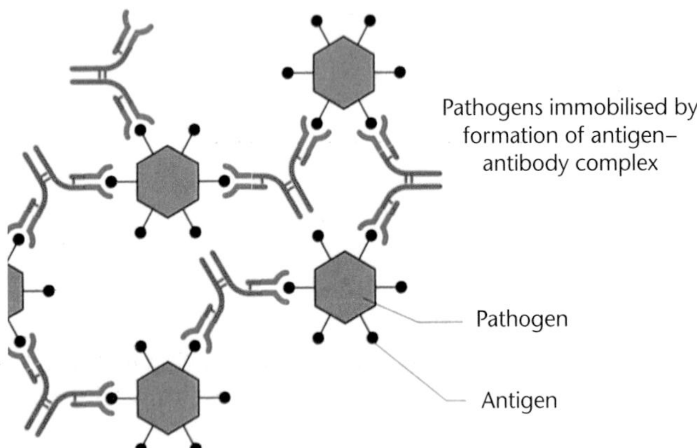

Antibody structure and function

Antibodies are found in plasma, tissue fluid and milk. They are proteins called immunoglobulins.

The variable region of the antibody is the part that combines with the antigen (Fig. 5).

There is a three-dimensional fit between the amino acid chain in the antibody and the antigen. The fit between the two molecules is similar to that between an enzyme and its substrate, although not as precise.

There are five different types of immunoglobulin in humans. The differences are in the constant region of the amino acid chains. The five types are known as IgG, IgM, IgA, IgD and IgE. IgG is usually referred to as **gamma globulin** (γ-globulin), and is the most abundant globulin found in blood. IgA is found in many of the body's secretions such as tears and mucus.

7 What type of cells produces antibodies?

How antibodies work

Antibodies have two binding sites, each of which can combine with a separate antigen molecule. This allows a number of antibody molecules to combine with a number of antigen molecules to form a lattice-like structure called an **antibody–antigen complex**. The antibody–antigen complex immobilises virus particles so they cannot latch on to host cells. The interlocking of antibody and antigen can render a toxic antigen harmless if its active region is blocked by an antibody molecule.

Phagocytes can more easily track down and ingest pathogens if they are immobilised in an antibody–antigen complex. The antibody–antigen complex also stimulates the activation of a number of plasma proteins. This leads to **complement** binding with the complex and destruction of the pathogens. Complement is a group of plasma proteins that work with (complement) antibody activity to eliminate pathogens. Sometimes, some of the proteins cover the outer membrane of the pathogens so that phagocytes can ingest them more easily. The preparation of pathogens for ingestion by phagocytes is called **opsonisation**.

8 Why do antibodies have two active sites on each molecule?

Primary response

The **primary response** occurs the first time an individual comes into contact with a particular antigen. It takes 3–14 days after infection for the body to produce antibodies. This period between infection and the onset of antibody production is the **latent period**. After the latent period, the amount of antibody in the blood rises rapidly and then begins to fall. During the immune response, memory cells are produced. These are **clones** of the lymphocytes that fought off the pathogen, and they remain in the body as a long-term defence against a second or subsequent infection by the bactera or virus.

Secondary response

The **secondary response** occurs if the body encounters the same antigen again. A much smaller amount of antigen will induce the

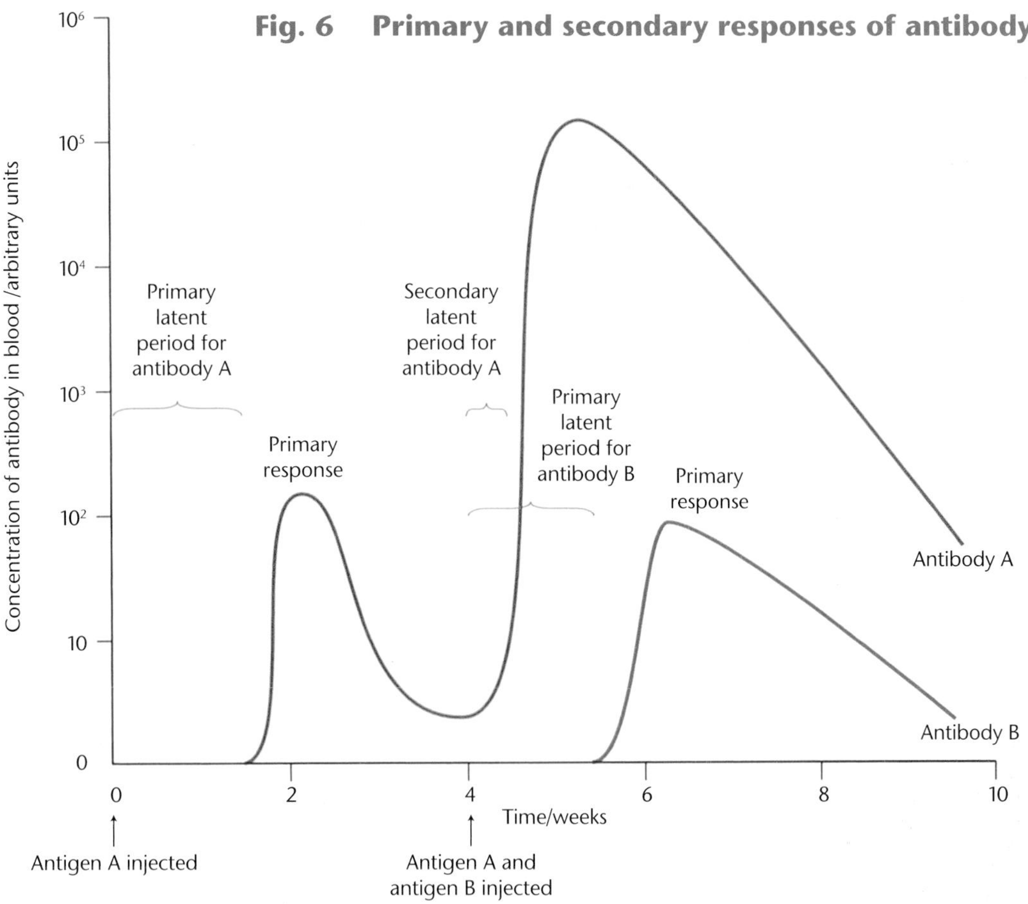

Fig. 6 Primary and secondary responses of antibody production

secondary response. The response is much more rapid than the primary response and more antibodies are produced. The speed of the secondary response means that the pathogen is destroyed before it fully infects the body and causes the symptoms of a disease. The individual has acquired immunity to the disease (Fig. 6).

Antigenic variability

So why do we suffer from many colds and bouts of influenza in our lives? Doesn't this contradict the description of primary and secondary responses outlined above? The problem is that the viruses that cause these two very common diseases occur in many forms. Each form contains a different antigen and therefore causes its own primary response. In addition, new mutations that the body doesn't recognise constantly arise.

Influenza outbreaks occur every year in Britain, and can reach epidemic proportions. In 1989, influenza is thought to have killed 25 000 people. High-risk individuals are offered a yearly vaccination against flu. These are the people who are most likely to develop serious complications that could lead to death.

Three groups of influenza virus are recognised: type A, type B and type C. However, there are many different strains in each group.

Influenza vaccines are prepared from highly purified inactivated viruses, grown in chicken **egg albumen**. GPs order the amount of vaccine they think they need, often a year in advance, and administer it in preparation for the oncoming winter. Most vaccines are cocktails of type A and type B antigens. Type C is rarely involved in UK outbreaks.

9 Which has the longer latent period, the primary response or the secondary response?

10 Why is the secondary response faster than the primary response?

11 Why doesn't an attack of influenza or cold give immunity to these diseases?

7.2 Vaccination

When the body is exposed to a pathogen, an impressive array of defence mechanisms contribute to maintaining health. However, some bacteria and viruses get past these mechanisms and cause disease. Some of these pathogens are so active that the disease ends in death. Examples of such diseases are cholera, smallpox and diphtheria. Diseases such as these were the targets of public health programmes for proper sewage treatment and the development of safe water supplies, and for the many vaccinations now available.

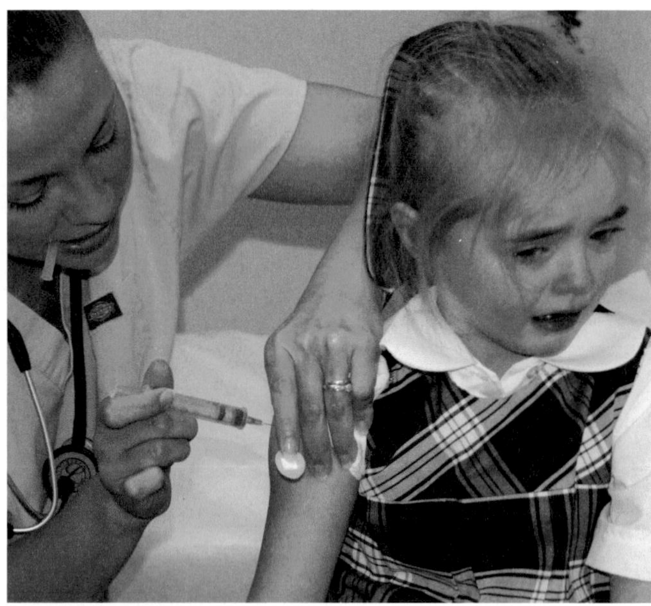

This girl is being vaccinated against rubella, a disease that can cause malformations in the developing baby if a woman gets it during pregnancy.

The term **vaccination** comes from the Latin word *vaccinia* (cowpox), which is derived from *vacca* (cow). A vaccine contains antigen derived from pathogenic organisms.

When injected into an individual, the antigen stimulates a primary response that leaves memory cells to generate the secondary response if the individual is subsequently infected by the relevant pathogen.

As adults we have immunity to most common diseases, apart from the common cold and influenza. This immunity is usually acquired through either contact with the pathogen at an earlier age or vaccination, and is classed as active immunity (Table 1). In this century, immunity against diseases such as polio, tuberculosis (TB) and diphtheria is likely to be the result of childhood vaccinations.

During the first few days of breastfeeding, the breasts produce a high-protein low-fat liquid called **colostrum**. This contains many antibodies, and provides the infant with immunity against a number of infectious diseases. Antibodies are also present in the milk that is produced after the colostrum. This is **naturally induced immunity** but no primary response takes place in the baby, because the infant simply receives antibodies from another individual. This is called **passive immunity**.

Artificially induced passive immunity arises when someone receives an injection containing antibody (as opposed to antigen). Whether naturally or artificially induced, passive immunity normally lasts for a few months. The body does not produce its own antibody or memory cells, since both these processes require the presence of an antigen.

Table 1 Types of immunity

Immunity type	How acquired	Duration
Active natural immunity	Immunity develops following natural exposure to antigen	Memory cells develop to produce long-lasting immunity
Active artificially induced immunity	Immunity develops after immunisation with a vaccine	Memory cells develop to produce long-lasting immunity
Passive natural immunity	Immunity develops through transfer of antibodies from mother to baby through the placenta and breast milk	No memory cells develop so the immunity is short term and lasts only a few months
Passive artificially induced immunity	Immunity develops after injection with antibodies	No memory cells develop so the immunity is short term and lasts only a few months

12 Why is passive immunity temporary?

Fig. 7 Global cases of diphtheria and immunisation coverage

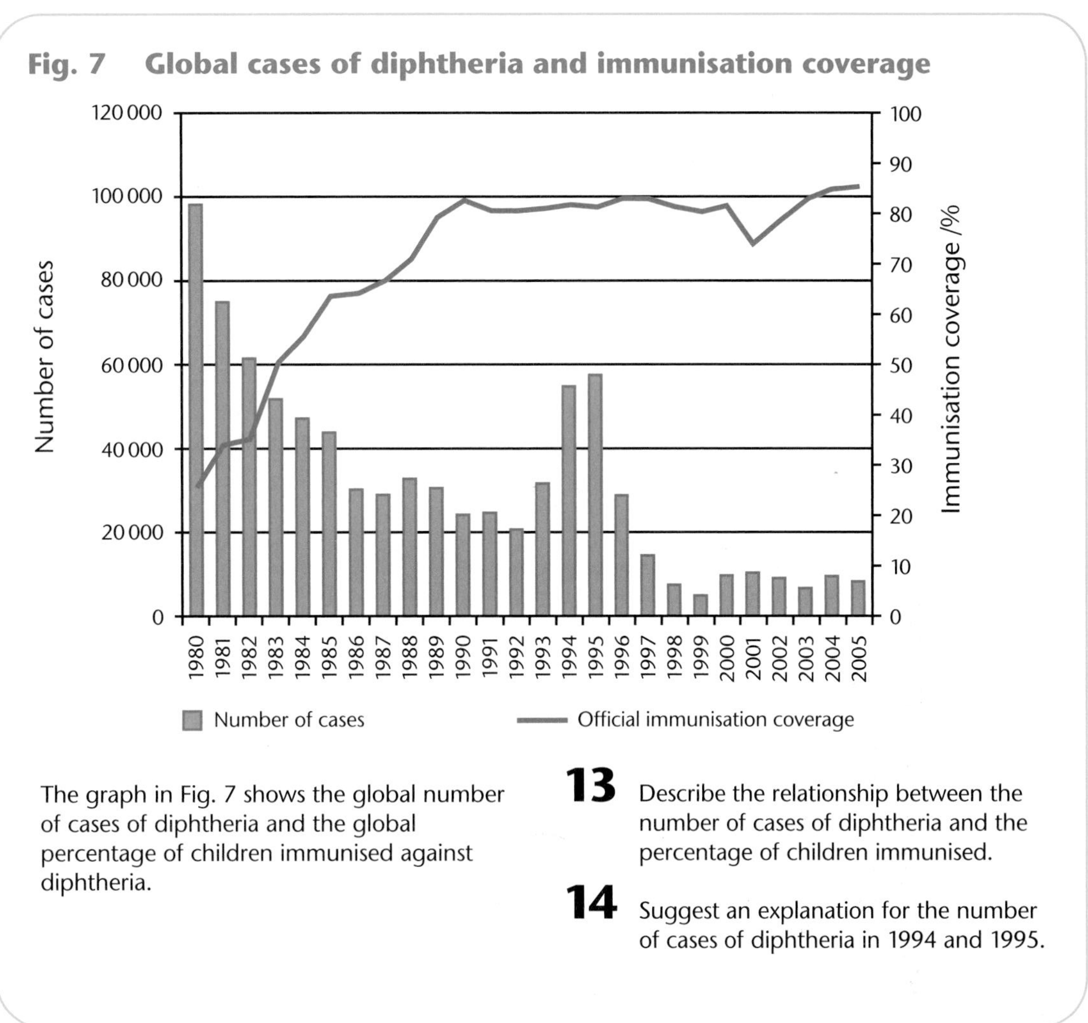

The graph in Fig. 7 shows the global number of cases of diphtheria and the global percentage of children immunised against diphtheria.

13 Describe the relationship between the number of cases of diphtheria and the percentage of children immunised.

14 Suggest an explanation for the number of cases of diphtheria in 1994 and 1995.

Vaccination schedules for children

It is important that children receive the necessary vaccinations in order to prevent large-scale outbreaks of disease. Artificially creating immunity is called immunisation. Table 2 shows the sequence of vaccinations for children in the UK. Children who go through this schedule are likely to be protected for life from some potentially lethal diseases, but it is not compulsory in the UK for children to have these vaccinations. Occasionally, booster vaccinations are needed if the original injection is not adequate. The possibility of a measles epidemic was so high in 1994–95 that a major publicity campaign was launched to ensure that all school-age children were immunised against the disease. School-age children can be very ill if they get measles; some cases are fatal. The symptoms are a high temperature, a rash, cough and sore eyes. Measles can also lead to

Table 2 Vaccination schedule for children in the UK

Age	Immunisations
2 months	DtaP: diphtheria, tetanus, whooping cough (pertussis) Hib: *Haemophiius influenzae* type b IPV: polio PCV: pneumococccal conjugate vaccine against pneumococcal infection
3 months	DtaP/IPV/Hib meningitis C (menC)
4 months	DtaP/IPV/Hib meningitis C (menC) PCV
12 months	Hib/menC
13 months	Measles, mumps, rubella (MMR) PCV
3 years 4 months–5 years	DtaP/IPV MMR
13–18 years	diptheria, tetanus and IPV

pneumonia, and may result in blindness, deafness and brain damage. Possible side effects of the measles vaccine include a mild fever or rash after about a week, but these should not last longer than 3–4 days. There is also a very small risk of brain damage. However, the dangers of measles greatly outweigh the dangers of the vaccine's side effects.

Parents took a very different view over the whooping cough vaccine. In the 1980s and 1990s there was a widespread refusal to immunise children against whooping cough because the vaccine was thought to have possibly fatal side effects. There was a rise in the occurrence of whooping cough but vaccination levels have gone up again.

Percentage cover

The proportion of individuals who must be immune to a disease in order to prevent an epidemic is called the **percentage cover**. Percentage cover varies from one disease to another. For instance, polio epidemics are prevented by 70% cover, but influenza requires 90–95% cover to prevent epidemics. This high level is never reached, so there are frequent influenza epidemics.

Monoclonal antibodies

Specific antibodies are produced by competent B cells in response to the presence of an antigen. During an infection, competent B cells undergo repeated **mitosis** (or cloning) to produce an array of cells capable of producing the particular antibody needed. Antibodies produced in this way, and specific to only one antigen, are termed **monoclonal antibodies**.

Monoclonal antibodies are causing a revolution in diagnosis. Techniques using monoclonal antibodies are usually quick and easy to perform.

Patients whose heart muscle is damaged in a heart attack are at risk of further heart problems. Such damage can be identified using radioactively labelled monoclonal antibodies specific to the protein **myosin**. This is found in all muscle cells but is only exposed when the muscle has been damaged. Damaged cardiac muscle is radioactively labelled by the anti-myosin. The label is easily detectable and doctors can then give appropriate therapy.

Growing and using monoclonal antibodies

Monoclonal antibodies are grown in cell cultures using tumour cells so that the B cells divide at a much faster rate than usual. This is known as the **hybridoma technique** (Fig. 8). In this technique, a B lymphocyte is fused with a tumour cell to form a 'hybrid' cell called a hybridoma. This cell divides rapidly, forming a clone of hybridoma cells, each of which produces the same antibody as the original B lymphocyte.

Each monoclonal antibody is specific to a particular antigen, and can be used to identify cells with that antigen. Monoclonal antibodies are also used in cancer treatment. The antigens on cancerous cells are different from those found on normal cells. By selecting a monoclonal antibody specific to a cancerous cell antigen and attaching a cytotoxic drug to the antibody, it is possible to send the drug specifically to the tumour cells.

15 Why do you think monoclonal antibodies are so called?

16 What does cytotoxic mean?

17 Why do you think the combination of a cytotoxic drug and a monoclonal antibody is sometimes called a 'magic bullet'?

18 Why should the radioactive isotopes used to label monoclonal antibodies have a very short half-life?

Fig. 8 Tumour cells and lymphocytes

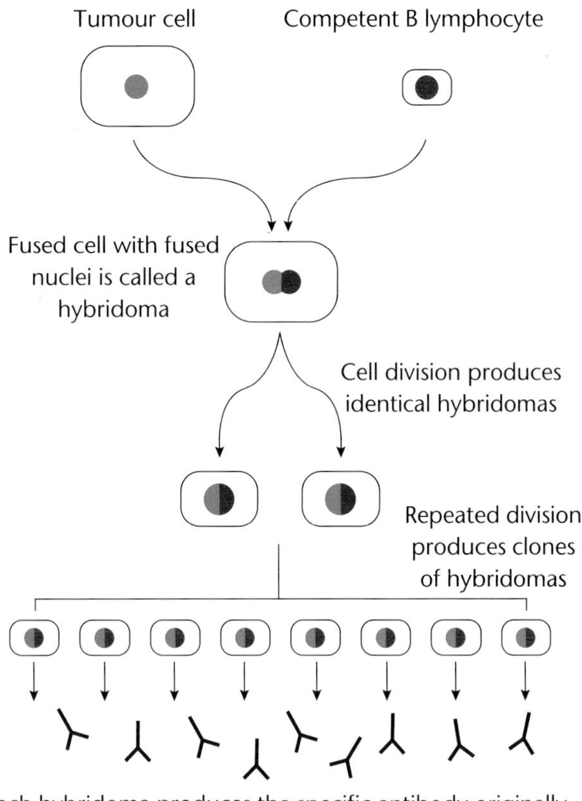
Tumour cell
Competent B lymphocyte
Fused cell with fused nuclei is called a hybridoma
Cell division produces identical hybridomas
Repeated division produces clones of hybridomas
Each hybridoma produces the specific antibody originally coded for by the competent lymphocyte

key facts

- Antibodies are proteins called **immunoglobulins**.
- Antibodies have two binding sites so that they can each attach to two antigen molecules and produce a lattice-like structure that immobilises the pathogens.
- Binding an antibody to the antigen can also render a toxic antigen harmless by blocking its active site.
- Immobilised pathogens are easier for phagocytes to ingest.
- The primary response (antibody production after first infection) takes 3–24 days. The secondary response (antibody production after a subsequent infection) is much faster.
- There are three groups of influenza virus – types A, B and C. Most influenza vaccines are against types A and B.
- A vaccine is a preparation of antigen from a pathogen that will generate a primary response when injected into an individual but will not cause the disease.
- Children in the UK are usually vaccinated against many diseases: diphtheria, tetanus, whooping cough, polio, measles, mumps, rubella and meningitis.
- Vaccination and natural exposure produce long-term active immunity that relies on the production of antibodies and memory cells.
- Passive immunity is the result of receiving antibody. It is short lived.
- During an infection, competent B cells undergo repeated mitosis (or cloning) to produce an array of cells capable of producing the particular antibody needed.
- Antibodies produced in this way, and specific to only one antigen, are termed monoclonal antibodies.
- Each monoclonal antibody is specific to a particular antigen, and can be used to identify cells with that antigen.

examination questions

1 There are two responses to infection, humoral and cellular.

a Describe the role of macrophages and lymphocytes in:
- **i** the humoral response; (3)
- **ii** the cellular response. (3)

b The diagram shows an antibody.

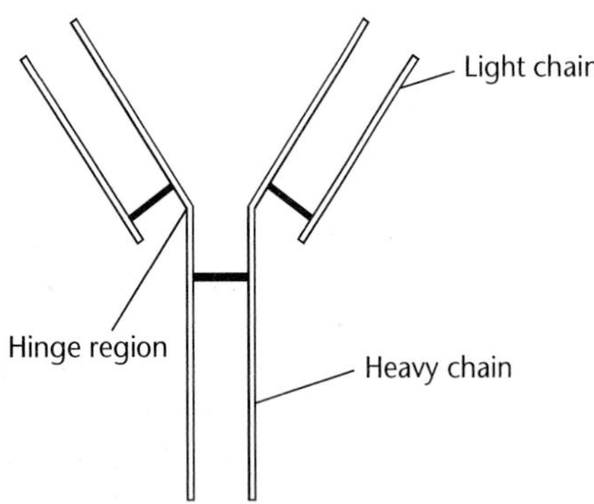

- **i** Explain how an antigen-antibody complex is formed. (3)
- **ii** Explain how the structure of an antibody is related to its function. (3)

c
- **i** Explain what is meant by 'monoclonal antibody' (2)
- **ii** Describe and explain the role of monoclonal antibodies in treatng one disease. (3)

Total 17

2

a Describe how each of the following parts of the body is protected to prevent microorganisms entering living cells.
- **i** Skin (1)
- **ii** Lungs (1)
- **iii** Eyes (1)

b Describe how macrophages help to prevent the spread of microorganisms that enter the blood and other tissues. (2)

Total 5

AQA, June 2005, Unit 7, Question 1

3 One blood group system is Rhesus. Rhesus positive people have rhesus antigens on their red blood cells. Rhesus negative people have no rhesus antigens. To find out whether a person is rhesus positive or negative, a sample of blood is mixed with rhesus antibodies.

a Explain why rhesus positive blood agglutinates when it is mixed with rhesus antibodies. (1)

b A person who is rhesus negative will produce rhesus antibodies if rhesus antigens get into the blood.
- **i** Describe how these antibodies are produced. (5)
- **ii** Explain how antibodies are produced more quickly if the same type of antigen gets into the blood on a second occasion. (2)

c In some pregnancies, rhesus antibodies move from the blood of the mother into the blood of a rhesus positive fetus. These antibodies destroy some red blood cells in the fetus. When the baby is born it will suffer from blue-baby syndrome.
Babies who suffer from blue-baby syndrome may breathe at an unusually high rate. Suggest an explanation for this. (2)

Total 10

AQA, June 2003, Unit 7, Question 6

4

a Whooping cough is a childhood respiratory disease caused by a bacterium. The graph shows the incidence of whooping cough in Central Europe from 1969 to 1999.

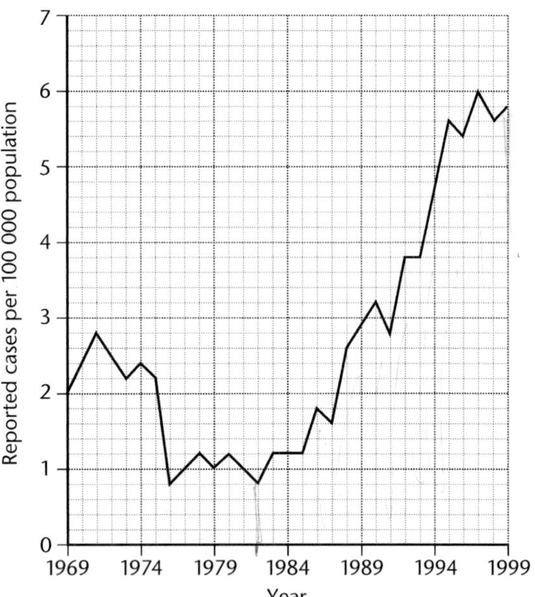

- **i** Calculate the percentage change in the incidence of whooping cough from 1982 to 1999. Show your working. (2)
- **ii** Suggest **one** reason for the trend in the number of cases of whooping cough since 1982. (1)

b Whooping cough bacteria prevent the normal functioning of cilia in the respiratory tract. Explain how this effect is linked to the persistent coughing associated with the disease. (2)

Total 5

AQA, June 2004, Unit 7, Question 3

5 The diagram shows an antibody.

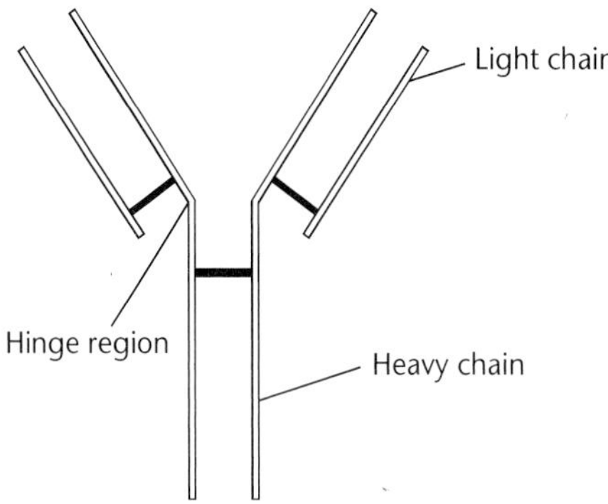

a Each heavy and light chain is made up from one type of monomer. Name the type of monomer in each chain. (1)

b Write **X** on the diagram to show where an antigen may form a complex with this antibody. (1)

c Each antibody can form a complex with only one type of antigen. Explain why. (1)

d The hinge region of the antibody allows both ends to pivot and rotate in relation to one another. Suggest how this action assists the role of antibodies in agglutination. (1)

Total 4

AQA, January 2003, Unit 7, Question 5

6 Wells **A, B** and **C** were cut into an agar plate and the following substances were put in them.

Well **A** – antigens from one strain of the influenza virus
Well **B** – a 'cocktail' vaccine containing antigens from several strains of influenza virus
Well **C** – blood plasma from a human volunteer

After ten hours, precipitation lines were observed in the agar. Precipitation lines appear where there is agglutination. The diagram shows where these lines occurred.

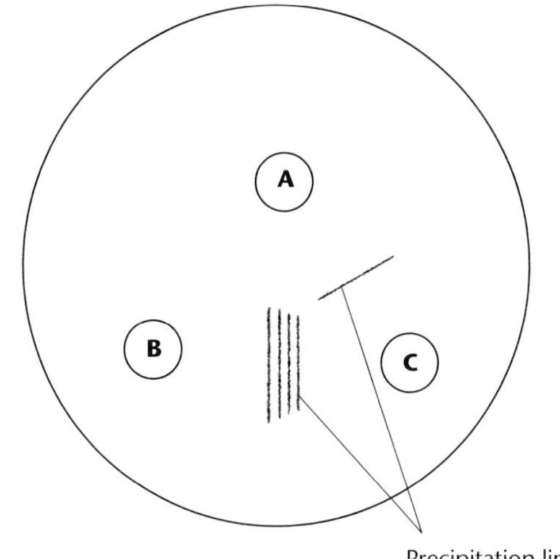

a Describe how the precipitation line between wells **A** and **C** was formed. (2)

b Explain what the precipitation lines suggest about the volunteer's plasma. (2)

Total 4

AQA, January 2004, Unit 7, Question 4

7 Measles is an infectious disease that can cause serious complications in children. In countries where measles is uncommon a combined measles, mumps and rubella vaccine (MMR) is given at 15 months. In a country where measles is common a single measles vaccine (MV) may be given at 9 months, followed by MMR at 15 months. In an investigation, the efficiency of the two vaccination programmes was compared in a country where measles is common. The amounts of measles antibody in the blood of children before vaccination and after completing vaccination were measured. The graph shows the results. All differences are statistically significant.

i What was the effect of vaccination in the MMR only group? Express your answer as the percentage increase in the amount of measles antibody in the MMR group after vaccination. Show your working. (2)

ii The MV + MMR group had more measles antibodies in their blood before vaccination than the MMR only group. Suggest an explanation for this. (1)

Total 3

AQA, January 2005, Unit 7, Question 4

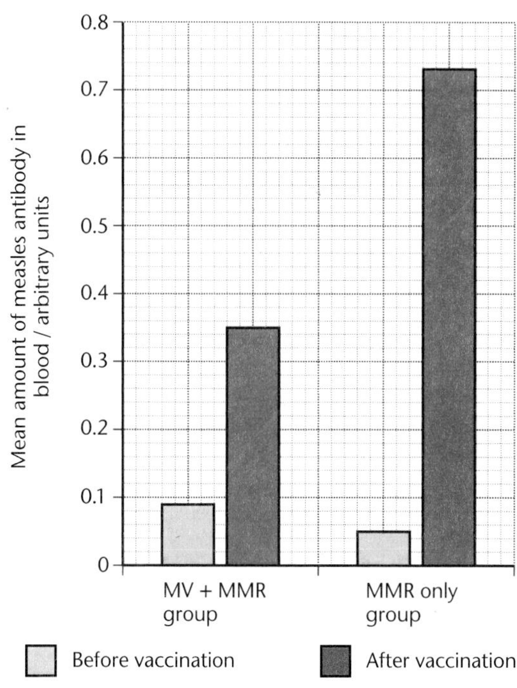

how science works **assignment**

The MMR controversy – timeline

February 1998

Research led by Dr Andrew Wakefield suggests that the MMR vaccine might be linked to an increased risk of autism and bowel disorders.

Dr Wakefield says he has evidence that children's behaviour changed drastically shortly after they received the MMR jab. He studied 12 children whose parents claimed they had developed autism after being vaccinated with the MMR vaccine. The work was partly funded by lawyers acting on behalf of these children.

Dr Wakefield theorises that the combination of the three virus strains contained in MMR may overload the body's immune system and cause the bowel disorder to develop.

March 1998

A panel of experts set up by the Medical Research Council says there is 'no evidence to indicate any link' between the MMR vaccine and bowel disease or autism in children.

Thirty-seven scientific experts gather to review all the evidence and conclude there is no reason to change the current MMR vaccination policy for children.

April 1998

A 14-year study by Finnish scientists finds no danger associated with the MMR vaccine.

Out of the three million children given the MMR vaccine, those who developed gastrointestinal side effects lasting 24 hours or more were traced. In all, 31 youngsters developed gastrointestinal symptoms such as diarrhoea and vomiting within 15 days of the injection. None of the 31 children developed any signs of autism or any similar syndrome.

January 2001

Dr Wakefield renews his concerns about MMR, saying the vaccine has never undergone proper safety tests.

Dr Wakefield says original safety checks on the vaccine were poorly conducted and lasted only for 4 weeks.

The claim is rejected by the Department of Health.

February 2001

A major statistical analysis published on the *British Medical Journal* website concludes that the soaring rate of autism in recent years is almost certainly not due to the MMR injection.

The study finds that the number of cases of autism has continued to rise even though MMR coverage has remained roughly the same.

If MMR were the cause of the illness, say the experts, the number of autism diagnoses would also have levelled off by now.

February 2002

A team from the Royal Free Hospital – where Dr Wakefield carried out his initial research – publishes a study on the *British Medical Journal* website saying there is no link between the MMR vaccine and autism.

The team looked at almost 500 children with autism born between 1979 and 1998. It found that the proportion of children with developmental regression (autism) or bowel disorders did not change significantly over that time.

September 2004

A Medical Research Council team looked at the vaccination records of 1294 children diagnosed with autism or related conditions between 1987 and 2001 in England and Wales.

These children were compared with 4469 children of the same sex and similar age who were registered with the same GP surgeries but did not have autism.

Overall, 78% of the children with autism had received the MMR vaccine. But 82% of the other children had also been given the MMR vaccine. The researchers say this 4% difference is not significant, and they argue that the sheer size of their study makes their finding very powerful.

March 2005

Japanese scientists say they have strong evidence that the MMR vaccination is not linked to a rise in autism after they found a rise in the incidence of autism after the withdrawal of the MMR vaccine in their country in 1993. Researchers looked at the incidence of autism in 31 426 children up to the age of 7 born from 1988 to 1996.

There were 48 cases of autism per 10 000 children born in 1988. The rate was seen to rise steadily to 117.2 per 10 000 for those born in 1996 – after the MMR vaccine had been withdrawn.

May 2006

US scientists report that they have found measles virus in the guts of autistic children with bowel disease. However, Dr Stephen Walker, the scientist behind the work, states that this finding did not show that the MMR vaccine caused the condition.

July 2007

The General Medical Council charges Dr Wakefield with 40 offences, mostly relating to professional misconduct in carrying out invasive procedures without the requisite ethical approval, and misleading an ethics committee and *The Lancet*.

Confidence in the MMR vaccine is now improving and more than 90% of parents have their children vaccinated.

A1 Evaluate the claim that MMR causes autism.

8 Variation and selection

People vary but we all take it for granted that we can recognise each other by our distinctive physical features. Some of the more obvious characteristics are used to group people into races, but closer analysis shows patterns of differences that do not match with generally accepted racial groups. Our genes suggest that we share many more similarities than differences.

Each human being has about 35 000 functional genes. Of these, about two-thirds seem to be identical in all individuals. The other third are genes that have two or more alleles. These alleles combine in a vast number of different ways, so no two fertilised ova are exactly the same. It is rather like doing the National Lottery with 30 000 numbers!

Two randomly chosen people of the same race have 85–88% of their genes in common. If the two people happen to be from different races, the proportion changes by less than 2%. Though small, this difference is finding an unexpected application – in archaeology. When a set of bones is unearthed, tests can be done to find out more about the person and his or her life. Analysis of carbon and nitrogen isotopes can reveal how much meat and fish he or she ate, and DNA analysis can reveal his or her racial origins. Key sequences in the human genome are known to be a 'fingerprint' of people whose ancestors lived in, say, the Middle East, or Roman Britain. These sophisticated techniques can supplement the more traditional techniques used in archaeology to build up an accurate picture of the past.

No two faces in a crowd are the same.

A reconstruction of a human face based on information provided by the shape of a skull found during an archaeological dig.

8.1 Introduction to our differences

In this chapter we look at how the differences between individuals of the same species arise. Both genetic and environmental factors contribute to variation. It is variation that has made evolution possible.

Discontinuous variation

Variation in a characteristic may be clear-cut, such as the distinction between being hairy or hairless. This is **discontinuous variation**. More often, especially in humans, there is a range of variation (e.g. from tall to short); this is called **continuous variation**.

A good example of discontinuous variation is the human blood grouping system. All human beings have one of four major blood groups: A, B, AB or O. Each of us must be one of these four possible **phenotypes**. You cannot be halfway

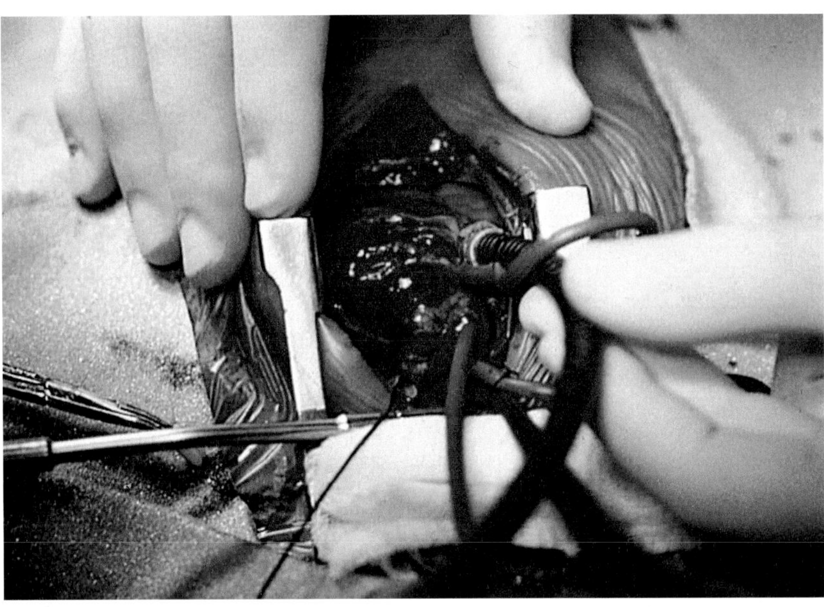

A heart transplant in progress. Donors for heart transplants must be matched carefully with the recipient to avoid the new heart being rejected. It is rare to find two people with similar HLA antigen patterns – which is why it can be so difficult to find an organ for someone who is desperately ill.

between groups A and O, or just slightly blood group B. Such variation is described as discontinuous because each of the phenotypes is quite distinct. Discontinuous variation is caused by differences in a single gene. Blood group is determined by a single gene, one that has three **alleles**, I^A, I^B and I^o. Alleles I^A and I^B code for the production of a cell surface protein – you either have it, or you don't. There are no states in between. Furthermore, your blood group depends entirely on which alleles you inherit from your parents. Your blood group is not affected by environmental factors such as the amount or type of food you eat. Knowledge of the ABO blood groups is vital for doctors performing blood transfusions. If the blood is not matched correctly between donor and recipient, the result can be fatal.

Other cell surface proteins are also important when doing tissue typing for organ transplants. These **antigens** are found on the surface of white blood cells. Two of the most important are known to have at least 23 and 47 alleles respectively. Each produces a slightly different antigen molecule, which can be detected by a technique called human leucocyte antigen (HLA) typing. If HLA antigens are not matched well between organ donor and recipient, the result is less dramatic than transfusing poorly-matched blood, but it is just as deadly. After a few days, a host-versus-graft reaction occurs in which cells from the immune system of the person who has been given the new organ go to work. The transplant is treated as a foreign organism. The same processes that destroy disease-causing bacteria and viruses come into play. The cells of the new organ are damaged and it starts to die.

HLA genes also show discontinuous variation, even though they have such a large repertoire of alleles. Each antigen produced is still distinctly different – you either have one or you don't. Any characteristic that shows two or more clear-cut phenotypes is described as showing discontinuous variation. Eye colour in humans is another good example.

However, do not be misled into thinking that discontinuous variation is limited to humans. There are, in fact, rather few examples in mammals; discontinuous variation is much more common in plants and other less complex organisms. The features of Mendel's garden peas, such as round and wrinkled peas, are also quotable examples of discontinuous variation for exam answers.

Continuous variation

In humans, most characteristics show continuous variation. This type of variation is called continuous because it is difficult to sort out all the possible types into distinct groups. It is not only physical characteristics, such as mass, length of big toe, hair colour and nose shape that have a range of continuous variation, but also metabolic characteristics such as heart rate, speed of reaction, muscle efficiency and the ability of the brain to process information (or 'intelligence', or whatever we choose to call it!).

Characteristics that vary continuously are usually the result of several different genes acting in such a complex way that it is not possible to distinguish separate phenotypes. A person's height depends on many different factors, including the growth rate of several different bones, hormone production and metabolic rate. Each factor may be controlled directly, or indirectly, by several genes. In most cases the interaction of different genes causes the majority of individuals to lie near the middle of the range, with many fewer at the extremes.

This range and distribution can be illustrated in a graph called a normal distribution curve (Fig. 1).

Fig. 1 Normal distribution curve

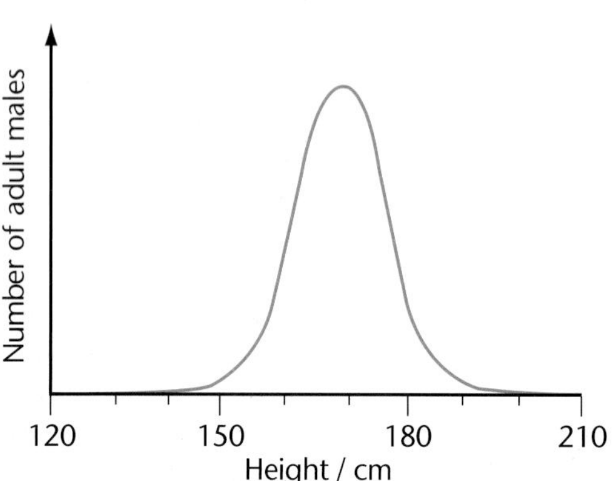

The Ukranian Leonid Stadnyt is the world's tallest man at 2.57m (8 feet 5 inches).

Mean and standard deviation

How many people in the world have more than the average number of hands? Most people who answer this question without thinking will say 'none' or 'very few', since most of us don't know anybody with three hands. The correct answer, however, is 'nearly everyone' because, although hardly anyone has more than two hands, there are unfortunately quite a lot of people with one or none, because of accidents or because they were born without them. The arithmetical average, or mean, is therefore rather less than 2.0, whereas, of course, most people do have exactly two. This illustrates the importance of being very careful when interpreting data. Common sense and statistics don't always go together!

The exact shape of the normal distribution depends on the **mean** and the **standard deviation (SD)** of the distribution. The SD is a measure of how much measurements are spread out from the mean.

Differences in SD affect the shape of the distribution. All the curves in Fig. 2 represent normal distributions. Although the distribution is symmetrical in all the curves shown, the distribution changes as the SD changes. When the SD increases, the curve becomes flatter; if the SD decreases, the curve is taller.

In a normal distribution:

- measurements greater than the mean and measurements less than the mean are equally common
- small deviations from the mean are much more common than large ones
- 68% of all the measurements fall within a range of ± 1 SD from the mean and 95% within ± 2 SDs.

Fig. 2 Normal distribution curves

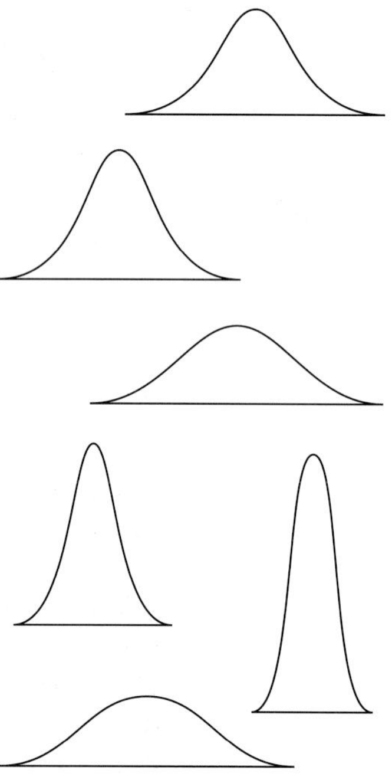

Fig. 3. Normal distribution of blood pressure

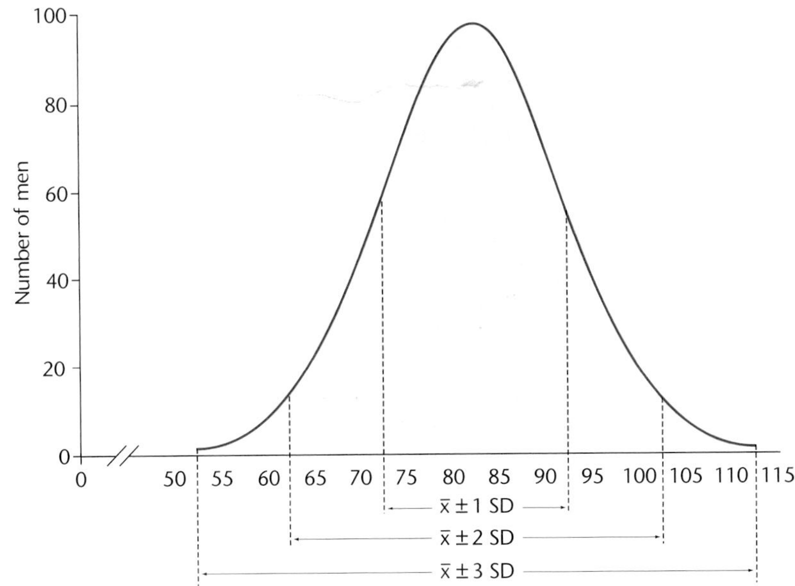

Fig. 3 shows the **normal distribution** of diastolic blood pressure for a sample of men.

The mean blood pressure is 82 mmHg and the SD is 10 mmHg. This means that 68% of the sample have blood pressure between 72 and 92 mmHg; 95% of the sample have blood pressure between 62 and 102 mmHg. It follows that if we sample, there is only a 0.05 (1 in 20 chance) of selecting a man with a blood pressure less than 62 or greater than 102 mmHg.

It is therefore important to take a random sample of a population. Simple random sampling is the basic sampling technique where we select a group of subjects (a sample) for study from a larger group (a population). Each individual is chosen entirely by chance and each member of the population has an equal chance of being included in the sample.

A practice ISA on the collection and analysis of data relating to variation can be found at www.collinseducation.co.uk/advancedscienceaqa

Continuous variation and the environment

Characteristics that vary continuously are also subject to interaction with the environment. A person's adult height may well be affected by diet or disease during childhood. Often it is very difficult to distinguish the contributions made by genes and those due to environmental factors. For instance, there is much disagreement between scientists about intelligence. Is it determined by a person's genes? Or is it solely to do with their environment? Probably the true answer lies somewhere between the two.

In cats a gene, C, is concerned with the colour of the coat. The normal allele, C^C, makes the cat's coat blackish and is dominant. Siamese cats have particular alleles of this gene. Siamese cats have a recessive allele, C^S, and they must be **homozygous** for this allele (C^SC^S). This mutant allele codes for an enzyme that synthesises black pigment, but only when below body temperature. This is because the mutant allele codes for an enzyme with a tertiary structure that is slightly different from that of the normal enzyme; molecules of this enzyme happen to unfold (denature) at about 37 °C. The Siamese cat therefore is black only in the cooler parts of the body, such as the tail, ears and lower legs. The rest of the coat is pale-cream coloured. This shows how both the genotype and the environment can affect the phenotype.

Alligators provide one of the most interesting illustrations of the interaction between genes and the environment. In these ancient reptiles, temperature affects the genes that determine sex. Eggs incubated at 30 °C all produce females; eggs incubated at 33 °C become males. The position and depth of a nest can therefore affect whether male or female young are produced. This means that alligators can produce more females than males, and in practice there may be as many as eight females for every male.

An alligator on its nest of eggs in Florida.

It is becoming clear that many people have alleles of genes that make them more likely to develop certain diseases, such as some forms of cancer. However, the disease may only occur if the genes are triggered by something, such as a chemical, in the environment. This may explain why many heavy smokers develop lung cancer in middle age whereas others carry on well into old age without being affected.

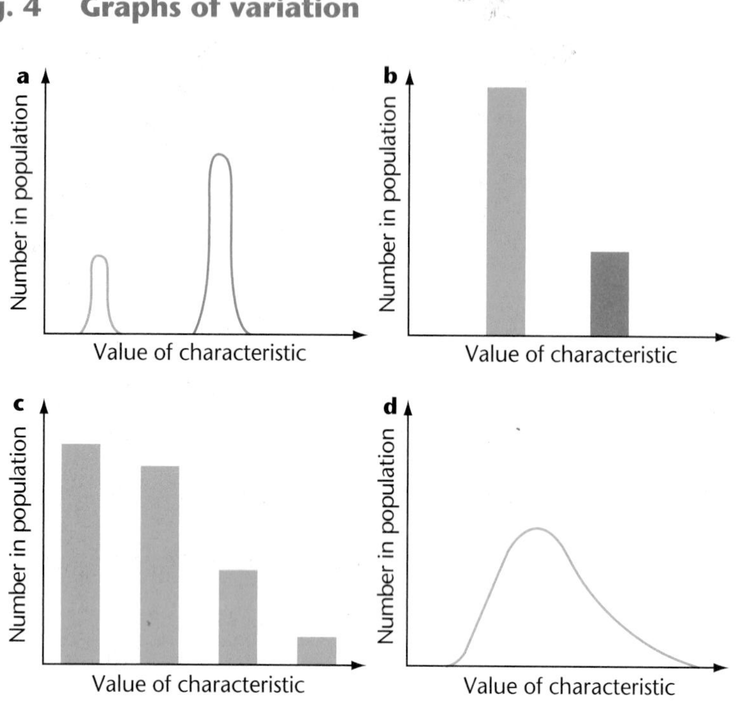

Fig. 4 Graphs of variation

1 Which of the graphs or bar charts in Fig. 4 shows the number of individuals in a population for each of the following characteristics? In each case give a reason for your answer.

a Human ABO blood groups

b Human earwax type, which is either wet or dry and is determined by a gene with two alleles; it is unaffected by environmental factors

c Human mass, which is determined by several genes and is affected by environmental factors

d Tall and dwarf peas, determined by a gene with two alleles and affected by environmental factors

2 What environmental factors might affect how tall a plant grows?

3 Explain how the phenotype of Siamese cats is affected by the effect of temperature on the pigment gene.

4 Suggest an advantage of alligators having a ratio of eight females to one male in a population.

5 How might rising environmental temperatures affect the alligator population?

key facts

- There are many differences between individuals of the same species. This is **variation**.
- Variation results from both genetic and environmental influences.
- Where only one or a small number of genes cause the variation in a characteristic, there are clear-cut differences. This is **discontinuous variation**.
- Where many genes and environmental factors are involved, there is a range of variation, without distinct types. This is **continuous variation**.
- Variation within a sample may be described by the mean and standard deviation.
- In a normal distribution, the majority of measurements lie near the middle of the range.
- The standard deviation is a measure of how much measurements are spread out from the mean.
- It is important to take a random sample of the population when investigating variation.

1

a Explain what is meant by:

i continuous variation (2)

ii discontinuous variation. (2)

b The bar chart below shows the heights of plants in a population of sweet peas growing in the wild.

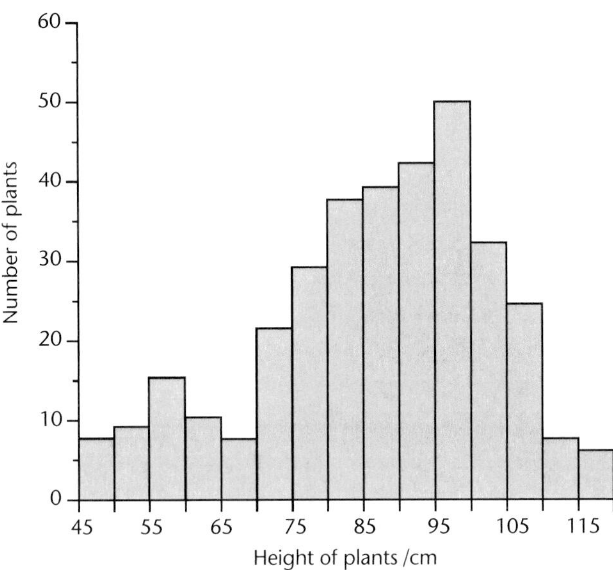

i List three factors that might affect the growth of sweet pea plants. (3)

ii What type of variation is exhibited by the height of sweet pea plants? (1)

iii What evidence is there that there are two strains of sweet pea plant in this population? (2)

Total 10

2 The bar chart below shows the distribution of blood group types in a population.

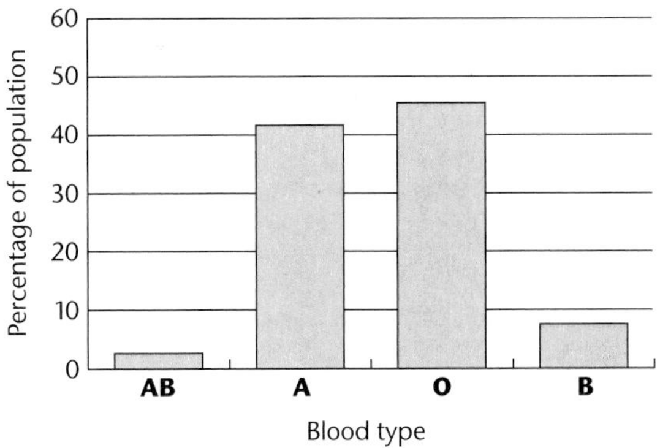

a What type of variation is shown by the distribution of blood types? (1)

b Using information from the graph, explain what determines the blood group of a person. (2)

Total 3

3 The chart below shows the variation of diastolic blood pressure in a population.

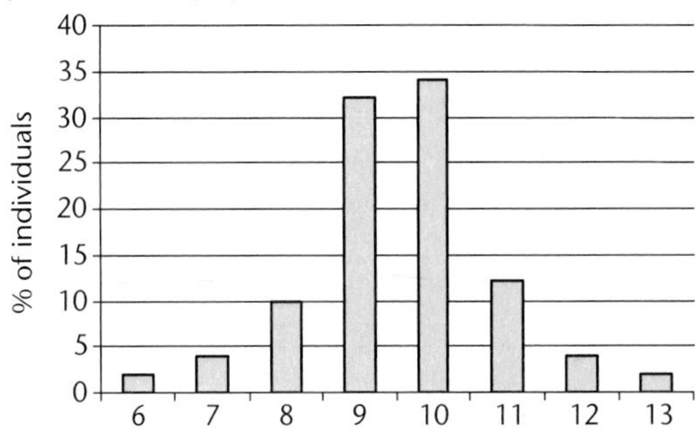

a High blood pressure is indicated by a diastolic pressure of 11 kilopascals or above.
What proportion of this population has high blood pressure? (1)

b Does the data indicate a normal distribution for blood pressure?
Explain the reason for your answer. (1)

c Using the data, suggest factors that affect the distribution of blood pressures.
Explain the reasons for your choices. (3)

Total 5

4 The table below shows the distribution of shells lengths in two populations of a marine mollusc.

	Shell length / mm	
	Population X	**Population Y**
	32	38
	31	43
	27	34
	34	40
	37	44
	38	39
	36	45
	22	46
	34	48
	23	39
	31.4	41.6
	5.7	4.3
Mean		
Standard deviation		

a Use information from the table to describe two differences between the two populations. (2)

b What type of factor is most likely to cause these differences?
Explain the reason for your answer. (2)

Total 4

5 IQ test scores have been used as a measure of intelligence. Genetic and environmental factors may both be involved in determining intelligence. In an investigation of families with adopted children, the mean IQ scores of the adopted children was closer to the mean IQ scores of their adoptive parents than to that of their biological parents.

a Explain what the results of this investigation suggest about the importance of genetic and environmental factors in determining intelligence. (1)

b Explain how data from studies of identical twins and non-identical twins could provide further evidence about the genetic control of intelligence. (4)

Total 5

AQA, June 2006, Unit 4, Question 2

6

a ABO blood groups in humans are an example of discontinuous variation, whereas height in humans is an example of continuous variation. Describe how discontinuous variation differs from continuous variation in terms of

- **i** genetic control
- **ii** the effect of the environment
- **iii** the range of phenotypes. (3)

b Genetically identical twins often show slight differences in their appearance at birth. Suggest one way in which these differences may have been caused. (1)

Total 4

AQA, January 2006, Unit 4, Question 4

7 Yarrow is a herbaceous plant which grows in California at altitudes from 1500 m to 3000 m. The mean height of the stems of plants growing at 3000 m is smaller than that of plants growing at 1500 m.

a The higher the altitude, the lower the mean temperature. Explain how the lower temperature at high altitude reduces the growth of plants. (4)

b The relative contribution of environmental and genetic factors on the growth of the plants was investigated. Samples of young plants were taken and grown outdoors in prepared plots at altitudes of 1500 m and 3000 m.

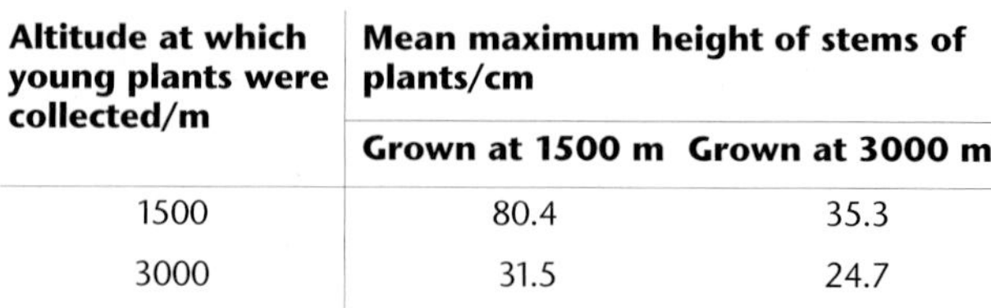

Altitude at which young plants were collected/m	Mean maximum height of stems of plants/cm	
	Grown at 1500 m	Grown at 3000 m
1500	80.4	35.3
3000	31.5	24.7

Describe the evidence from the table above that the variation in height is

- **i** partly genetically determined (1)
- **ii** partly environmentally determined. (1)

Total 6

AQA, January 2006, Unit 5, Question 5

8

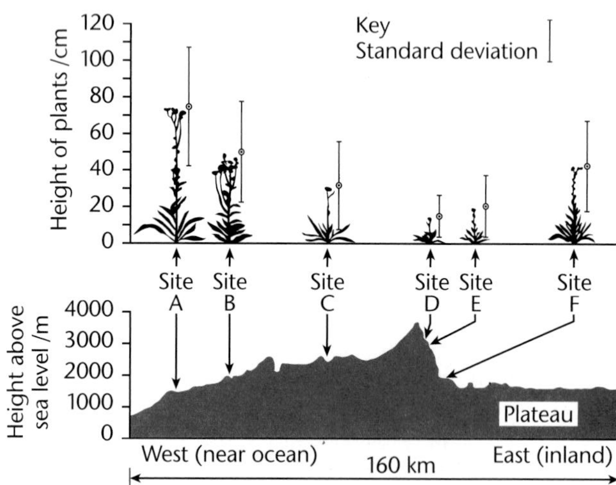

a What information does the bar representing standard deviation give about the plants in a sample? (1)

b Describe what the results show about the variation of the height of the plants in relation to altitude. (2)

c There was a significant difference between the mean heights of the plants grown from seeds taken from sites **A** and **D**. Describe the evidence from the information given which shows that this is likely to be due to genetic differences between the two populations. (1)

Total 4

AQA, June 2005, Unit 5, Question 6

Variation and human history

Fig. 5 The different proportions of people with blood group B in some populations in different parts of the world

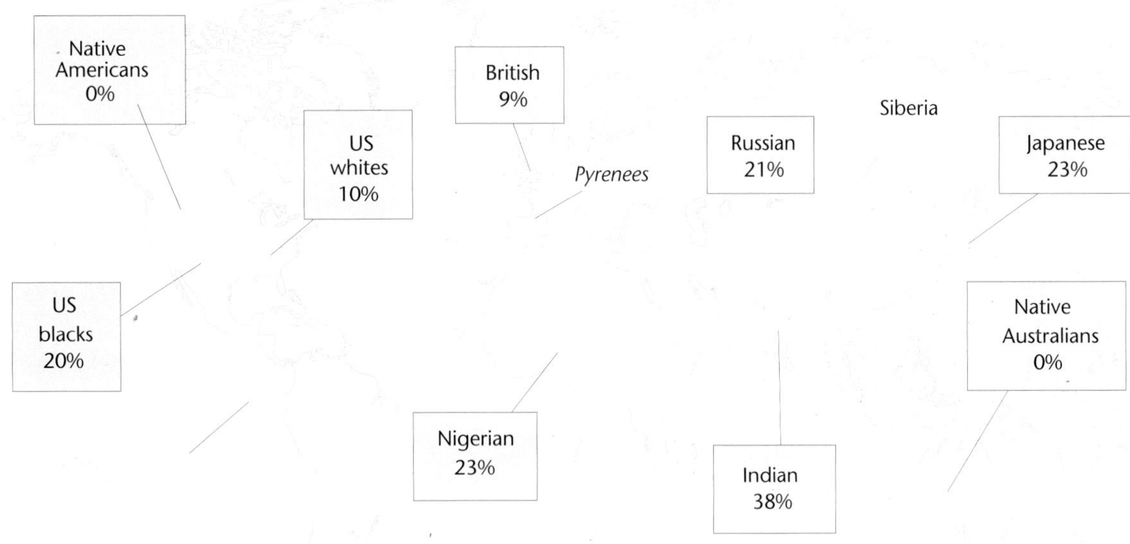

By studying the distribution of blood groups and HLA types it is possible to build up a picture of the complex patterns of migration that have taken place in human history. It seems that the first humans originated in Africa and then spread around the world. Studies of the genes of the native (original) inhabitants of North and South America suggest that a small group of people, with a limited range of alleles of some genes, reached North America from Siberia and then gradually populated the continent.

A1 It is suggested that none of the group of people that reached North America from Siberia carried the blood group allele, I^B.

a What evidence from the map supports this suggestion?

b Which two blood groups would you expect to be absent in these people. Which genotypes would you expect to be present?

A2

a Is the occurrence of blood group B correlated with skin colour? Use evidence from the map to explain your answer.

b How might the different proportions of group B in US blacks and US whites be explained?

A3 Most West European populations have about 9 or 10% with blood group B. However, this percentage is much lower in the Basque people, who live in the Pyrenees.

a What does this suggest about the origins of the Basque people?

b What additional information might now be available to help confirm relationships between peoples from different parts of the world?

9 The genetic code

For hundreds of thousands of years, human genes have passed from generation to generation, their presence shown only by the features that they have conferred on the human beings. At the start of the 21st century, our knowledge of genes is changing rapidly. The recent massive expansion of communications technology and the advances made in genetics mean that genes are now visible as sequences of letters (ACTG) recorded on CD ROM and published on the Internet. The consequences of this knowledge are yet to be explored, but run way beyond any thoughts that Crick and Watson could even have dreamt about in the 1950s, when they first deciphered the structure of DNA. The Human Genome Project has determined and published the sequence of all the nucleotides in the human genome. This information is available for any researcher to use. The function of large parts of the genome has yet to be determined.

Whilst this is good news in some respects – medical research will inevitably benefit – it also raises some ethical issues that we have never had to face before. Does someone who has a mutation that gives them a very 'unusual' gene have any rights over that sequence? Can they say what the sequence can be used for? Do they 'own' it? Should they be paid for it? Can a pharmaceutical company or a biotechnology company apply for a patent on a gene? Should companies refuse to employ someone because they know the person has a particular gene? Will our knowledge of genetics create a new 'genetically challenged underclass'?

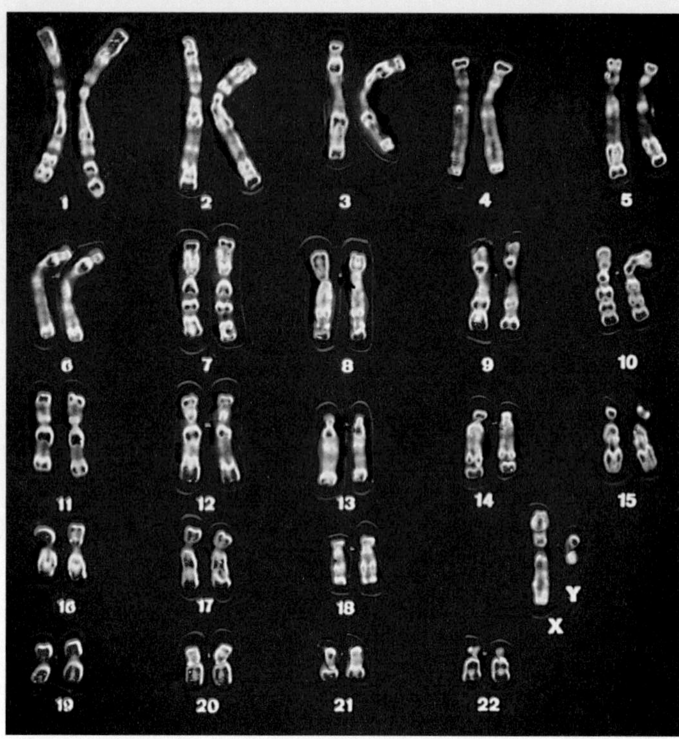

The full complement of human chromosomes (male) arranged in numbered homologous pairs. The female karyotype is similar, but has two X chromosomes instead of an X and a Y.

9.1 The genetic code

The collection of all genetic information within an organism has two remarkable properties:

- it carries the information that codes for the characteristics of that organism
- it can copy itself exactly and pass a complete copy on to every new cell.

Every human being starts life as a fertilised egg. This single cell contains two sets of coded information, one set from the mother and one set from the father. All the information is copied every time a cell divides, so the nucleus of every one of the billions of cells in the body has a full complement of **genes**. We do not know for sure exactly how many human genes there are – latest estimates say 50 000–100 000. The term **genome** is used to refer to all the different genes in a single individual.

In this chapter, we see how genes determine the nature and development of humans and other organisms

What are genes?

A gene is a chemical code that contains the instructions for making a complete protein, or, more usually, a polypeptide. Often two or more polypeptide chains must be joined together to produce a functional protein. For example, a haemoglobin molecule contains two copies each of two different types of polypeptide. Genes are important because the proteins they code for determine the characteristics of an organism. The huge varieties of different proteins act as enzymes, structural components, carriers and hormones. Our genes contribute to the development of every human feature that we recognise, such as the colour of our eyes or the shape of our nose. Usually many genes are involved in shaping a particular feature; human beings are so complex that it is rare to find a single gene that has one clear-cut effect. Of course, the environment also has an effect; it modifies the actions of individual genes and groups of genes. For example, a serious illness during childhood in a person whose genes code for a tall, athletic build could lead to a shorter, less muscular adult. This means that discovering the sequence of all human genes is just the beginning – the huge task of finding out what they all do and how they interact with each other and the environment still lies ahead.

Deoxyribonucleic acid

Deoxyribonucleic acid (DNA) is the **nucleic acid** that carries the genetic code. DNA is a remarkable substance. Its properties make it the key to all life on Earth and it has essentially the same structure in bacteria, plants and mammals. It has survived throughout evolution as the one substance that can store blueprints for each of the millions of species that have existed. DNA molecules:

- are huge, and able to store vast amounts of information in a small volume
- have small variations in structure that act as a simple code
- are stable, so that the information is not easily corrupted
- can reproduce themselves and so copy the information.

Just before cells divide, their DNA is copied so that the information it contains can be passed on. The DNA then contracts into **chromosomes** and remains in this condensed form until cell division is complete. When the cell is not dividing, the DNA exists in its uncondensed form as chromatin and is used as a guide for making proteins that the cell needs. In humans, DNA is organised into 46 chromosomes. Each chromosome consists of a single very long DNA molecule surrounded by proteins. The DNA contains sections that code for particular proteins – these are the genes. There are also sections in-between the genes that do not code for proteins. In prokaryotic cells the DNA molecules are smaller and circular, and are not associated with proteins.

As the photograph opposite shows, the 46 chromosomes in a body cell consist of 23 pairs; each pair consists of a copy of a chromosome from the egg and one from the sperm. The members of a pair are not identical. The genes they carry occur at the same position on each chromosome in the pair. This position is called the locus (plural; loci) of the gene (Fig. 1).

Fig. 1 The gene locus

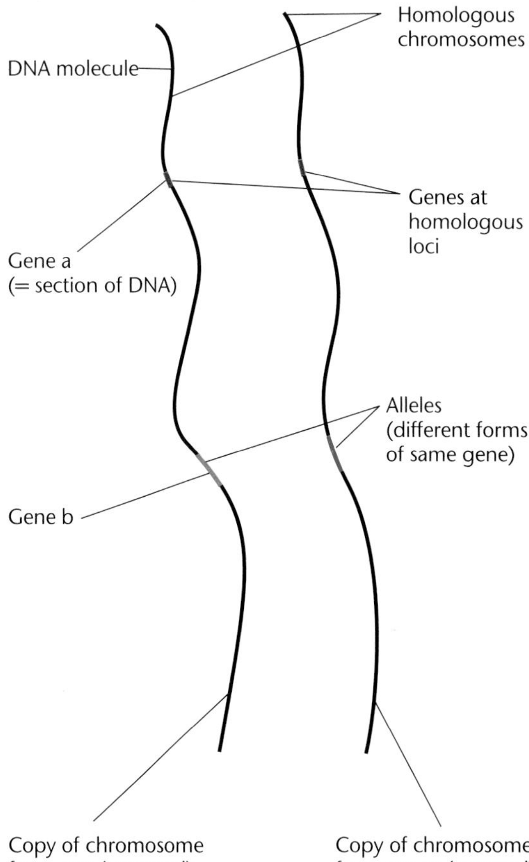

Each cell therefore has two copies of the gene that codes for each protein. The chromosomes themselves are called **homologous chromosomes**. The genes that code for a particular polypeptide are normally found at the same position in the DNA molecule of their chromosome in every cell of the body. You might expect the two genes that code for a particular polypeptide to have exactly the same DNA codes. In some cases this is true; both copies of the gene are identical. In other cases the two genes have slightly different codes, although they still occur at the matching locus on their chromosome. These different forms of the same gene are called **alleles**. A variant allele can produce a polypeptide with a different structure and function to the one produced by the normal allele. Sometimes the variant polypeptide does not function at all. Although each individual person can have only two alleles of the gene, a wide variety of alleles for that gene can occur within a population. These different alleles are one of the sources of variation between individuals of the same species.

The photographs show a blue eye and a brown eye in close-up. It is difficult to see from photographs but the iris has no blue or brown pigment. Its colour is due to reflections from black and white patches. There are two alleles of the gene for eye colour. One allele causes extra black pigment to be produced in the iris. This forms a black layer in front of the white, and makes the eye look brown. Usually the white and black layers are streaky because they are incomplete. Other genes also affect the structure of the iris, producing other shades of colour.

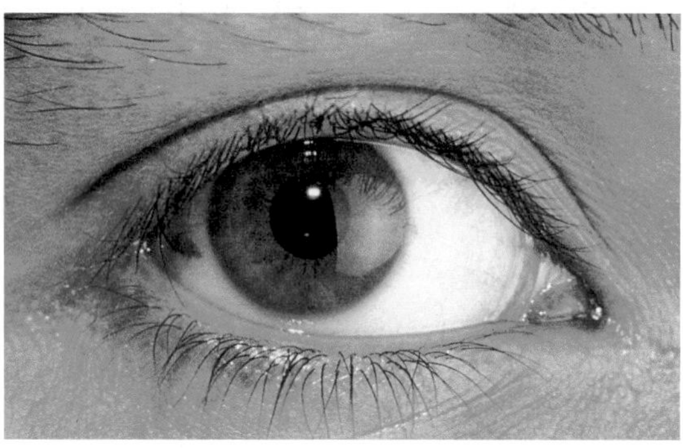

In blue eyes, the iris has a patchy white layer on a black background. In brown eyes, there is an extra black layer in front of the white one.

1 Assume that human chromosome 22 carries 450 genes.

a How many DNA molecules does one copy of chromosome 22 contain?

b How many copies of chromosome 22 are contained in a sperm, a fertilised egg and a young embryo with 16 cells?

c For how many polypetides or proteins does a single chromosome 22 code?

key facts

- Genes are sections of DNA that contain coded information for making polypeptides. These make the proteins that determine the characteristics of organisms.
- Chromosomes contain one very long molecule of DNA. Each molecule carries many genes.
- The DNA molecules in prokaryotic cells are smaller and circular and are not associated with proteins.
- In body cells, chromosomes occur in homologous pairs. Each pair consists of a copy of one maternal and one paternal chromosome.
- Genes that code for the same polypeptide occupy the same relative position on homologous chromosomes. This position is called the gene locus.
- Genes can have different forms, called alleles. The coded information in alleles differs, so the polypeptides they code for also differ.

9.2 The structure of DNA

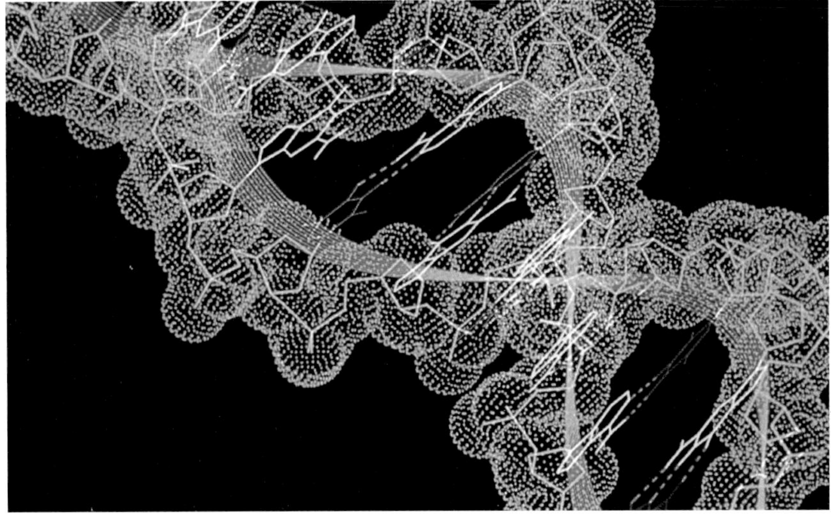

This computer representation of a small piece of DNA may look complicated, but its basic structure is very simple.

People have selected for favourable features in animals and plants for roughly 10 000 years – really since the start of agriculture. For most of that time they were unaware of the rules of genetics that controlled the inheritance of those features. Genetics as a science began roughly 100 years ago. Molecular genetics, which allows us to explain the reactions of the chemicals that control the inheritance of genes, is even younger. It all started in 1953 with Francis Crick and James Watson, at Cambridge University, when they worked out the molecular structure of DNA. Crick and Watson unlocked the puzzle of how the components of DNA fit together into a complex three-dimensional structure.

DNA is made up of monomers called **nucleotides**. Each nucleotide has three parts: a sugar, a phosphate group and a base.

The sugar in DNA is called deoxyribose. There are four different bases, all of which contain nitrogen. These four bases are called **adenine, thymine, cytosine** and **guanine**, and are often referred to by their initial letters. This means that there are four types of nucleotide in a DNA molecule. The nucleotide monomers link together to make long strands, forming a polymer called a **polynucleotide** (Fig. 2).

A DNA molecule consists of two polynucleotide strands joined together to make a structure rather like a twisted ladder. Weak hydrogen bonds form between the bases to produce the 'rungs' of the ladder. The hydrogen atoms on one base are attracted to oxygen and nitrogen atoms on another. The shapes and sizes of the bases mean that the correct distance between the two sugar–phosphate backbones can only be maintained by adenine–thymine and cytosine–guanine bonding. This produces a regular and stable DNA molecule with two sugar–phosphate sides joined by pairs of bases (see Fig. 2).

Fig. 2 A DNA molecule

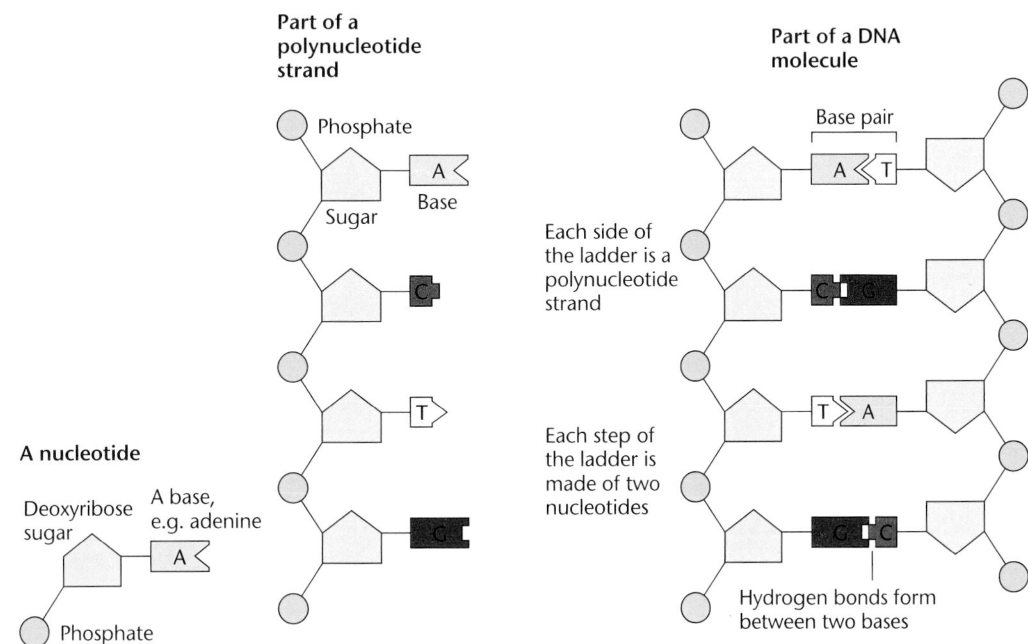

On a molecular scale, DNA molecules are huge. This is an advantage to an organism because it means that a vast number of different genes are confined to a fairly small number of DNA molecules. This makes it more likely that all the information is passed on during cell division and from one generation to the next. It would be disastrous for an organism if some of its cells did not have a full complement of genes, since it would have no instructions for synthesising some of its proteins.

A photograph of a fossilised midge insect embedded in Baltic amber. This specimen is approximately 40 million years old.

The twisting of the two strands, like two long springs plaited together, earns DNA its famous nickname – the **double helix**. Twisting the strands into a helical structure ensures that the weak hydrogen bonds linking the bases are protected in the centre of the molecule, which prevents the code from being corrupted by other chemicals present in the nucleus. DNA is a very stable molecule and can withstand relatively high temperatures. Samples of intact DNA have been found in centuries-old woolly mammoths frozen in the Arctic permafrost, and even from 20-million-year-old fossils of insects preserved in amber – the basis of the book *Jurassic Park*.

The way in which the bases pair up enables stored information to be copied quickly and accurately. When the DNA molecules untwist, the hydrogen bonds break so that the strands can be separated like the sides of a zip. Exact copies can then be produced, since each exposed base will combine with only one of the four types of nucleotide.

Different molecules of DNA differ in the number and order of the pairs of bases that join the two sugar–phosphate backbones. Different

2 Draw a diagram of a polynucleotide strand with the bases in the following order: thymine, thymine, adenine, guanine, cytosine and adenine, using the shapes shown in Fig. 2.

3 Complete the other strand of the DNA molecule you have just drawn.

Table 1 shows the proportions of the four bases in DNA from four organisms.

4 hsw Use your understanding of the structure of DNA to explain the pattern in these proportions.

5 In an organism, 26% of the bases in the DNA are found to be adenine. What percentage would be cytosine? Explain how you worked it out.

Table 1 Bases in DNA

Source of DNA	Adenine %	Guanine %	Thymine %	Cytosine %
Human	30	20	30	20
Rat	28	22	28	22
Yeast	31	19	31	19
Turtle	28	22	28	22

genes have different sequences of base pairs. Only the four base-pair combinations shown in Fig. 2 are possible. In effect, the instructions in genes are written in an alphabet with only four letters. But, by having long sequences, DNA can code vast amounts of information. After all, computers can store massive amounts of information with only a two-letter 'binary' code. A human fertilised egg has about a billion (10^9) pairs of nucleotides. These carry all the coded instructions for the development of the adult body and for the maintenance of all the metabolic processes.

Copying the DNA

Every time a cell divides, it makes a complete copy of its DNA; it copies every single one of its genes. The copy must be exactly the same as the original in order to preserve the information. But how does a molecule with such a complex structure make a perfect copy of itself? Fig. 3 shows what happens. The hydrogen bonds that connect the bases are broken by an enzyme called DNA helicase. The two strands separate easily, exposing the bases. The bonds between the sugar and phosphate groups in the polynucleotide strands are relatively strong, and they keep the separate strands intact.

Another enzyme, called **DNA polymerase**, attaches free nucleotides to the exposed bases on each strand. Only the complementary bases will fit together. A new strand is built on each of the original strands, so that the two new DNA molecules are exactly the same as the original. This process of making perfect copies of DNA is called **replication**. As you can see from the diagram, each of the two new molecules of DNA has one of the original polynucleotide strands and one new one made from the supply of nucleotides in the cell. The system is therefore called **semi-conservative replication**, because one strand in each molecule is conserved (Fig. 4).

Fig. 3 DNA replication

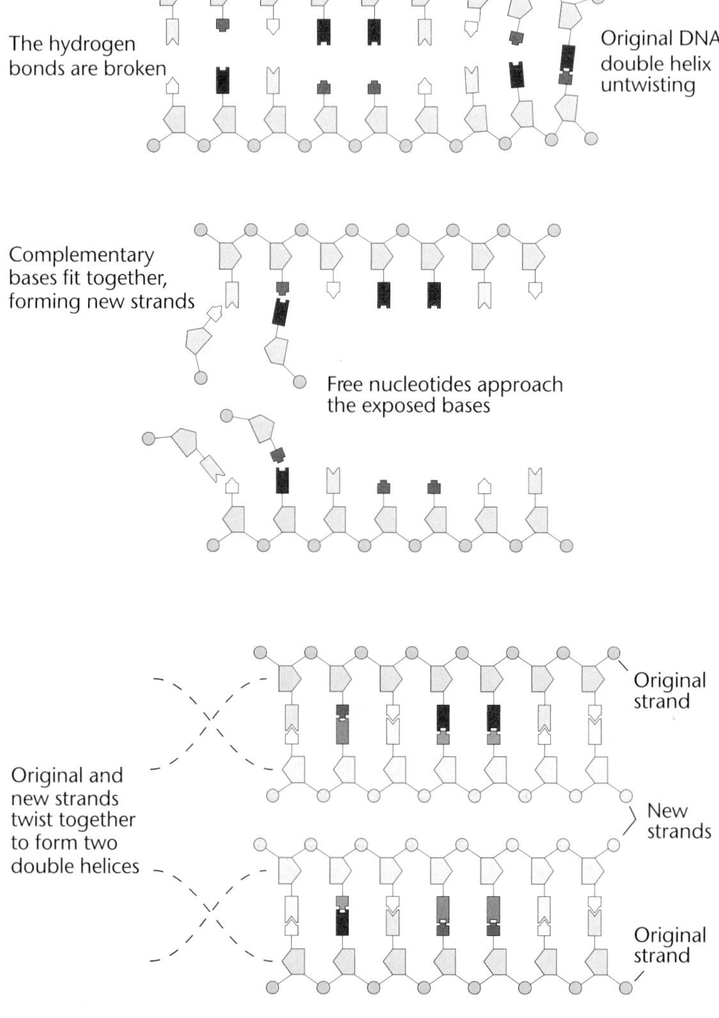
The hydrogen bonds are broken
Original DNA double helix untwisting
Complementary bases fit together, forming new strands
Free nucleotides approach the exposed bases
Original strand
New strands
Original and new strands twist together to form two double helices
Original strand

Fig 4. Semi-conservative replication

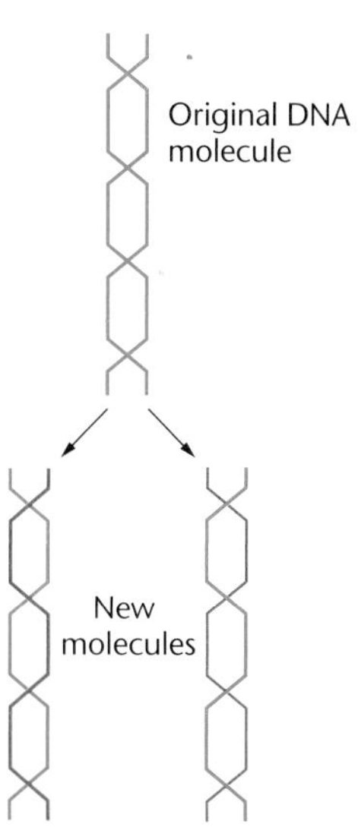
Original DNA molecule
New molecules

6 Draw diagrams to show how the section of DNA in the diagram below would be replicated.

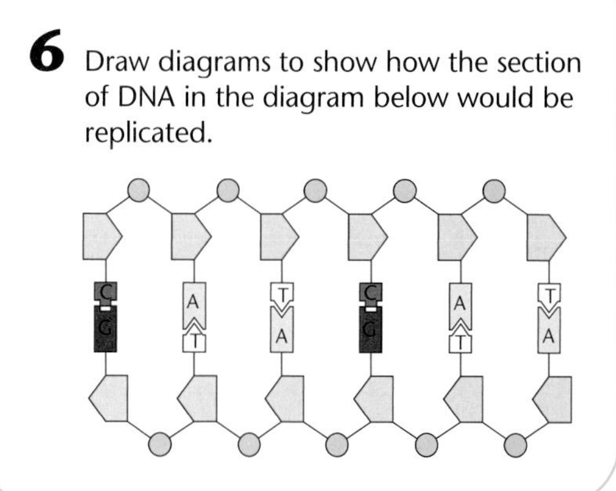
C
G
A
T
T
A
C
G
A
T
T
A

how science works

Evidence for semi-conservative replication

Fig. 5 The Meselon and Stahl experiments

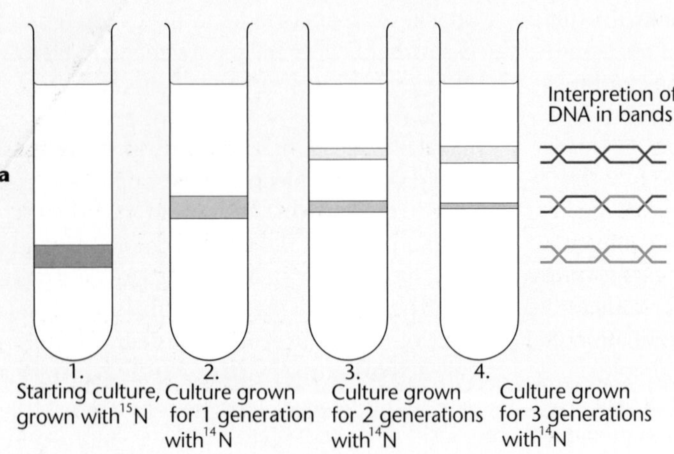

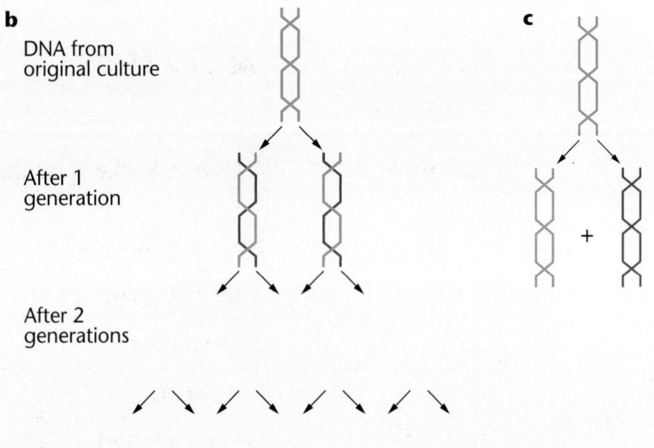

Shortly after Watson and Crick published their theories and suggested semi-conservative replication, Matthew Meselson and Franklin Stahl set out to investigate whether DNA did replicate like this.

Part a of Fig.5 summarises the experiment. Meselson and Stahl grew bacteria in a culture medium containing 'heavy nitrogen' (the isotope 15N). Some of the nitrogen in the nucleotides of the bacterial DNA was therefore 15N. The researchers extracted DNA from these bacteria and centrifuged them in caesium chloride solution. The DNA molecules settled at a point in the centrifuge tube depending on the mass of the molecule. (You can read about centrifugation on page 26).

After centrifuging, the concentration of caesium chloride varies uniformly from the top to the bottom of the tube, with the highest concentration at the bottom. The density changes slowly from the top to the bottom and the DNA extract settles as a band at a particular level.

As the diagram shows, Meselson and Stahl then took bacteria from the 15N medium and grew them on medium containing the common 'light' isotope 14N.

7 Copy and complete diagram B to show Meselson and Stahl's prediction. Use different colours for the 15N and the 14N strands for the second and third generations.

8 Suppose that DNA replicated by producing a new molecule made completely of new nucleotides, as suggested in diagram C. What results would you expect to find in tube 2 after one generation?

Coding for proteins

A protein is a polymer built up from units called amino acids. The order of amino acids in a protein determines its three-dimensional structure and therefore its function. The first protein to be sequenced was insulin, the hormone that regulates blood glucose concentration. In 1959, Frederick Sanger published the order of the 51 amino acids that make up the insulin molecule.

How does the cell put the amino acids together in the correct order? There are only four different bases in the nucleotides of DNA, but there are 20 naturally occurring amino acids. So how does the code work? After many years of research we now know that a sequence of bases, rather than a single base, codes for each different amino acid. A single-base code could code for only four amino acids. A two-base code could code for more amino acids but a sequence of three bases (triplets) is necessary to provide a code for all 20 amino acids. In fact, by using three bases, there are plenty of sequences to spare.

The triplet sequences of bases on the sense strand of the DNA molecule code for different amino acids are called **codons**. Since there are surplus codes, some amino acids have more than one codon. Some amino acids have six different codons. For others there is only one codon. Other codons act as 'start' and 'stop' signals to indicate where a gene begins and ends. Table 2 shows some examples of the codons in DNA that correspond to specific amino acids (with their standard abbreviations in brackets).

Fig. 6 shows a section of the gene that codes for the first four amino acids of insulin. The first triplet of bases, AAA, is a codon for phenylalanine, the next for valine and so on.

Introns and multiple repeats

Only about 3% of the DNA actually codes for amino acids, which in turn make proteins. The remaining 97% of the DNA, originally known as 'junk DNA', are introns. One function of introns is to enable several different proteins that share some sections in common to be produced from a single gene.

At some loci there are **multiple repeats** of certain triplets of letters in the DNA sequence like GAT, GAT, GAT etc. For example, the number of times that this triplet of letters is repeated in the sequence of DNA at a locus on chromosome 3 where a marker called D3 is located could be 12, 13, 14, 15, 16, 17, 18 or 19, which means that there are eight different possible forms or alleles at the D3 locus.

9 Fig. 6 shows the section of the insulin gene that codes for the first four amino acids of the insulin molecule. Using Table 2, decipher the code and list these amino acids.

10 The nucleotide sequence in a DNA strand is: A C G T T G G T G C A C G T G. What sequence of amino acids will this section of DNA add to a protein?

Fig. 6 Part of the insulin sequence

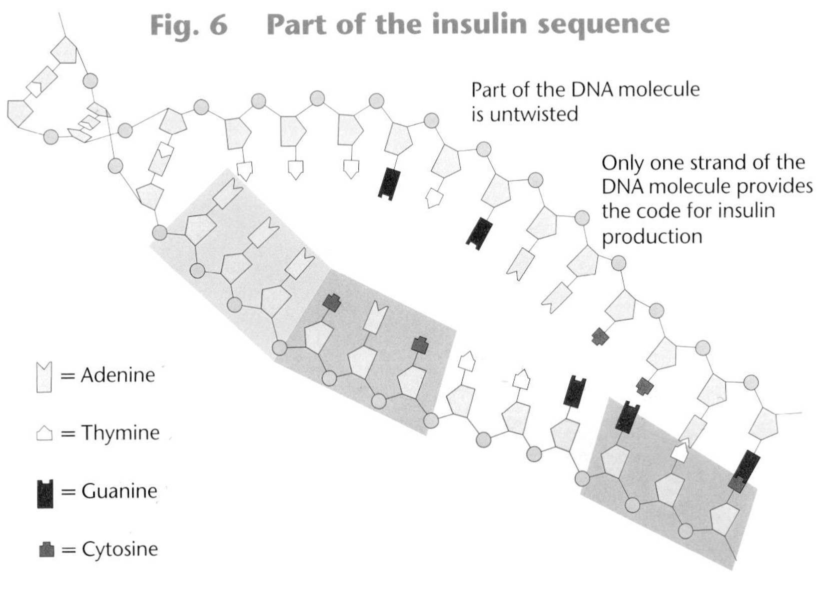
Part of the DNA molecule is untwisted
Only one strand of the DNA molecule provides the code for insulin production
= Adenine
= Thymine
= Guanine
= Cytosine

Table 2 Condons and their amino acids

Codon in DNA	Corresponding amino acid
AAA	phenylalanine (Phe)
GTC	glutamine (Gln)
ACG	cysteine (Cys)
GTG	histidine (His)
TTG	asparagine (Asn)
GAG	leucine (Leu)
CAC	valine (Val)

key facts

- DNA molecules consist of two polynucleotide strands linked together.
- The sequence of bases in the nucleotides enables the DNA to store information.
- The double-stranded structure of DNA and the way in which the bases pair up enable this stored information to be copied precisely and with a high degree of accuracy.
- The large size of the DNA molecules allows a great deal of information to be held in one molecule. This makes it easier to ensure that all the information is passed from generation to generation.
- DNA replicates by a semi-conservative mechanism, which means that half of each new molecule comes from the original molecule.
- Introns are regions of DNA that enable several different proteins that share some sections in common to be produced from a single gene.
- Some parts of DNA contain multiple repeats of triplets.

1 The diagram shows part of a DNA molecule.

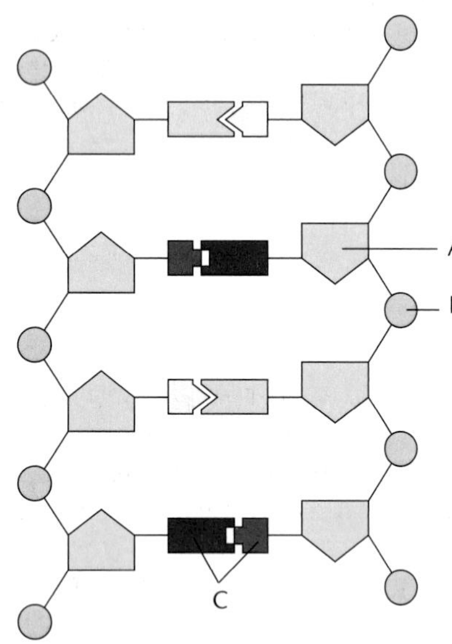

a Name the structures labelled **A**, **B** and **C**. (3)

b The sequence of bases on one strand of a DNA molecule is

ACTGCATCC

How many amino acids could be coded for by this section of DNA? (1)

c Explain what is meant by

i intron (2)

ii multiple repeat. (2)

Total 8

2

a A nucleotide of DNA contains a sugar molecule and two other molecules.
Name the other two molecules. (2)

b Draw a simple diagram to show the structure of a DNA nucleotide. (2)
A piece of DNA contained 16 base pairs.
Complete the table below to give the numbers of the bases in this piece of DNA.

Number of bases

	Adenine	Thymine	Cytosine	Guanine
Strand X	6			
Strand Y	2			4

(3)

Total 7

3 The diagram shows the structure of a DNA molecule.

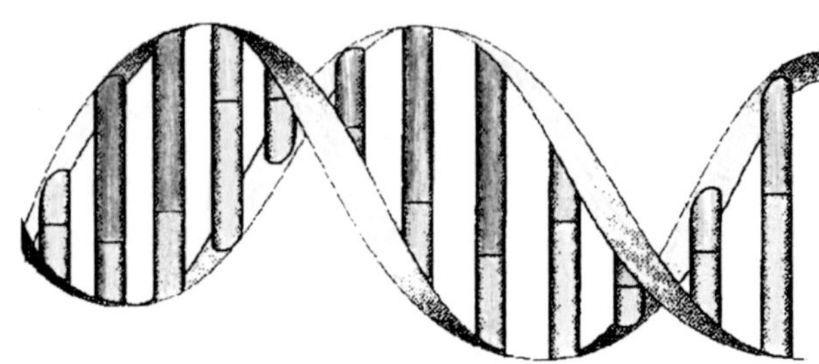

a What term describes the shape of a DNA molecule? (1)

b How are the two nucleotide chains joined to each other? (1)

c Explain three ways in which the structure of the DNA molecule is related to its function. (3)

Total 5

4 The diagram shows how a DNA molecule is replicated.

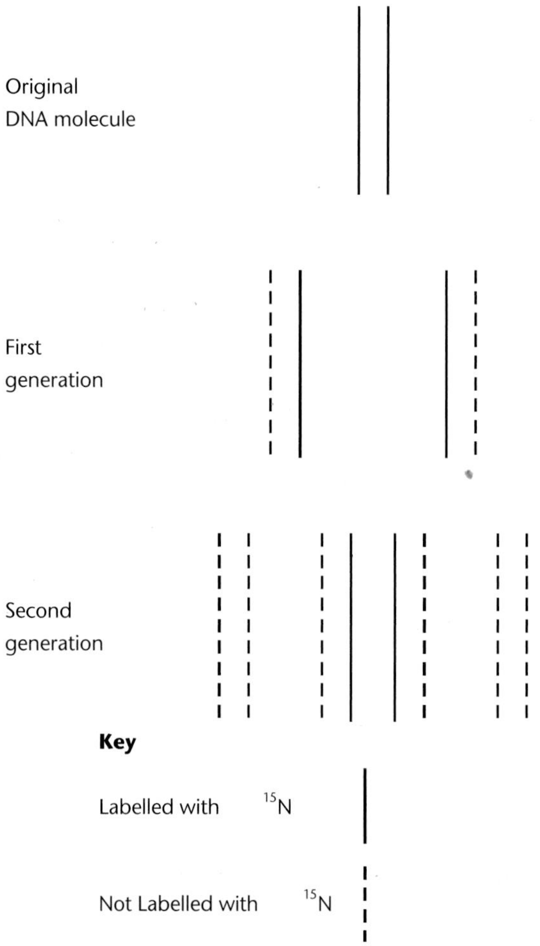

Original DNA molecule

First generation

Second generation

Key

Labelled with ^{15}N

Not Labelled with ^{15}N

a What is ^{15}N? (1)
b Which part of the DNA molecule is labelled with ^{15}N? (1)
c Explain how molecules of DNA containing ^{15}N can be distinguished from those which do not contain ^{15}N. (3)
d Explain as fully as you can:
 i Why all the first generation DNA molecules contained ^{15}N (2)
 ii Why only half of the second generation DNA molecules contained ^{15}N. (2)

Total 9

5 The diagram shows the replication of a molecule of DNA.

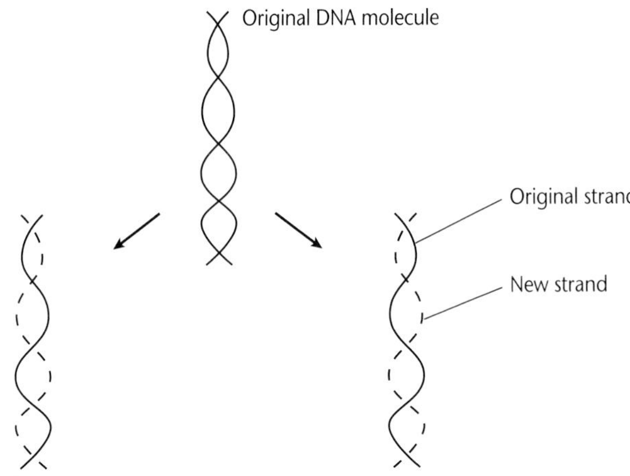

a Explain why DNA replication is described as *semi-conservative.* (1)
b **i** What is meant by *specific base pairing*?
 ii Explain why specific base pairing is important in DNA replication. (3)
c Describe **two** features of DNA which make it a stable molecule. (2)

Total 6

AQA, June 2003, Unit 2, Question 2

6 The diagram shows one nucleotide pair of a DNA molecule.

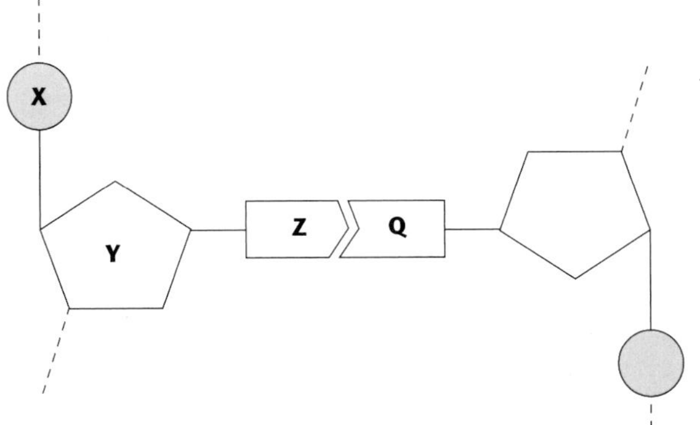

a Name the parts of the nucleotide labelled **X**, **Y** and **Z** (3)
b What type of bond holds **Z** and **Q** together? (1)
c A sample of DNA was analysed. 28% of the nucleotides contained thymine. Calculate the percentage of nucleotides which contained cytosine. Show your working. (2)

Total 6

AQA, June 2004, Unit 2, Question 1

7

a Explain why the replication of DNA is described as semi-conservative. (2)
b Bacteria require a source of nitrogen to make the bases needed for DNA replication. In an investigation of DNA replication some bacteria were grown for many cell divisions in a medium containing ^{14}N, a light form of nitrogen. Others were grown in a medium containing ^{15}N, a heavy form of nitrogen. Some of the bacteria grown in a ^{15}N medium were then transferred to a ^{14}N medium and left to divide once. DNA was isolated from the bacteria and centrifuged.
The DNA samples formed bands at different levels, as shown in the diagram below.

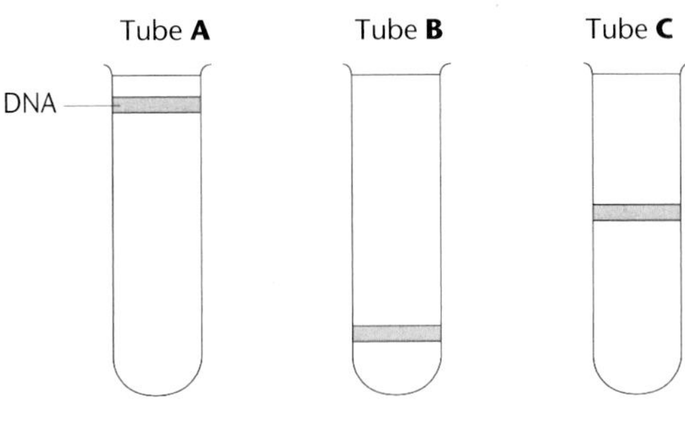

Tube A	Tube B	Tube C
DNA from bacteria grown in a ^{14}N medium	DNA from bacteria grown in a ^{15}N medium	DNA from bacteria grown originally in a ^{15}N medium, but then transferred for one cell division to a medium containing ^{14}N

i What do tubes **A** and **B** show about the density of the DNA formed using the two different forms of nitrogen? (1)
ii Explain the position of the band in tube **C**. (2)

Percentage of base present

DNA sample	Adenine	Cytosine	Guanine	Thymine
Strand 1	26		28	14
Strand 2	14			

c In a further investigation, the DNA of the bacterium was isolated and separated into single strands. The percentage of each nitrogenous base in each strand was found. The table shows some of the results.
Use your knowledge of base pairing to complete the table. (2)

Total 7

AQA, June 2006, Unit 2, Question 4

8 The diagram shows a short section of a DNA molecule.

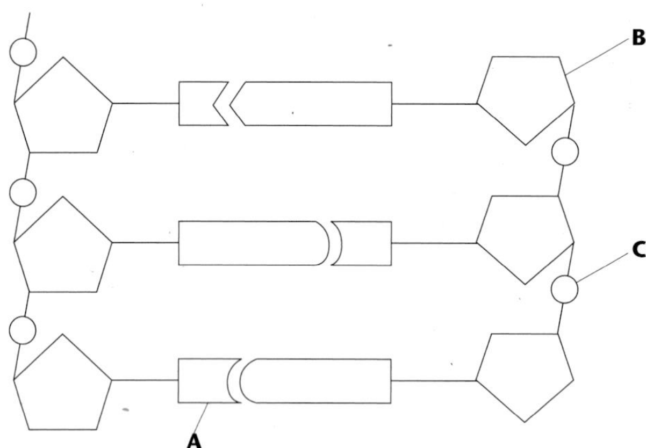

a On the diagram draw a box round **one** nucleotide. (1)

b **i** The sequence of bases on one strand of DNA is important for protein synthesis. What is its role? (1)

ii How are the two strands of the DNA molecule held together? (1)

iii Give **one** advantage of DNA molecules having two strands. (1)

Total 4

AQA, January 2005, Unit 2, Question 1

hsw

how science works **assignment**

Could dinosaurs live again?

The novel, *Jurassic Park*, by Michael Crichton, is based on the idea that DNA from dinosaurs could be used to recreate them. The following text from the book illustrates the principles being suggested.

'Regis introduced Henry Wu, a slender man in his thirties. "Dr Wu is our chief geneticist. I'll let him explain what we do here."

Henry Wu smiled. "At least I'll try," he said. "Genetics is a bit complicated. But you are probably wondering where our dinosaur DNA comes from."

"As a matter of fact," Wu said, "there are two possible sources. Using the Loy antibody extraction technique, we can sometimes get DNA directly from dinosaur bones."

"What kind of a yield?" Grant asked.

"Well, most soluble protein is leached out during fossilisation, but twenty percent of the proteins are still recoverable by grinding up the bones and using Loy's procedure. As you can imagine, a twenty percent yield is insufficient for our work. We need the entire dinosaur DNA strand in order to clone. And we get it here." He held up one of the yellow stones. "From amber – the fossilised resin of prehistoric tree sap." …

"Tree sap," Wu explained, "often flows over insects and traps them. The insects are then perfectly preserved within the fossil. One finds all kinds of insects in amber – including biting insects that have sucked blood from larger animals."

"Sucked the blood," Grant repeated. His mouth fell open. "You mean sucked the blood of dinosaurs … ".'

A1

a Explain why you would need a sample of its entire DNA in order to produce a dinosaur.

b Wu suggests that only 20% of protein is recovered by Loy's procedure, and that this would not yield enough DNA. Explain the mistake.

c The second method of obtaining dinosaur DNA was to extract it from the nuclei of the red blood cells found in the insects preserved in amber. This method could, in theory, work for reptiles, but not for mammals. Explain why not.

A2 After extracting the DNA it would be necessary to make copies of the DNA strands. Which enzyme could be used to do this?

A3 In *Jurassic Park*, the dinosaur DNA that was extracted was placed into a crocodile egg from which the nucleus had been removed.

a Explain why the egg nucleus would be removed.

b Explain what the egg would provide that could make it possible for a young dinosaur to develop.

A4 The ideas in the book are not as far-fetched as they might seem. Similar techniques are being proposed as ways of re-creating more recently extinct animals, such as the dodo. Suggest the practical difficulties that would be likely to arise when trying to use these methods to re-create dinosaurs.

10 The cell cycle

Researchers in Texas have cloned a domestic cat, producing a 2-month-old kitten called CopyCat. CopyCat is a copy of her genetic mother, not of the cat that actually gave birth to her. Mark Westhusin, a member of the cloning team, has explained serious scientific reasons for cloning a cat: 'Cats have a feline AIDS [aquired immune deficiency syndrome] that is a good model for studying human AIDS.'

Animal welfare groups have voiced concern over the experiments. Cats Protection, a UK feline welfare charity, said cloning was not the answer to replacing a lost pet. Chief Executive Derek Conway said: 'the cloning of cats interferes with nature and raises serious questions concerning whether a pet can ever be truly replaced.'

CopyCat is the only surviving animal of 87 kitten embryos created by cloning. The success rate will have to improve if pet cloning is to become a reality. The cloning experiments were funded by an 81-year-old financier called John Sperling, who wants to charge wealthy pet owners to clone their animals. He is quoted as saying that he would also like to see cloning used for socially useful animals such as rescue dogs.

The first ever clone of an adult mammal was Dolly the sheep, in 1997. Dolly was grown from a single udder cell from a 6-year old sheep. As you learned in Chapter 9, every cell has a full set of genes in its nucleus. Before Dolly's birth it was thought that once a cell had been specialised in an organ such as the heart or udder, it was impossible to reprogram it to go through the same full cycle of development as a fertilised egg.

Dolly was euthanised in February 2003 because she was suffering from a lung disease and arthritis. Scientists think this was unrelated to her being a clone, as the lung disease is fairly common in sheep kept indoors. There is less certainty about the arthritis, however.

Although widely publicised as a scientific success story, Dolly's birth was achieved only after many failures. Would it be acceptable for deformed or handicapped babies to be born in the pursuit of a perfect clone? What will be the legal status of a child born by cloning one individual, and will such children have problems accepting their own identity? Will cloning only be available to the super-rich who can afford the expense? These issues need debate and legislation, and the future is uncertain. The only thing that is sure is that there is no possibility of reversing the scientific understanding and technological advances that have made cloning a reality.

Copy Cat – the first cloned cat.

10.1 Mitosis

The key to producing genetically identical offspring is the process of cell division called **mitosis** (Fig. 1). Every time a cell divides by mitosis, the instructions contained in its DNA are copied faithfully. Each new cell therefore receives a complete set of the genetic code from the parent cell and can make all the same proteins.

When a cell is not dividing the individual chromosomes are not visible. When stained, the whole nucleus appears as a dark mass because the chromosome material, called **chromatin**, is spread out (Fig. 1). This makes it easier for mRNA molecules to move away from the DNA molecules and out of the nucleus.

In a dividing cell, there is a clear sequence of events and, for much of the process, the chromosomes exist in a condensed state and are easier to distinguish. Replication of the DNA (see Chapter 9) takes place just before the visible

Fig. 1 Interphase and the stages of mitosis

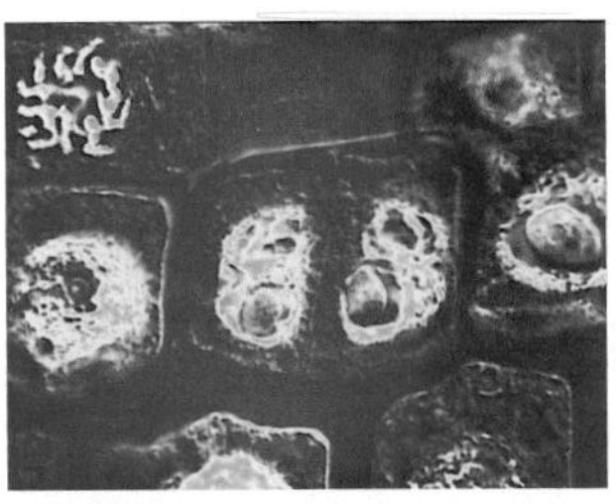

Interphase
The period of the cell cycle between mitosis
For most of the time the chromosomes in a nucleus cannot be seen. The DNA molecules are stretched out and busy synthesising proteins. At the onset of mitosis the DNA replicates. This happens during interphase, before any sign of cell division can be seen.

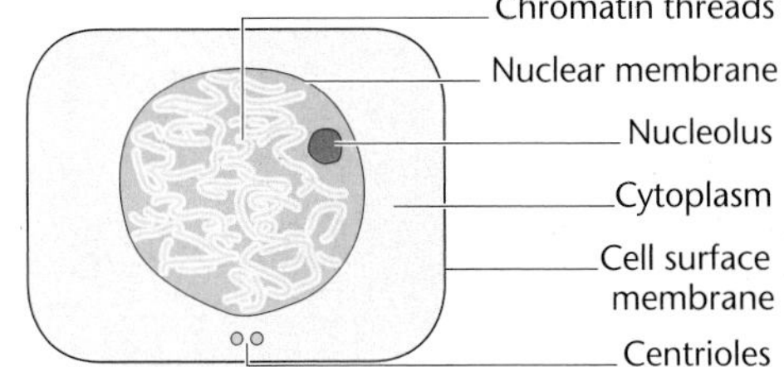

These light micrographs show the stages of mitosis in cells of a hyacinth root.

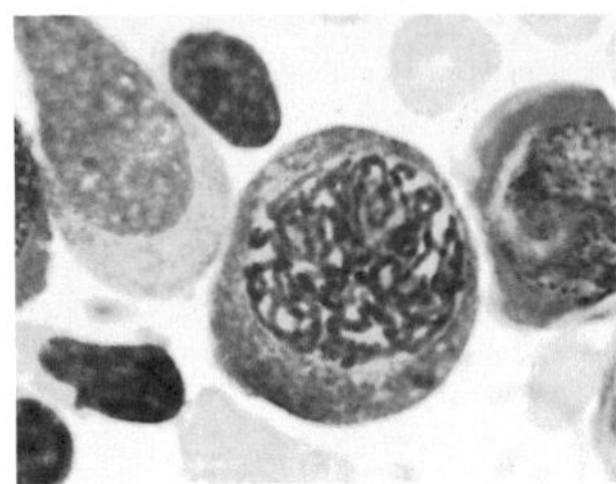

Prophase
After the chromosomes have replicated, they coil up and contract. They then become visible. The replicated chromosomes appear as double strands. In fact they consist of the two new chromosomes, at this stage called chromatids, still firmly joined together at the centromere.

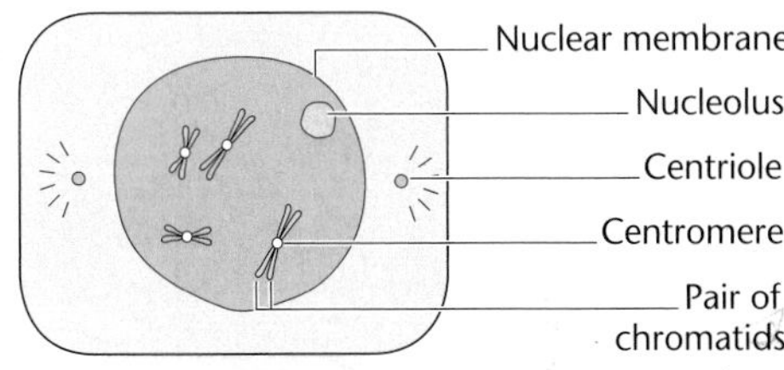

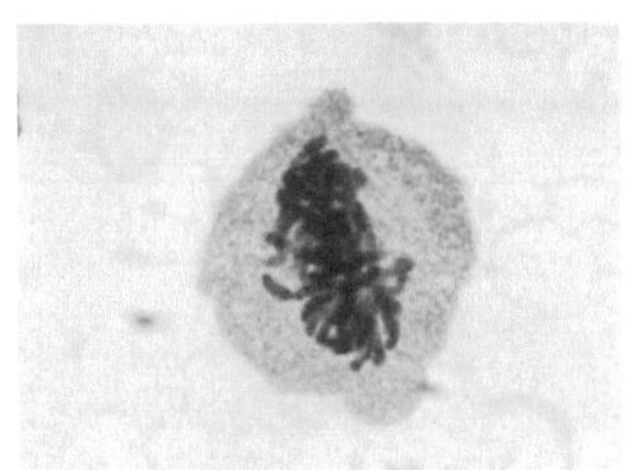

Metaphase
The membrane of the nucleus breaks down and a web of protein fibres called spindle fibres (microtubules) forms from from one end of the cell to the other. The centromeres attach attach to the spindle fibres in the middle of the cell.

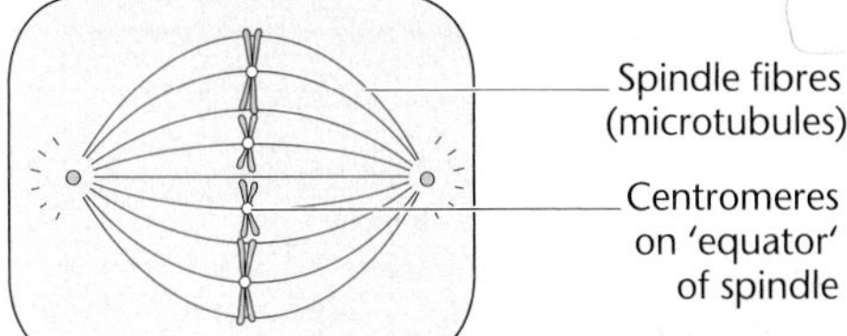

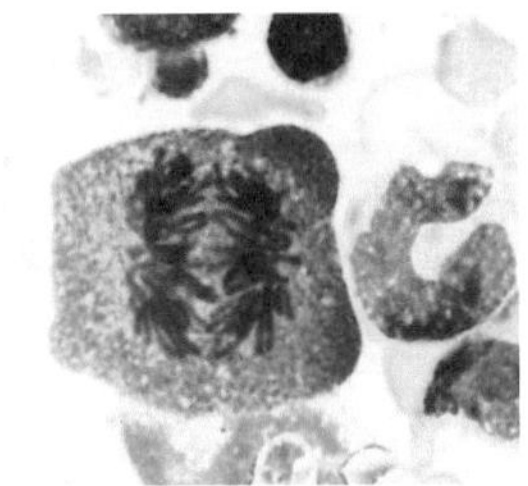

Anaphase
The centromeres now split and the chromatids separate. The chromatids move along the spindle fibres to opposite ends (poles) of the cell.

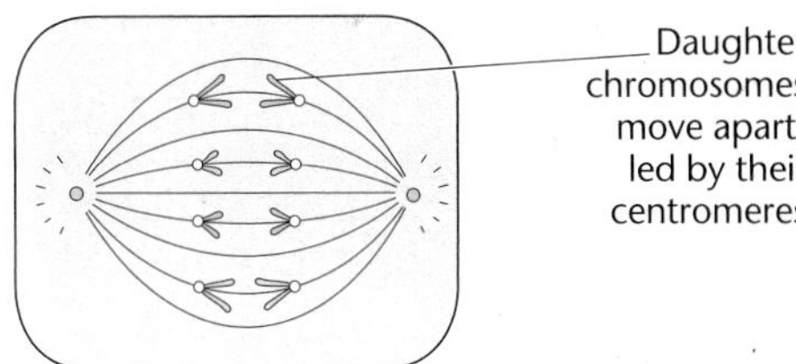

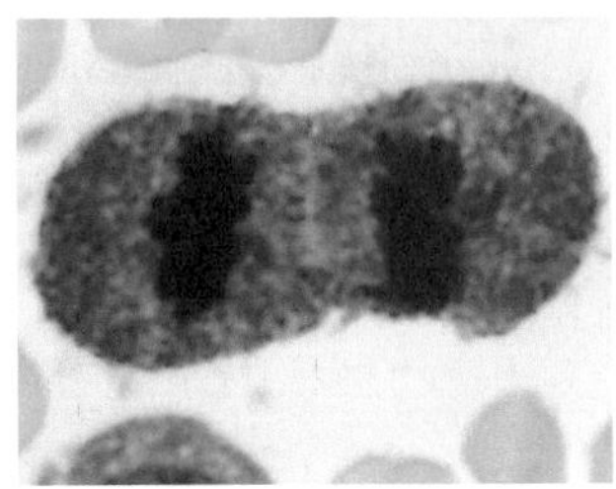

Telophase
The separated chromatids, which are exact copies of the original chromosomes, group together at opposite ends of the cell. New nuclear membranes develop and the chromosomes uncoil.
Mitosis is complete when the cytoplasm divides and new cell membranes form. Plant cells also form new cell walls.

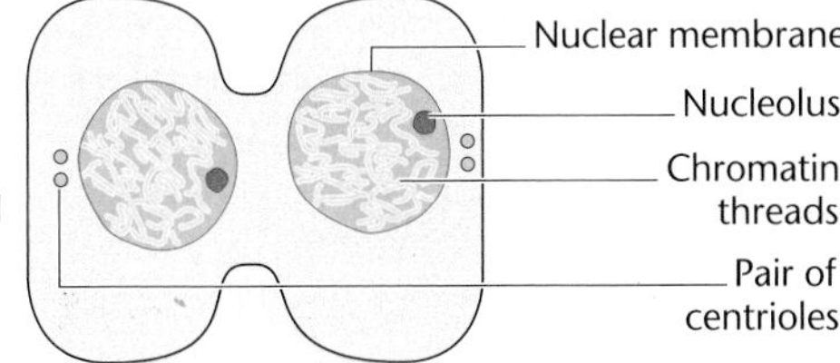

stages of mitosis begin. At this point, each chromosome has two molecules of DNA. The chromosomes shorten and thicken, and the DNA forms very tightly packed coils of coils, called supercoils. This shrinkage of chromosomes into smaller packages makes it less likely that sections of DNA will break away and be lost during cell division.

After the chromosomes have contracted, they become visible when stained and it can be seen that each chromosome consists of two separate threads, called **chromatids**. The chromatids are identical and have been produced from the original chromosome DNA by replication. The two chromatids remain attached to each other at the **centromere** until the nucleus divides. Each dividing cell has two sets of chromatids

Fig. 2 The cell cycle

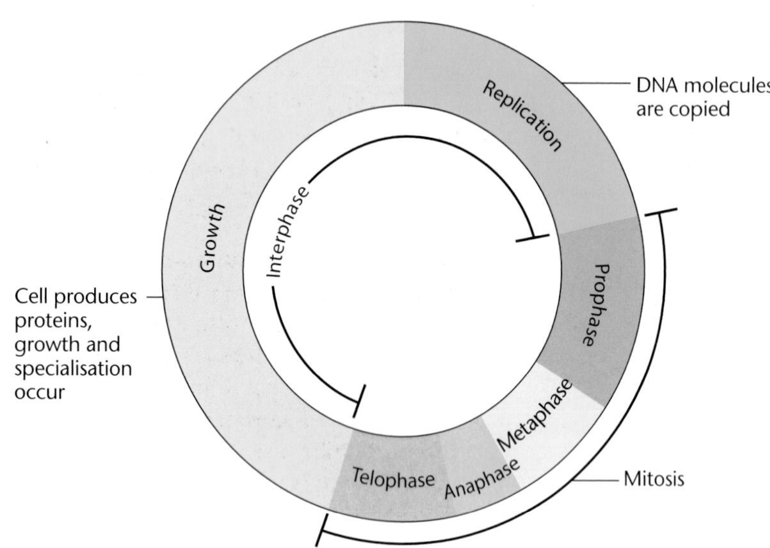

because it has two of each chromosome. When the cell divides by mitosis, the chromatids separate and each new cell nucleus gets a full complement of all of the chromosome pairs in the original cell.

The cell cycle

Mitosis is a continuous process, but for convenience the cell cycle is described as having five stages. Fig. 2 summarises the complete cell cycle.

The cell cycle and cancer

The cell cycle is affected by two groups of genes.

- Proto-oncogenes produce proteins that stimulate cell division. These genes also inhibit cell death.
- Tumour suppressor genes prevent cell division or lead to cell death.

1

a How does the amount of DNA in a nucleus differ between the start and the end of interphase?

b There are 46 chromosomes in a human body cell. How many DNA molecules are there in a cell nucleus at the start of prophase?

c Suggest why it is an advantage for the chromatids to contract independently during prophase.

d Suggest why it is important that the chromatids remain attached at the centromere until anaphase.

2 Draw diagrams of metaphase, anaphase and telophase for the cell shown below.

Fig. 3 Animal cell with six chromosomes

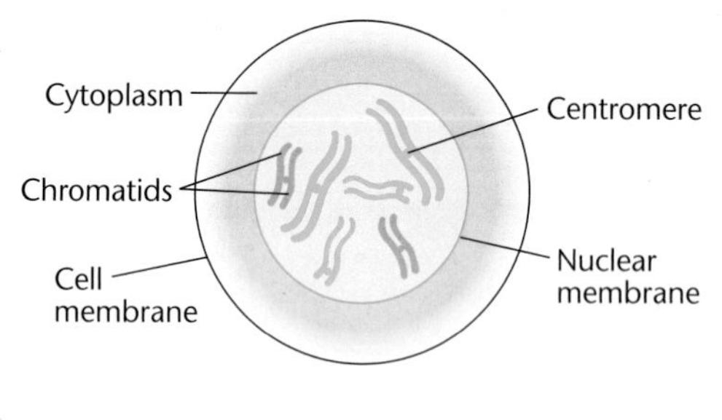

Mutated forms of proto-oncogenes are called **oncogenes**. Oncogenes cannot be controlled by tumour suppressor genes, so they continue to stimulate cell division. This results in a tumour. Mutated forms of tumour suppressor genes cannot control proto-oncogenes, again resulting in uncontrolled cell division.

key facts

- Clones are genetically identical organisms. The offspring of plants and other organisms that reproduce asexually are clones.
- Mitosis is a type of cell division. When the cell's DNA is replicated in mitosis, each new cell produced receives an exact copy of the DNA in the parent cell.
- Replication of the DNA in the chromosomes occurs during interphase, before the chromosomes contract and become visible.
- Replication produces two identical chromatids from each chromosome. The chromatids are separated during mitosis, each daughter nucleus receiving one of each pair.
- Proto-oncogenes and tumour suppressor genes control the cell cycle. Mutations of these genes may result in uncontrolled cell division, leading to tumour formation.

10.2 Meiosis

Humans, like other sexually reproducing organisms, inherit their genes from their parents' sex cells. **Haploid** sex cells have only one of each pair of homologous chromosomes. When a sperm (a haploid cell) fertilises an egg (also a haploid cell) to form a **zygote** (a diploid cell), each contributes one set of chromosomes, and so one set of genes. It is important that one copy of each chromosome, and hence of each gene, is passed on from each parent, otherwise there would be an incomplete set of instructions in the zygote.

Gametes that do not have one copy, and one copy only, of each chromosome will not produce a viable zygote. If a chromosome or part of a chromosome is missing, there will be an incomplete set of instructions and some essential proteins will not be made. Extra chromosomes or fragments of chromosomes are usually harmful to the development of an embryo, which usually dies before birth. The survival of sexually reproducing species depends on accurate meiosis; and not surprisingly, the process is quite complex. It involves two cell divisions.

In humans, a cell about to begin meiosis has 46 chromosomes, each consisting of two chromatids. At the end of the first division, one chromosome from each homologous pair has been separated into a new nucleus. These two new nuclei have 23 chromosomes, but each chromosome still consists of a pair of chromatids. By the end of the second division, the chromatids have separated and are now independent chromosomes, so there are four nuclei, each with 23 chromosomes.

Fig. 4 An overview of meiosis

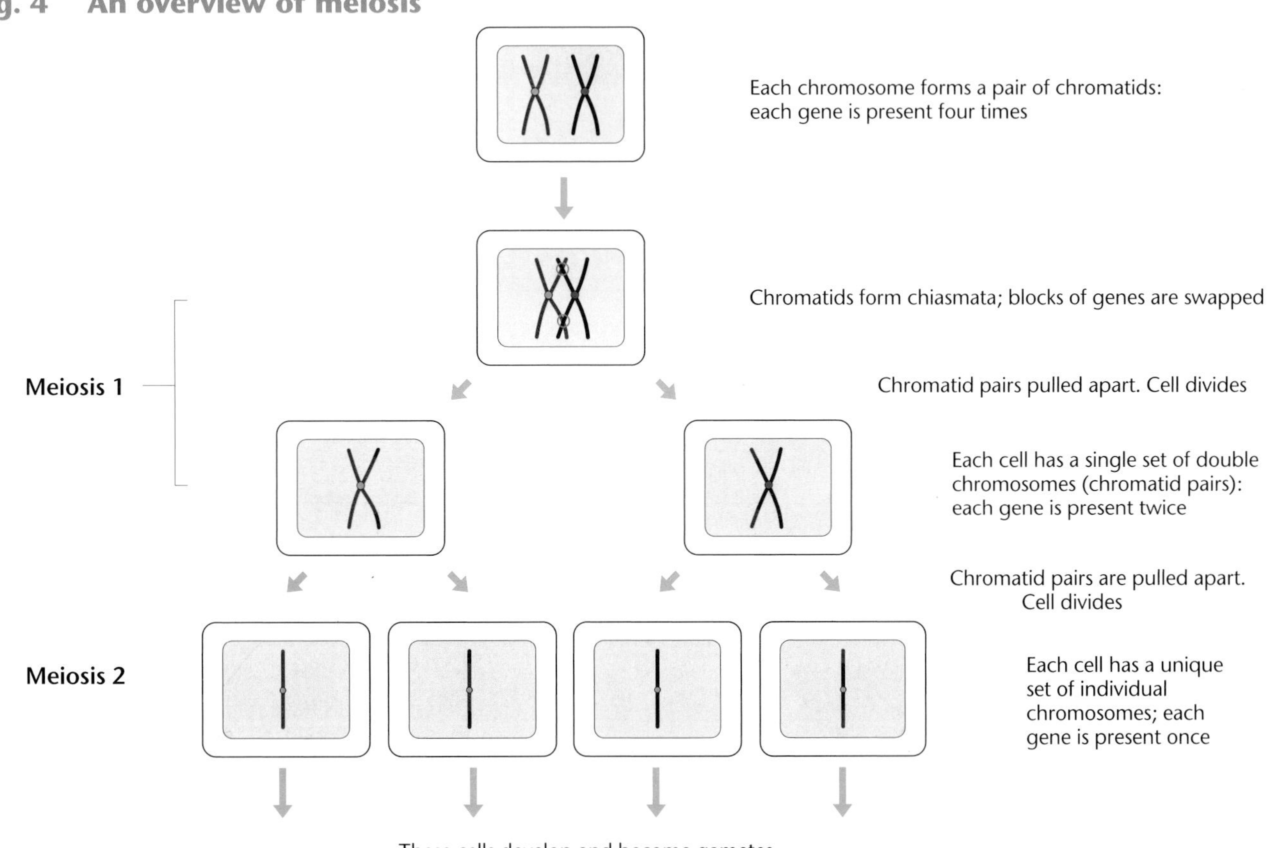
Each chromosome forms a pair of chromatids: each gene is present four times
Chromatids form chiasmata; blocks of genes are swapped
Meiosis 1
Chromatid pairs pulled apart. Cell divides
Each cell has a single set of double chromosomes (chromatid pairs): each gene is present twice
Chromatid pairs are pulled apart. Cell divides
Meiosis 2
Each cell has a unique set of individual chromosomes; each gene is present once
These cells develop and become gametes

The advantages of meiosis

A female ocean sunfish can produce 50 million eggs. If every one was fertilised, there could be 50 million baby sunfish, and every one would be different! Not surprisingly, since the mother makes no effort to look after them, very few of these babies survive. But, because of the differences, some may be slightly better adapted for survival than others. That's the great advantage of sex. Meiosis followed by fertilisation greatly increases the amount of genetic variation. New combinations of alleles are produced, and these may give rise to individuals with new features that favour their survival. The survivors then pass on their successful gene combinations to their offspring.

The ocean sunfish *Mola mola* is a notoriously neglectful parent, but mass production of offspring ensures that some survive and keep the species going.

Fig. 5 Independent assortment

A cell from an organism with two pairs of homologous chromosomes

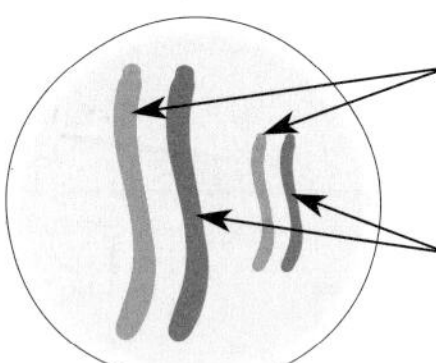
The pink chromosomes are originally from the female parent
The purple chromosomes are originally from the male parent

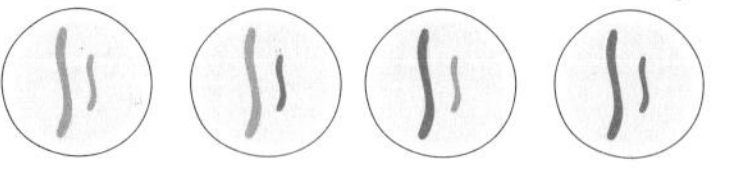
There are four possible way the chromosomes could be reshuffled in the gametes

A cell from an organism with three pairs of homologous chromosomes

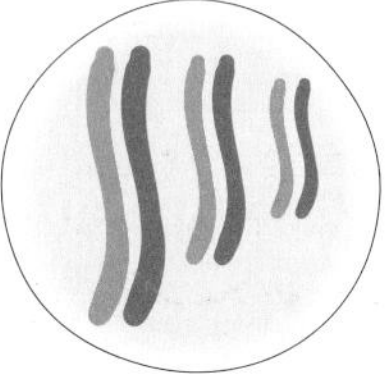

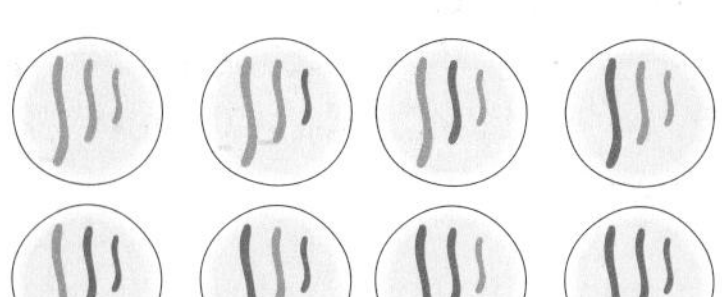
There are eight ways the chromosomes could be reshuffled in the gametes

Sources of variation

Combinations of alleles may be produced in three ways:

- **independent assortment**
- **random fertilisation**
- **crossing over**.

Independent assortment. In each homologous pair, one chromosome is derived from the mother and one from the father. During meiosis these maternal and paternal chromosomes can be reshuffled in any combination – they are independently assorted. This process occurs at anaphase of the first division of meiosis, because the maternal and paternal chromosomes move apart entirely randomly.

3 Use the pattern shown in Fig. 5 to work out how many different combinations are possible if there are:

a 4 pairs

b 23 pairs
of homologous chromosomes.

4 Look at Table 1. Describe three ways in which mitosis is similar to meiosis.

Table 1 Differences between mitosis and meiosis

Mitosis	Meiosis
One cell division only	Two stages of cell division
Two cells produced	Four cells produced
Daughter cells have the same number of chromosomes as parent cell	Daughter cells have half the number of chromosomes
Homologous chromosomes do not pair up	Homologous chromosomes pair in prophase I
No bivalents or crossing over	Crossing over occurs in bivalents

Random fertilisation. Fertilisation is random – any female gamete can join with any male gamete. The gametes are from different individuals, each with chromosomes carrying different alleles of many of the genes. In humans there are about 30 000 genes, many of which have several alleles, so the number of possible combinations is astronomical.

Crossing over. Even more variation is introduced during prophase I by a process called crossing over, as shown in Fig. 6. Once a homologous pair of chromosomes has formed a bivalent, two of the chromatids coil together like mating snakes. At the points where the chromatids cross over each other, they join together. These links are called **chiasmata** (singular – chiasma). When the chromosomes separate again during anaphase, parts of the chromatids are swapped from one chromatid to another. As a result, the new chromatids have some sections that have been copied from the maternal chromosome and some sections that have been copied from the paternal chromosome. Fig. 7 shows a close up of this process.

Fig. 6 Crossing over in meiosis

A homologous pair of chromosomes has a gene for hair colour and a gene for hair structure. The maternal and paternal chromosomes carry different alleles for these genes.

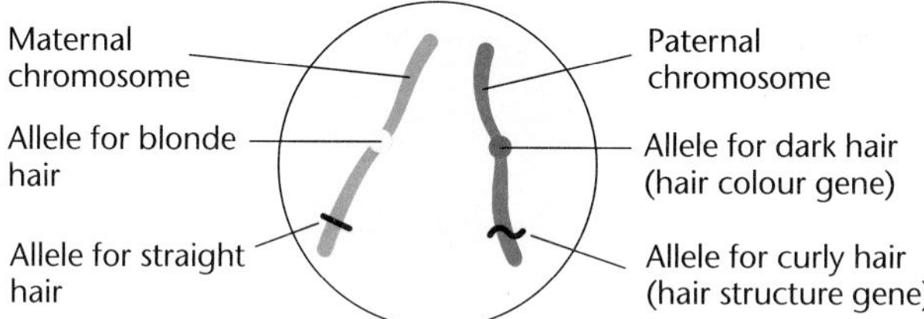

The chromosomes replicate, forming two chromatids joined at the centromere

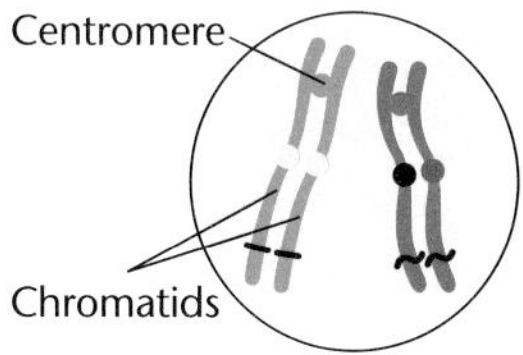

The chromosomes come together as a pair, and two of the chromatids cross over

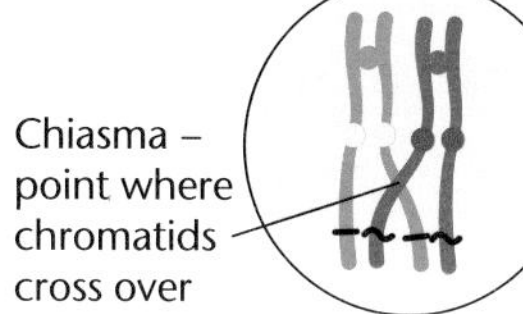

Four possible gametes can be produced by meiosis

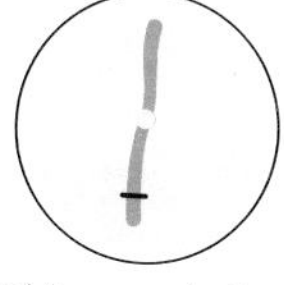

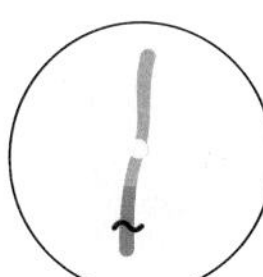

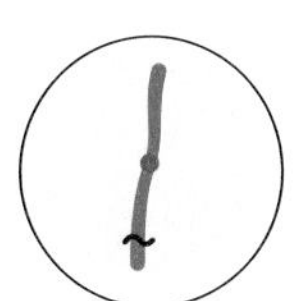

This gamete has the same alleles as the maternal chromosome

These two gametes have new combinations of alleles – one for blonde and curly hair, the other for dark and straight hair

This gamete has the same alleles as the paternal chromosome

Fig. 7 A possible crossing over

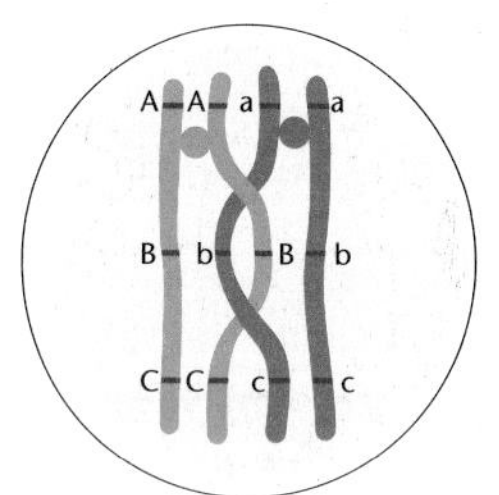

The great advantage of crossing over is that alleles of genes that occur on the same chromosome can be combined in new ways. If only whole chromosomes were passed on, much less genetic variation would be possible.

5 A chromosome has three genes on it. Each gene has two alleles, i.e. A and a, B and b, C and c. Crossing over occurs at two chiasmata, as shown in Fig. 7. Use coloured pens to draw diagrams to show the chromosomes in the gametes that would be produced.

key facts

- Meiosis is involved in the production of **gametes** (sex cells). This type of cell division halves the number of chromosomes in a cell. The gametes are haploid, having only one of each homologous pair of chromosomes.
- In the first stage of meiosis the homologous pairs of chromosomes pair up and form **bivalents**.
- Meiosis continues in two stages. In the first, the two chromosomes of each pair are separated, in the second, the chromatids are split apart.
- New combinations of alleles result from independent assortment of the maternal and paternal chromosomes; from crossing over of sections of chromatids; and from random fertilisation.

examination questions

1 The diagram shows the cell cycle.

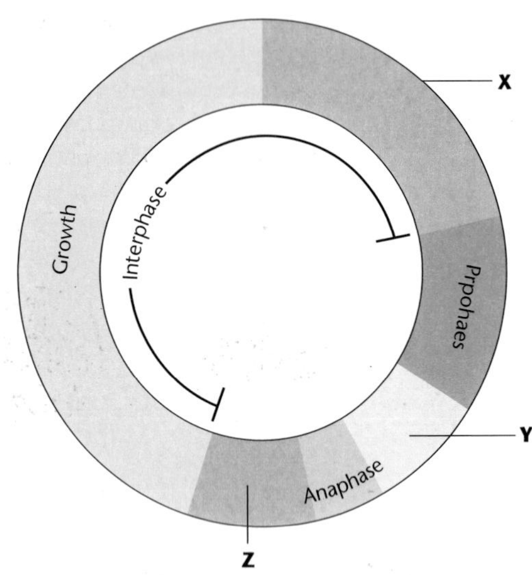

a **i** Name the stage of mitosis labelled **Y**. (1)
ii Descibe what happens during this stage. (2)
b **i** Name the stage of mitosis labelled **Z**. (1)
ii Descibe what happens during this stage. (2)
c **i** Describe what happens during the stage of the cell cycle labelled **X**. (2)
ii Explain the significance of the this stage in the development of an organism. (2)
d Explain the relationship between oncogenes and cancer. (3)

Total 13

2 The diagram shows the life cycle of an alga.

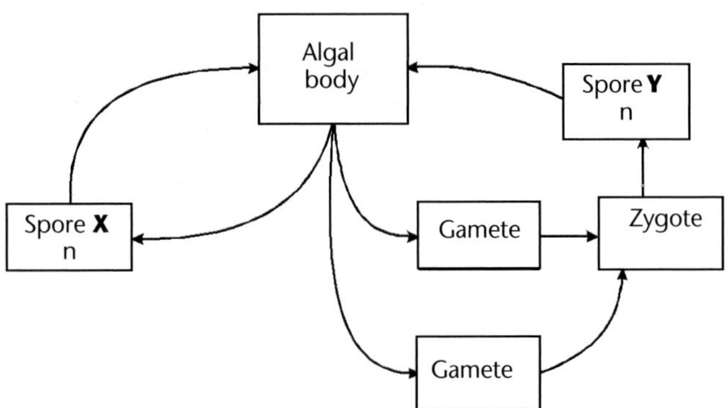

a Copy the diagram, then on the diagram
i Mark with an **H** a haploid cell (1)
ii Mark with a **D** a diploid cell (1)
iii Mark with an **M** where meiosis occurs. (1)
b **i** The algal body grown from spore **X** will be genetically identical to its parent. Explain why. (2)
ii The algal body grown from spore **Y** may be different from its parent. Explain why. (2)

Total 7

3

a Boxes **A** to **E** show some of the events of the cell cycle.

A Chromatids separate

B Nuclear envelope disappears

C Cytoplasm divides

D Chromosomes condense and become visible

E Chromosomes on the equator of the spindle

i List these events in the correct order, starting with **D**

D _____ _____ _____ _____ (1)

ii Name the stage described in box **E**. (1)
b Name the phase during which DNA replication occurs. (1)
c Bone marrow cells divide rapidly. As a result of a mutation during DNA replication, a bone marrow cell may become a cancer cell and start to divide in an uncontrolled way. A chemotherapy drug that kills cells when they are dividing was given to a cancer patient. It was given once every three weeks, starting at time 0. The graph shows the changes in the number of healthy bone marrow cells and cancer cells during twelve weeks of treatment.

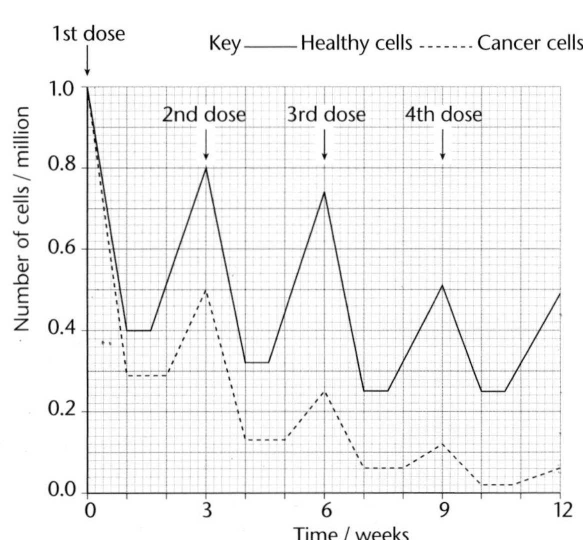

i Using the graph calculate the number of cancer cells present at week 12 as a percentage of the original number of cancer cells. Show your working.
_______________ % (2)

ii Suggest **one** reason for the lower number of cancer cells compared to healthy cells at the end of the first week. (1)

iii Describe **two** differences in the effect of the drug on the cancer cells, compared with healthy cells in the following weeks. (2)

Total 8

AQA, June 2006, Unit 2, Question 2

4

a The drawing shows a stage of mitosis in an animal cell.

i Name this stage of mitosis. (1)

ii Describe and explain what happens during this stage which ensures that two genetically identical cells are produced. (2)

b The graph shows the relative amounts of DNA per cell during two successive cell divisions in an animal.

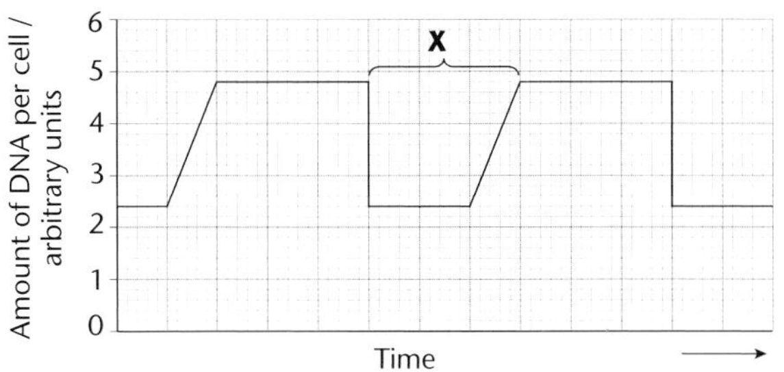

i What stage of the cell cycle is shown by **X**? (1)

ii Apart from an increase in the amount of DNA, give **one** process which occurs during stage **X** which enables nuclear division to occur. (1)

iii How many units of DNA would you expect to be present in a gamete formed in this animal as a result of meiosis? (1)

c The table shows the average duration of each stage of the cell cycle in the cells of a mammalian embryo.

Stage	Mean duration /minutes
Interphase	12
Prophase	50
Metaphase	15
Anaphase	10
Telophase	42

Give **one** piece of evidence from the table which indicates that these cells are multiplying rapidly. (1)

Total 7

AQA, June 2005, Unit 2, Question 2

5

a The drawing shows stages in mitosis in a cell.

A

B

C

D

E

a Using the letters **A** – **E**, place the drawings in the correct sequence starting with stage B. (1)

b Describe what happens during:

i stage **B** (2)

ii stage **E**. (2)

c Name two occasions in the life of an organism when mitosis occurs rapidly. (2)

Total 7

6

Two pairs of alleles **A** and **a**, and **B** and **b** are found on one pair of homologous chromosomes. A person has the genotype **AaBb**. **Figure 1** shows the chromosomes at an early stage of meiosis. The position of two of the alleles is shown.

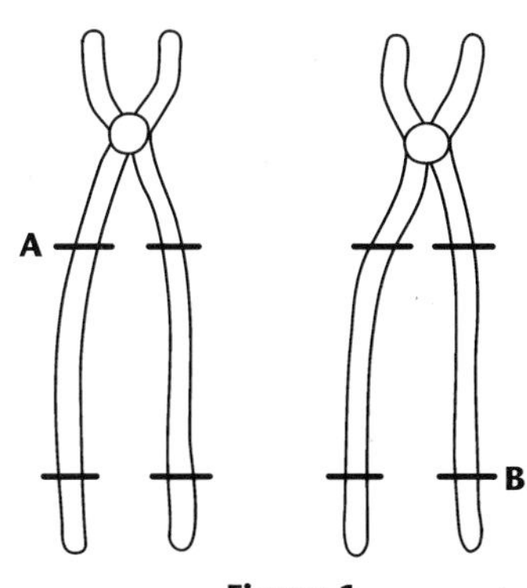

Figure 1

a Complete **Figure 1** to show the alleles present at the other marked positions. (1)

Crossing over occurs as shown in **Figure 2**.

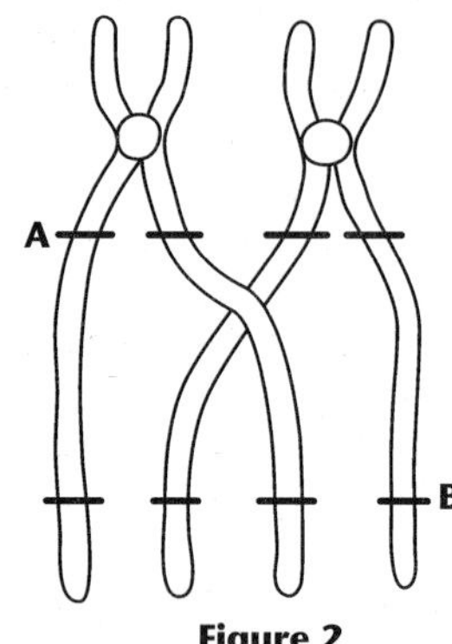

Figure 2

b What term is used to describe the pair of homologous chromosomes shown in **Figure 2**? (1)

c From **Figure 2**, give the genotypes of the gametes produced containing the chromatids
 i that have **not** crossed over;
 ii that have crossed over. (2)

d Give **two** processes, other than crossing over, which result in genetic variation. Explain how each process contributes to genetic variation.
Process
Explanation (2)
Process
Explanation (2)

Total 8

AQA, January 2005, Unit 4, Question 8

7 **Figure 4** and **Figure 5** show the chromosomes from a single cell at different stages of meiosis.

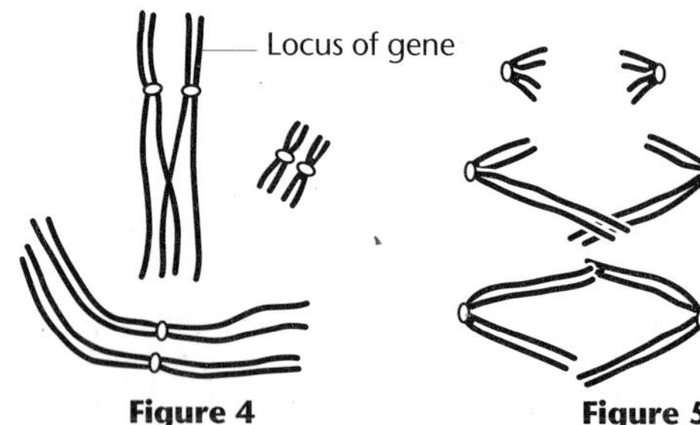

Figure 4 **Figure 5**

a What is the diploid number of chromosomes in the organism from which this cell was taken? (1)

b Describe what is happening to the chromosomes at the stage shown in
 i **Figure 4**; (2)
 ii **Figure 5**. (2)

c **i** The genotype of this organism is **Bb**. The locus of this pair of alleles is shown in **Figure 4**. Label **two** chromosomes on **Figure 5** to show the location of the **B** allele and the location of the **b** allele. (1)
 ii How many genetically different gametes can be produced by meiosis from a cell with the genotype, **Bb Cc Dd**? Assume these genes are located on different pairs of homologous chromosomes. Show your working. (2)

Total 8

AQA, January 2004, Unit 4, Question 2

8 The graph shows the changes in the DNA content of cells during the cell cycle.

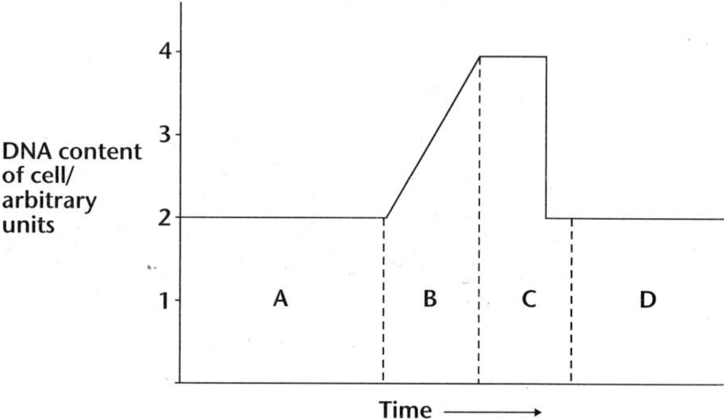

a In which of the stages, **A** to **D**, does each of the following take place?
 i DNA replicates
 ii The chromosomes become visible. (2)

b Describe and explain how the amount of DNA in the cell changes during stage **C**. (3)

c **i** Cytarabine is a drug used to treat cancer. It inhibits an enzyme needed to synthesise new DNA. Suggest how the graph would be different if cytarabine was present during the cell cycle. (1)
 ii Explain why cytarabine is effective in treating cancer. (2)

Total 8

Controlling the cell cycle

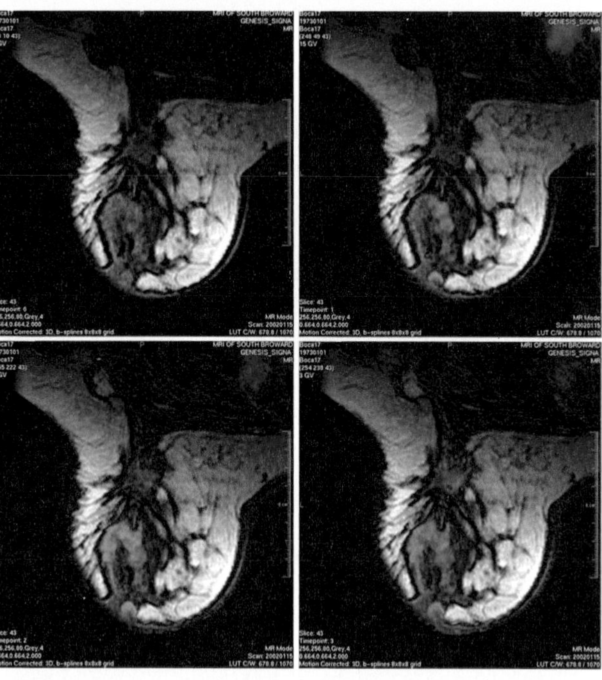

X-ray photograph of the developement of breast cancer.

What makes your fingers grow to a certain size and then stop? How is that some cells in your fingers make bone, while others produce skin, nerves, muscles, blood vessels and so on? As the skin on your fingertips wears away, cells underneath produce new layers, following exactly the same pattern of fingerprints. How do these cells 'know' what to do and when to stop doing it?

Cells produced by mitosis may grow and become specialised, or they may have a relatively short growth period before dividing again. Some, for example most nerve cells, may never divide again. They nevertheless remain active and their genes continue to produce the proteins necessary for their function as nerve cells. Until recently it was thought that no new brain cells are made in humans after about 16 years of age, and that as cells die they are never replaced. However, even with a loss of several thousand per day, there are still plenty to spare. More recently this idea has been challenged, and there is evidence that even specialised nerve cells may be stimulated to divide again.

Cells in embryos, and in tissues that have a high cell turnover, such as the skin, gut lining and bone marrow, have quite a short interphase. During the first part of interphase the genes are actively involved in protein synthesis and growth occurs. New organelles are formed, and some of these, such as mitochondria and chloroplasts, contain their own small sections of DNA that enable these organelles to reproduce independently. After a time the protein cyclin builds up in the cell. This seems to stimulate production of another protein that in turn stimulates the replication of the DNA and initiates mitosis. This second protein also breaks down cyclin.

Normal levels of a form of cyclin called cyclin D1 are known to be critical for normal growth of human mammary (breast) cells. However, the majority of human breast cancer cells have higher than normal levels of cyclin D1, which may lead to uncontrolled, cancerous growth.

Scientists have created a new strain of mice that lack the protein cyclin D1. They found that knocking out this important cog causes surprisingly little damage. These results have implications for treating human breast cancer and should lead to a better understanding of cancer. The study was led by Dr Robert Weinberg, a cancer research pioneer.

'Cyclin D1 is overproduced in more than half of human breast cancer biopsies, but treatment strategies have not targeted this protein because of the damage they might cause to normal cells in other tissues,' Dr Weinberg said. 'These new results suggest, however, that breast cancer therapies designed to block cyclin D1 action may prevent the growth of tumour cells without harming normal tissues.'

Researchers noticed a striking difference in the mutant mice when females delivered pups. The mother mice, whose breast tissue was otherwise normal, were unable to suckle their babies after giving birth. Without cyclin D1 their mammary cells failed to undergo the rapid growth that normally occurs during pregnancy and they produced no milk.

The researchers concluded that in adult female mice the cyclin D1 molecule seems critical only for a specialised process in breast tissue – the rapid growth of mammary cells during pregnancy.

A1 What is the role of cyclin in normal cell division?

A2 What limits the production of cyclin during normal cell division?

A3 What happens to cyclin production in most breast cancers?

A4 What is the role of cyclin D1 during pregnancy in mice?

A5 How might the research on cyclin D1 in mice be applicable to treating breast cancer in humans?

A6 Evaluate the ethical issues involved in using mice in these experiments.

11 Plant cells

Strombolites at Shark Bay, Western Australia. Strombolites are living fossils, composed of sticky mats of cyanobacteria that cement sand and sediments. They are found in rocks that are 3.5 billion years old.

The earliest prokaryotic cells probably had no internal organelles. They stored energy in long-chain hydrocarbons and perhaps used ATP from their surroundings. One thing is certain – since there was no oxygen in the atmosphere, these early cells respired anaerobically. This process released carbon dioxide into the atmosphere and carbon dioxide levels rose steadily over the next 500 million years. This gas reduced the amount of infrared and ultraviolet radiation that reached the Earth, causing the atmosphere to cool. This reduced the rate of production of organic compounds in the atmosphere. The prokaryotes that depended on these organic molecules for energy would probably have died out had it not been for the development of eukaryotes that could produce their own organic compounds. These eukaryotes used the energy from light to convert carbon dioxide into carbohydrate.

One of the earliest groups of Eubacteria to do this were the Cyanobacteria (also called blue-green algae). These organisms still exist in large numbers and fossil records show that they have changed very little over the last 3 billion years. They contain molecules of chlorophylla, the pigment found in all chloroplasts. In fact some Cyanobacteria look remarkably similar to the chloroplasts found inside the cells of modern green plants. Cyanobacteria use energy from sunlight to split water into hydrogen and oxygen. The oxygen and some of the hydrogen is given off, but some of the hydrogen is used to reduce carbon dioxide to carbohydrates.

About 2.5 billion years ago, as oxygen levels increased, some prokaryotes evolved the ability to use oxygen to generate ATP using a mechanism that is very similar to that used by mitochondria today. So, by this time there were three types of prokaryote:

- those that could photosynthesise
- those that could respire aerobically
- those that could respire anaerobically.

The first eukaryotic cells appeared about 1.5 billion years ago. The 'Endosymbiotic Theory of Eukaryote Evolution', first proposed in the 1960s, tries to explain how they evolved. This theory states that the ancestors of modern eukaryotic cells were 'symbiotic consortiums' of prokaryotic cells. For example, aerobic bacteria might have invaded larger amoeba-like anaerobic bacteria. Both organisms would benefit – the anaerobic bacteria would ingest organic material, and aerobic bacteria would oxidise this to provide ATP for both organisms. The embedded aerobic bacteria assumed the role of what we now call mitochondria. These cells were the ancestors of eukaryotic animal cells. Some amoeba-like bacteria also formed symbiotic consortia with cyanobacteria. These cells were the ancestors of eukaryotic plant cells. If this had not happened, life on Earth as we know it today, could not have evolved. Without plants there would be no animals.

At first this theory was ridiculed, but there was an important test. If chloroplasts and mitochondria were prokaryotic symbionts, they would have their own DNA – and this turns out to be the case. The division of mitochondria and chloroplasts is not under the control of the nucleus of the cell, but is controlled by DNA in the organelles.

You studied the structure of animal cells in chapter 1. In this chapter you will study the additional features of plant cells.

11.1 Structure of plant cells

Fig. 1 What you can see with a light microscope

Cell wall
Cell membrane
Nucleus
Chloroplast
Vacuole

A light micrograph of mesophyll cells from a leaf; the diagram shows the main features visible

Fig. 1 shows a cell from a plant leaf as seen through a light microscope.

1 List the differences you can see between this cell and the animal cell on page 25.

Plant cell organelles

Fig. 2 shows the ultrastructure of a plant leaf cell. If you compare this drawing with the drawing of the animal cell on page 25, you will see that plant cells have two additional organelles – a cell wall and chloroplasts.

The cell wall

The cell wall of a young plant cell is called the primary cell wall and is mainly **cellulose**. Cellulose is a **polysaccharide**, or carbohydrate **polymer**. Polysaccharides are giant molecules made up from many single sugar molecules joined together by **condensation reactions**. Starch, glycogen and cellulose are all polysaccharides. Part of a starch molecule is shown in Fig. 3. Starch and glycogen molecules have compact, coiled and branched molecules, making them ideal 'energy' stores. Cellulose molecules have long straight molecules, perfect for forming structural fibres.

Fig. 2 Plant cell

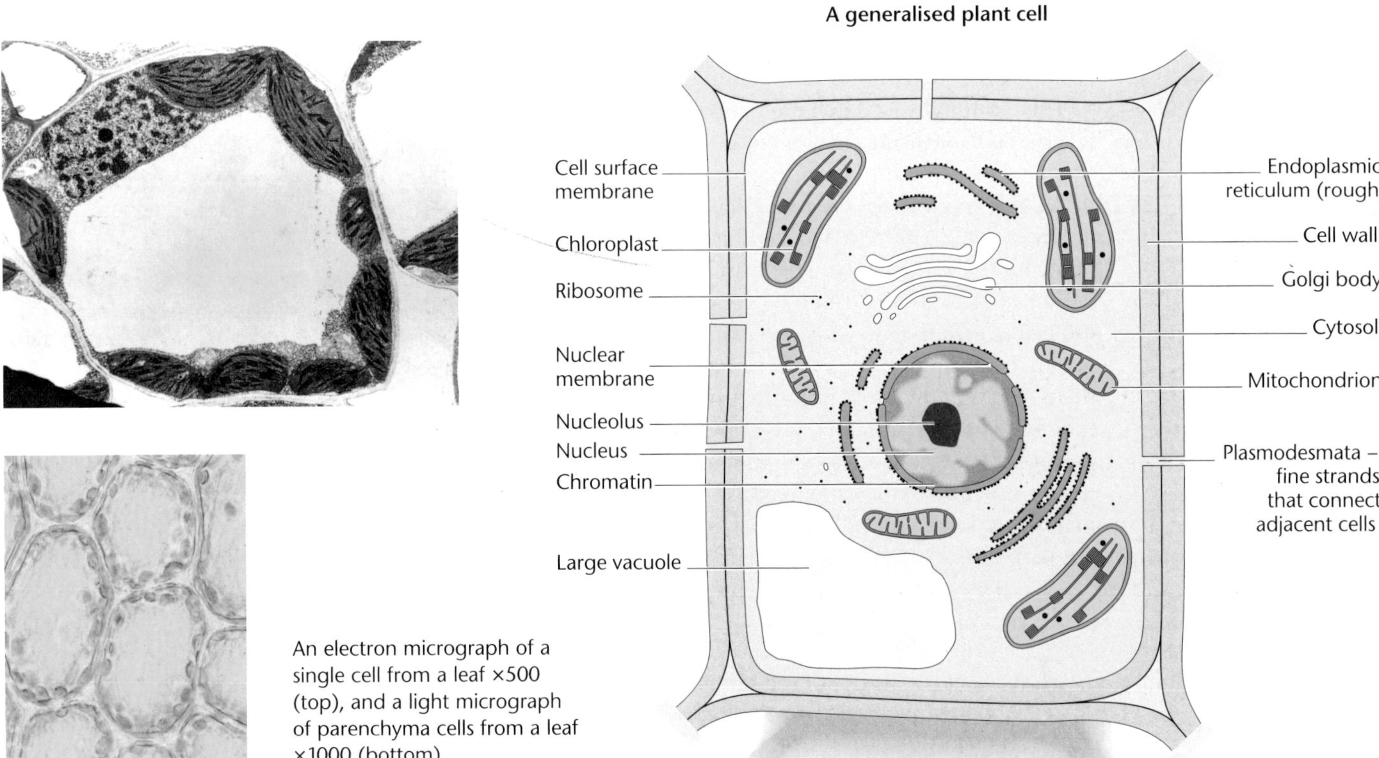

An electron micrograph of a single cell from a leaf ×500 (top), and a light micrograph of parenchyma cells from a leaf ×1000 (bottom).

Fig. 3 Starch and glycogen

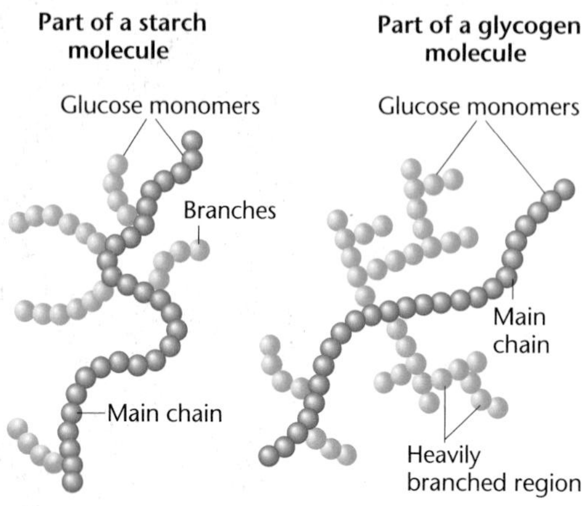

Starch and cellulose are both polymers. A polymer is a molecule made up of repeating units, rather like links in a chain. The individual units are called **monomers**, and the same monomer can be used to build different kinds of chains – long, short, straight or branched. For example, the same monomer, glucose, is linked in different ways in the polymers starch and cellulose found in plants and the polymer glycogen found in animals.

The glucose monomer in starch is called α-glucose. Cellulose is made from a slightly different glucose monomer, β-glucose. Because their monomer molecules differ, the polymer molecules have different shapes. The shape of each polymer determines its function.

The monomers in starch are joined by **α–glycosidic bonds**. These bonds produce twisted chains of monomers that form branched molecules. In glycogen, α-glycosidic bonds again link the monomers, but glycogen has even more branches than starch chains. The coiled and branched chains of starch and glycogen molecules give them a compact shape. This, together with their insolubility in water, makes them ideal 'energy' storage compounds. Starch is the major storage carbohydrate in plants; glycogen does the same job in animals.

The structure of the cellulose molecule, the major component of the cell wall, is shown in Fig. 4. In this polymer, the glucose monomers are joined by **β-glycosidic bonds**. These bonds result in straight chains. Individual cellulose molecules are therefore long, unbranched chains with many β-glycosidic bonds. The straight molecules lie side by side, forming **microfibrils** that are strengthened by many **hydrogen bonds**. This makes cellulose fibres very strong and stable – ideal for a structural material. These fibres are part of most cell walls and give them strength.

> **2** Look at the shapes of the starch and glycogen molecules. Suggest why animals store carbohydrates as glycogen rather than starch.

Cellulose and the cell wall

Cellulose microfibrils are embedded in a framework of other substances, the most common being complex molecules called **hemicelluloses** and **pectins**. The arrangement of microfibrils determines the shape of the primary cell wall. As well as being flexible enough to allow the cell to grow, the cell wall needs to be strong enough to resist the force exerted by water that has entered the cell by osmosis (see Chapter 3). This is particularly important in young plants, where the strength of the cell wall is the only thing that supports stems and leaves. If water is lost, the cells lose their shape. Think how quickly a young plant droops if it loses water.

Fig. 4 Cellulose

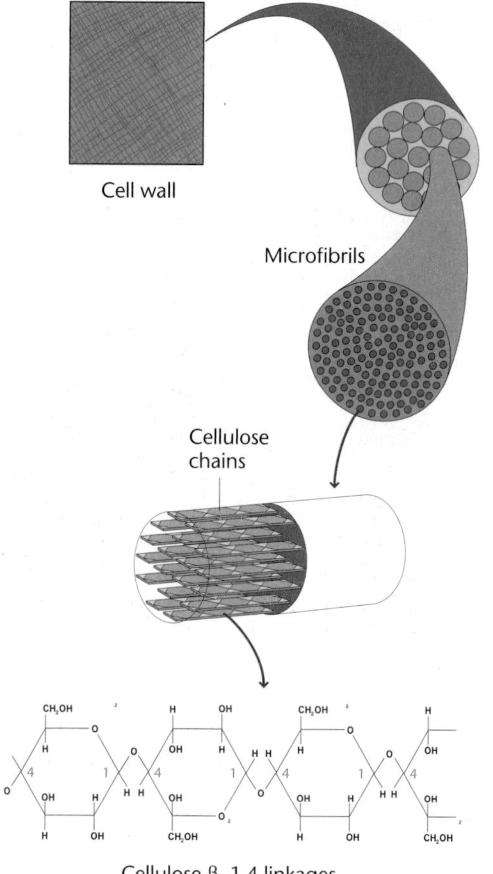

Specialised plant cells

Older cell walls often have extra thickening called a **secondary wall**, which gives a permanent shape and more strength. The trunks of trees contain a large proportion of **xylem** cells. The secondary walls of these cells become impregnated with a tough waterproof material called **lignin**. This waterproofing layer means that the cells cannot exchange substances with their environment and so the cells die.

Cork cells that form the outer layer of tree bark become impregnated with a fatty material called **suberin**, which is also waterproof. It may seem strange when you look at a large oak tree covered in green leaves in the middle of summer that most of what you see consists of dead cells. Only the leaves and thin layers of cells in the trunk and roots are actually alive.

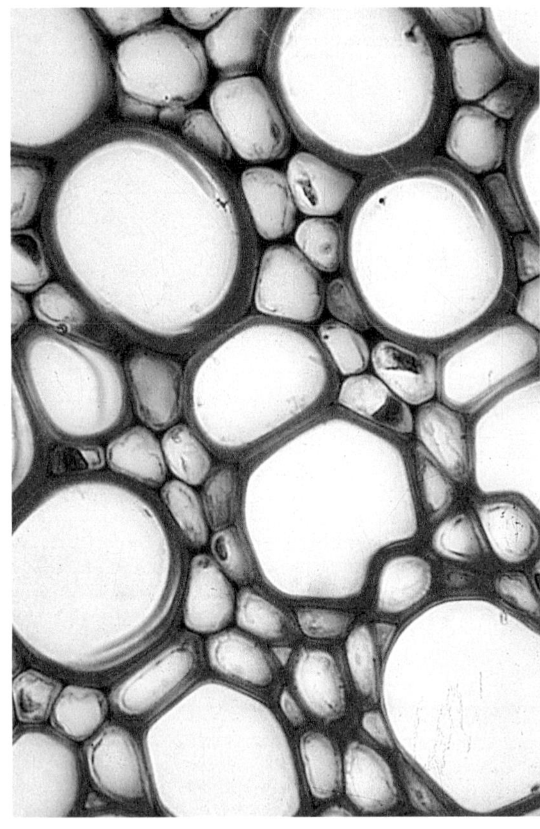

This photograph (above) shows a section through xylem tissue. The lignin that provides secondary thickening in the walls of the xylem vessels is stained red.

3 Describe two ways in which the cell wall of plant cells may be waterproofed.

This scanning electon micrograph (below) shows the rings of lignin in the walls of the xylem vessels.

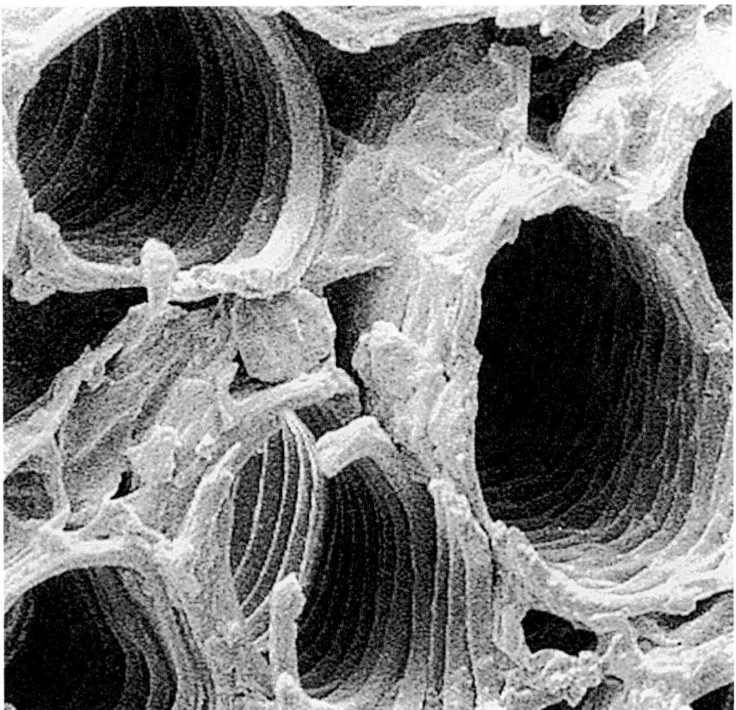

Chloroplasts

Chloroplasts (Fig. 5) are organelles that occur only in plant cells. They are found in many of the cells in the green parts of a plant (mainly the leaves and young stem). The primary function of the chloroplasts is to make the sugars that form the basis of all carbohydrates by the process of **photosynthesis**. Different parts of the chloroplast carry out the different stages of photosynthesis. First, light energy is absorbed by pigments such as **chlorophyll** and is transformed to chemical energy, mainly in the form of **ATP**. Energy from ATP is then used to fix atmospheric carbon dioxide into carbohydrate.

key facts

- Cellulose, the major component of the **cell wall** in plant cells is a **carbohydrate**.
- Polysaccharides (carbohydrate polymers) include starch and glycogen, as well as cellulose.
- The properties of these polymers depend on the shape of their molecules.
- Starch and glycogen molecules have compact, coiled and branched molecules, making them ideal 'energy' stores.
- Cellulose molecules have long straight molecules, perfect for forming structural fibres.

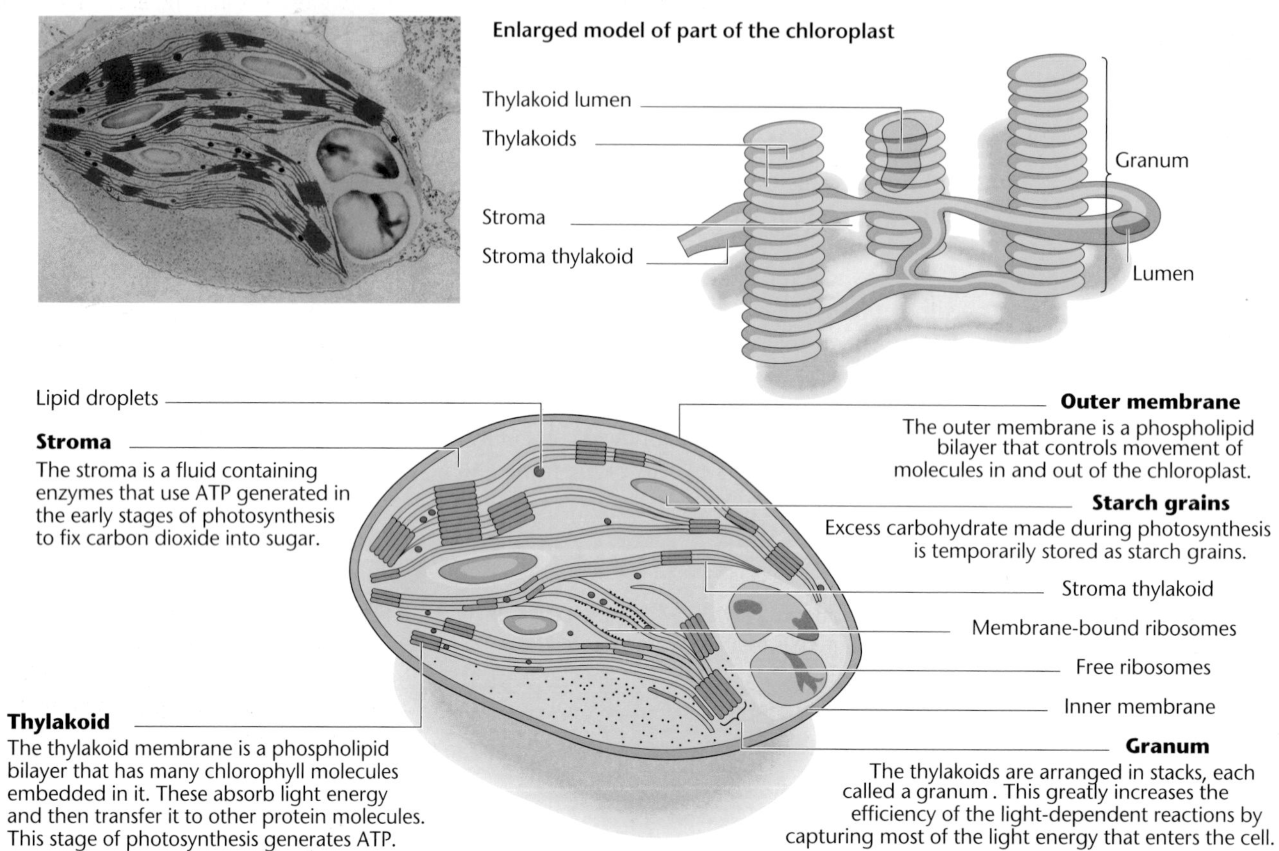

Enlarged model of part of the chloroplast
Thylakoid lumen
Thylakoids
Stroma
Stroma thylakoid
Granum
Lumen
Lipid droplets
Stroma
The stroma is a fluid containing enzymes that use ATP generated in the early stages of photosynthesis to fix carbon dioxide into sugar.
Thylakoid
The thylakoid membrane is a phospholipid bilayer that has many chlorophyll molecules embedded in it. These absorb light energy and then transfer it to other protein molecules. This stage of photosynthesis generates ATP.
Outer membrane
The outer membrane is a phospholipid bilayer that controls movement of molecules in and out of the chloroplast.
Starch grains
Excess carbohydrate made during photosynthesis is temporarily stored as starch grains.
Stroma thylakoid
Membrane-bound ribosomes
Free ribosomes
Inner membrane
Granum
The thylakoids are arranged in stacks, each called a granum . This greatly increases the efficiency of the light-dependent reactions by capturing most of the light energy that enters the cell.

Fig. 5 The chloroplast

Like the **mitochondria**, the chloroplasts also have two bilayered **phospholipid membranes**. The outer one surrounds the organelle; the inner membrane is highly folded, creating **thylakoids**, in between which is a fluid **stroma**. In places, the thylakoid membranes are arranged into structures called **grana**, which look a little like piles of green coins.

The thylakoids contain chlorophyll molecules that capture light energy from sunlight. The arrangement of the thylakoids into grana, and the distribution of the grana, maximises the absorption of light energy. In the first stage of photosynthesis, chemical reactions in the grana split water molecules into hydrogen ions and oxygen. The oxygen is given off as a 'waste product'. The hydrogen ions are used, along with energy from ATP, to reduce carbon dioxide to form sugars. Enzymes in the fluid-filled stroma control this second series of reactions.

4 Describe how chloroplasts and mitochondria are:

a similar in structure and function;

b different in structure and function.

key facts

- **Chloroplasts** contain light-absorbing pigments such as **chlorophyll**.
- These pigments are located in double **phospholipid membranes** called **thylakoids**.
- In places the thylakoids form stacks called **grana**.
- In the grana, light energy is transformed into chemical energy.
- In the fluid **stroma**, enzymes catalyse reactions that use **ATP** and hydrogen ions produced in the grana to fix carbon dioxide into sugars.

examination questions

1 The drawing shows an electron micrograph of part of a plant cell.

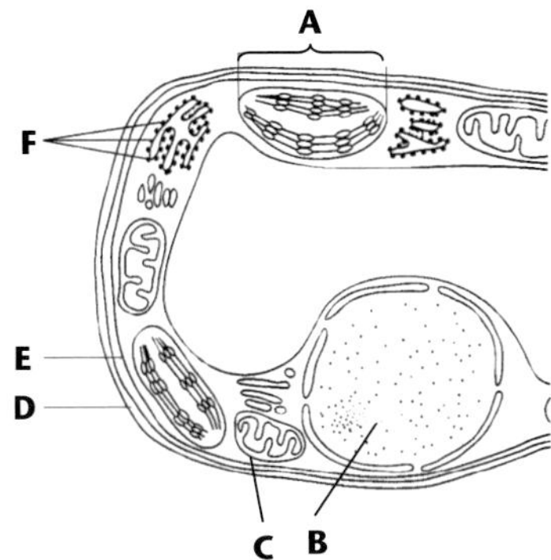

a Name the structures labelled **A–F**. (6)
b Name the main carbohydrate found in part **D**. (1)
c Give the functions of the parts labelled **C–F**. (4)
d Give three differences between the structure of this cell and an animal cell. (3)

Total 14

2 The drawing shows the structure of a β glucose molecule.

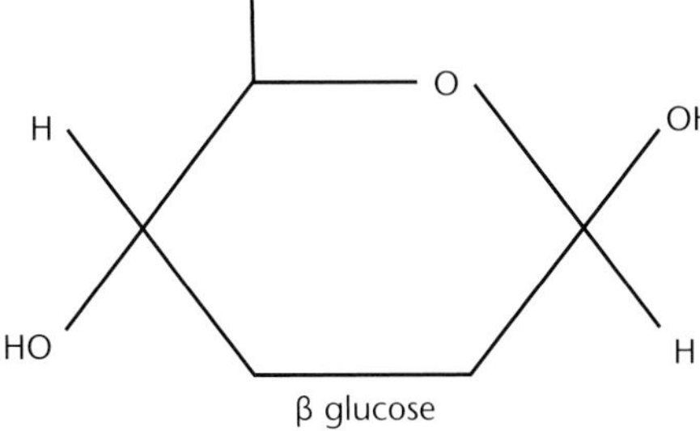

β glucose

a i Copy the drawing, then show how two β glucose molecules can be joined by a condensation reaction. (2)
ii Name the bond that joins two β glucose molecules together. (1)
b Explain how the structure of a cellulose molecule is related to its function. (3)

Total 6

3
a i Describe the structure of the cell wall in a palisade mesophyll cell. (3)
ii Explain how the structure of this cell wall is related to its function. (4)
b Describe and explain the ways in which the cell walls of the following cells are adapted to the function of the cells:
i xylem vessels (3)
ii cork cells. (3)

Total 13

4 The drawings show a starch molecule and a glucose molecule.

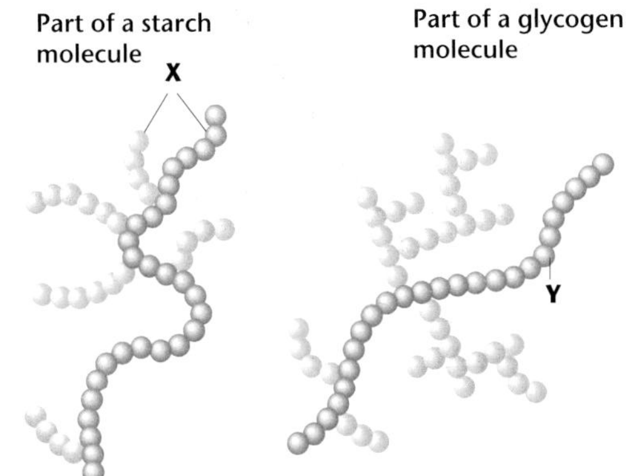

a Name one type of cell containing
i starch (1)
ii glycogen (1)
b i Name the molecules labelled **X**. (1)
ii Give the major structural difference between a starch molecule and a glycogen molecule. (1)
c Explain how the structure of a glycogen molecule is adapted for its function. (3)

Total 7

5 The drawing shows the structure of a chloroplast.
a Name the structures labelled **X, Y** and **Z**. (3)

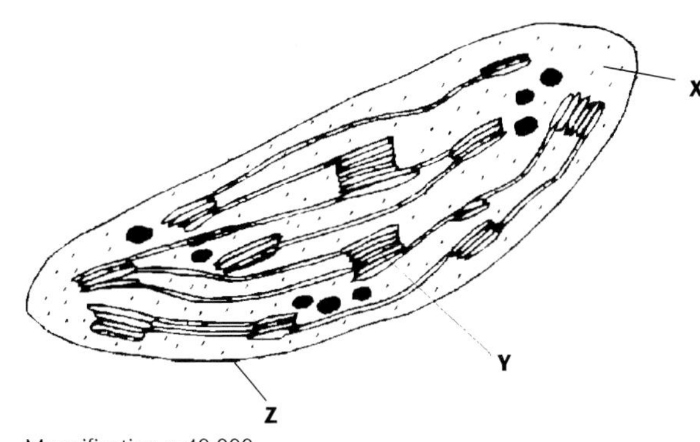

Magnification × 40 000

b Calculate the actual length of the chloroplast in μm. (2)
c Explain how the structure of a chloroplast is related to its function. (3)

Total 8

6 **Figure 1** shows a section through a palisade cell in a leaf as seen with a light microscope. The palisade cell has been magnified × 2000.

Figure 1

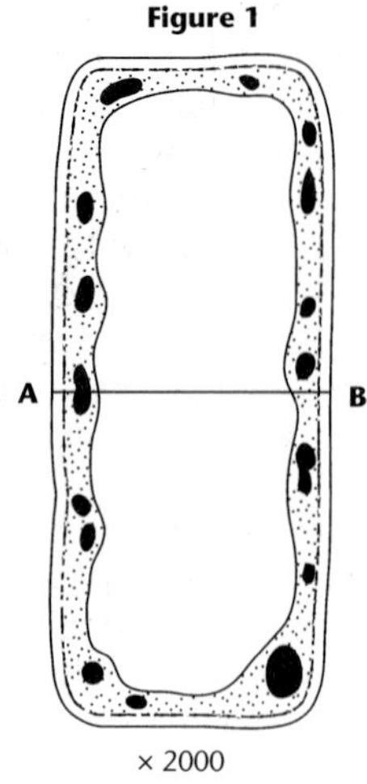

× 2000

a Calculate the actual width of the cell, measured from **A** to **B**, in μm. Show your working. (2)

b Palisade cells are the main site of photosynthesis. Explain **one** way in which a palisade cell is adapted for photosynthesis. (2)

Total 4

AQA, January 2006, Unit 1, Question 2

7

a Cells of multicellular organisms may undergo differentiation. What is meant by differentiation? (1)

b The drawing shows part of a plant cell as seen with an electron microscope.

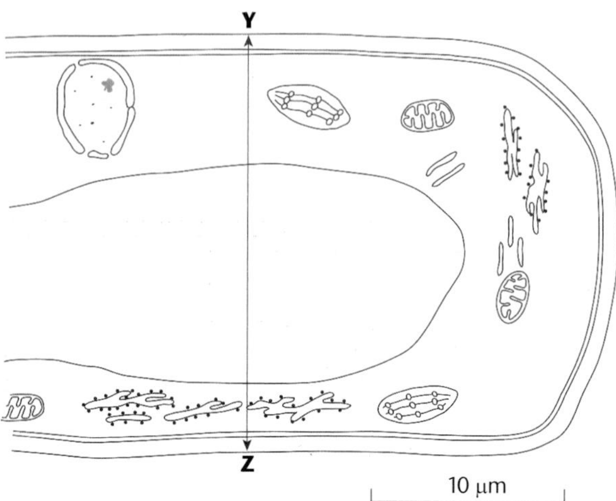

i Give **two** features shown in the drawing which are evidence that this cell is eukaryotic. (2)

ii Calculate the actual width of the cell from **Y** to **Z**. Give your answer in micrometres (μm) and show your working. (2)

iii Give **one** way in which a typical animal cell differs from the cell shown in the drawing. (1)

Total 6

AQA, June 2004, Unit 1, Question 1

how science works **assignment**

How did life on Earth begin?

Fig. 6 The Miller–Urey experiment

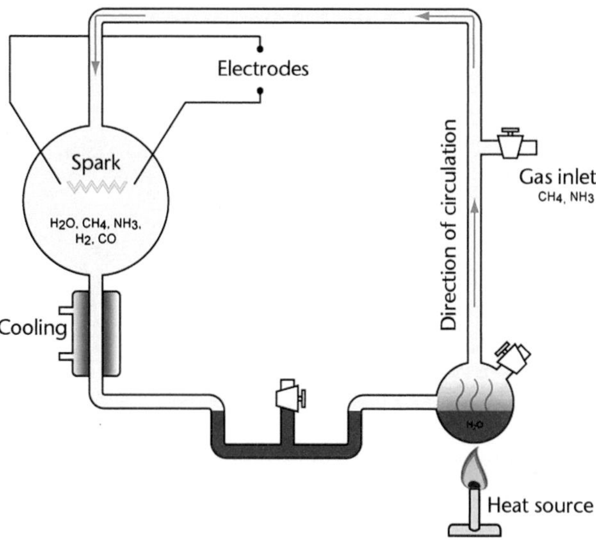

We can never be certain how life on Earth began, but many scientists believe that the first cells arose spontaneously about 3.5 billion years ago. Conditions were very different then. The atmosphere contained hydrogen, methane, ammonia and water, with very little or no oxygen – just right for the formation of the chemicals we find in living organisms.

In the early 20th century, Aleksandr Oparin and John Haldane independently suggested that if the primitive atmosphere was reducing (as opposed to oxygen-rich), and if there was an appropriate supply of energy, such as lightning or ultraviolet light, then a wide range of organic compounds might be synthesised.

Oparin suggested that the organic compounds could have undergone a series of reactions, leading to more and more complex molecules. He proposed that the molecules formed colloid aggregates, or 'coacervates'. These coacervates were able to assimilate organic compounds from the environment. They would have taken part in evolutionary processes, eventually leading to the first life forms.

Haldane's ideas about the origin of life were very similar to Oparin's. Haldane proposed that the primordial ocean served as a vast chemical laboratory powered by solar energy. The atmosphere was oxygen free, and the combination of carbon dioxide, ammonia and ultraviolet radiation gave rise to a host of organic compounds. The sea became a 'hot dilute soup' containing large populations of organic monomers and polymers. Haldane hypothesised that groups of monomers and polymers then acquired lipid membranes, eventually leading to the first living cells.

In the early 1950s, Miller and Urey experimented to test the Oparin–Haldane hypothesis. They passed electric sparks through a mixture of hydrogen, methane, ammonia and water. The sparks simulated lightning. After several days they detected newly formed organic compounds.

A1 How far did the results of the Miller–Urey experiment support the Oparin–Haldane hypothesis?

A2 Do the results of this Miller–Urey experiment prove that life on Earth started in this way? Explain the reasons for your answer.

A3 Do you think that cells could arise on a planet where the atmosphere and climate were similar to those on Earth at the moment? Support your answer with information from the text.

A4 Using the information above and information from the passage at the beginning of this chapter,

a draw a time line for eukaryotic cell evolution

b describe the endosymbiont theory.

12 Exchanges

Insects are an incredibly successful group of organisms that occupy every conceivable habitat. The largest insect that ever lived was probably a prehistoric dragonfly with a wingspan of over 60 cm. Today there are some large insects; the goliath beetle, for example, can grow as large as a man's fist, but it is an exception.

Why has no insect ever grown as big as an elephant? The answer lies in the way organisms exchange materials and heat with their surroundings. Insects 'breathe' through a system of tubes that carry air into the body, allowing oxygen to diffuse the short distance to every individual cell. This system works well for an ant, but imagine using air tubes to get oxygen to every cell of an elephant! To grow to this size, mammals like the elephant have evolved a complex blood system and lungs, which provide a large surface area for gas exchange.

Insects do not maintain an internal body temperature; their body temperature changes with the surroundings. So they don't have to worry too much about losing heat. Elephants, like humans, maintain a high constant body temperature. Their large size means they produce a lot of heat from the food they eat, which they cannot lose very quickly. This can be a problem but if the elephant were the size of an ant, its problems would be far worse. An elephant this small would need to spend 24 hours a day eating to maintain this high constant body temperature – there would be no time for wallowing in mud!

Why can't an ant grow as big as an elephant or a polar bear? What would life be like for an elephant or polar bear that was as small as an ant?

12.1 Cells, tissues and organs

Large multicellular organisms develop systems for exchanging materials. During the development of an animal, cells differentiate so that they can perform specific functions. Groups of cells with similar specialisations that form a common function are referred to as a **tissue**. **Organs** are groups of several tissues that function together. Examples of organs include the stomach, heart and lungs. An organ system is a group of organs that work together to perform a function. Examples of organ systems include digestive, circulatory and gas exchange systems. Plant organs include roots, stems and leaves.

12.2 Heat exchange in mammals

In the first part of this chapter you will find out more about how animals of different sizes deal with the challenges of heat exchange. The rest of the chapter concentrates on the mechanisms that organisms use to exchange gases with the environment. First we will look at the structure and function of lungs in humans and then we will take a briefer look at gas exchange in fish and in plants.

The elephant and the polar bear are both large mammals. The larger the mammal, the greater its volume and the more heat it generates by respiration in its tissues. The two animals have opposite problems because of their environments. The polar bear needs to retain heat to keep its body temperature constant despite the cold ice and snow of the Arctic; the elephant needs to keep cool during the hot days on the African savannah.

Mammals transfer heat to the environment by conduction, convection and radiation. The effectiveness of all of these processes depends on the surface area of the animal. An animal with a very large surface area can lose more heat in a minute than an animal with a smaller surface area. Heat transfer to the environment is also affected by the temperature of the animal's surroundings. On the African savannah, the air is hot during the day and so there is not much of a temperature gradient between the elephant's skin and the air. This makes it nearly impossible for the elephant to lose heat by conduction or convection. But having extremely large ears helps; the diameter of one ear can be as large as 1.5 metres. The earflaps are rich in blood vessels and the elephant flaps them frequently during the hottest part of the day. In the Arctic winter, the polar bear has the opposite problem. There is a steep temperature gradient between the polar bear's skin and the cold air and the bear loses heat quickly.

A shrew has a very large surface area to volume ratio. It therefore loses heat very rapidly. It is only able to maintain its body temperature by eating constantly, using large amounts of food in respiration to generate plenty of body heat.

1 How does flapping its large ears help an African elephant to reduce its body temperature?

2 Explain how the size and shape of the different parts of the polar bear's body help it to conserve heat.

Surface area to volume ratio

An elephant has a much larger surface area than a shrew. But the elephant's volume is also much bigger. If we calculate the ratio of surface area to volume in each animal, we find that the surface area to volume ratio for the elephant is much smaller than the value for the shrew. This is a general rule: the larger the animal, the smaller its surface area to volume ratio. The smaller the surface area to volume ratio, the more difficult it is for the animal to gain or lose heat. Large mammals need to compensate for this by having body features that help them to control the way they exchange heat with the environment.

3 A shrew and an elephant in the hot African savannah both need to cool down. Which animal is likely to be more successful? Explain why.

4 In Britain, shrews need to eat their own mass of food every day to survive. Suggest why.

Heat exchange in water

Water and air have very different properties with respect to heat transfer. In air, gas molecules are comparatively far apart – so air is a poor conductor of heat. This is why double glazing works: very little heat is conducted across the air gap between the two glass panes. Molecules in water are very close together, so water is a much better heat conductor than air. Animals tend to lose heat to cold water much faster than they do to cold air. A person who falls into very cold water loses heat so rapidly that their life is put in serious danger. Animals that live in cold water have developed adaptations that help them to survive.

how science works

The walrus and the manatee

The manatee and the walrus have many features in common but they live in completely different environments. The manatee lives in warm seas near the West Indies, while the walrus lives on the Arctic ice shelf off the coast of Alaska. Both must maintain a constant body temperature. Let's look at the way each animal does this. The table shows the relative body sizes of the manatee and the walrus, and a large land mammal, the elephant, for comparison.

	Manatee	Walrus	Elephant
Approximate body length/metres	4	3.5	3
Approximate body width/metres	0.75	1	2
Approximate body mass/kilograms	1000	1700	5000

5

a Assume the body of each animal has a regular cylindrical shape. Calculate the surface area of each animal.

Surface area of a cylinder = circumference × length

b Animals have approximately the same density as water – 1000 kg m^{-3}. Estimate the volume of each animal.

$$\text{Density} = \frac{\text{Mass}}{\text{Volume}}$$

c Calculate the surface area to volume ratio of each animal.

6 Do mammals lose heat more quickly in air or in water at the same temperature? Explain the reason for your answer.

7 How is the surface area to volume ratio of the two marine mammals related to the temperature of water they can live in?

key facts

- Mammals produce large amounts of heat because of the high rate of respiration in their body cells; much of this heat is lost to the environment.
- Mammals control the rate of heat loss to the environment in order to maintain a constant body temperature.
- As animals increase in size, their surface area [to] volume ratio decreases.
- In cold habitats it is an advantage to a mamm[al] to reduce the surface area of its body to redu[ce] the rate of heat loss.
- In hot habitats it is an advantage to a mamm[al] to increase the surface area of the body to increase the rate of heat loss.

12.3 Gaseous exchanges

Living organisms must exchange substances with the environment. The single-celled *Amoeba* exchanges gases through its surface by diffusion. Diffusion is efficient only over small distances; small, multicellular organisms have therefore become adapted to increase their surface area and to decrease the distance that substances need to diffuse across to maintain their body function. Tapeworms and roundworms have very different habitats. Tapeworms are parasites that live in the guts of animals; roundworms live in the soil. But they both exchange gases with the surroundings through their skin and, as they grow, they both reach their maximum width very quickly. Any further growth is confined to an increase in the length. The flattened shape of the tapeworm ensures that no point in its body is more than 0.1 mm from a supply of oxygen, food and water.

8

a What substances does a tapeworm exchange with the gut contents of its host?

b What mechanism does the tapeworm use to exchange each of these materials?

9 How does the body shape of a tapeworm increase the rate at which it can exchange materials?

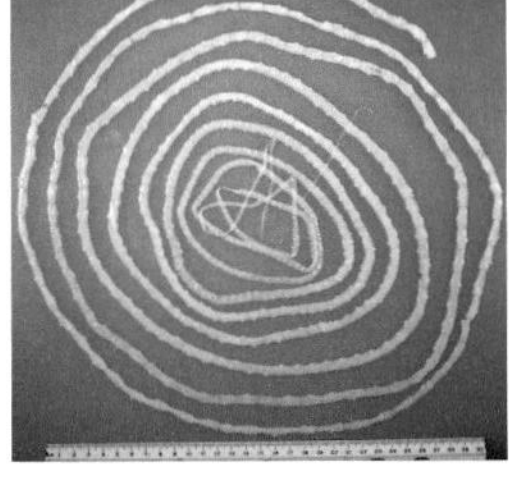

A coiled beef tapeworm, *Taenia saginata*. This tapeworm is several metres long, as you can see from the ruler. However, its body is thin and flattened.

In larger organisms, diffusion cannot supply the cells at the centre of the body with the oxygen they need, and waste such as carbon dioxide cannot escape quickly enough. No adaptations to body shape can make much difference; the problem of a low surface area to volume ratio needs more drastic solutions. Mammals and other animals have overcome this problem because they have evolved special internal exchange surfaces such as the **alveoli** in lungs and **villi** in the intestines. Alveoli and villi have a very large surface area that maximises the amount of materials that can be exchanged across them. Animals with internal exchange surfaces usually have a complex blood system to transport substances between the exchange surfaces and the rest of the body tissues.

Gas exchange in insects

Insects were one of the first groups of organisms to adapt to terrestrial life. One adaptation that contributed to this success was the development of a **cuticle** covering the body. The cuticle is impermeable to water so its helps the insect to conserve water. However, the cuticle is also impermeable to oxygen and carbon dioxide.

Insects did not develop lungs, nor did they develop a blood system to transport respiratory gases. Instead, they have small openings in the cuticle called **spiracles**. These are connected to the inner organs by a system of highly branched, gas-filled tubes called **tracheae**. Oxygen uptake and carbon dioxide release by the cells mainly occur at the tips of the smallest branches, which are called **tracheoles**. In highly active organs, such as flight muscle, the tracheoles enter the cells and reach the **mitochondria** directly.

Fig. 1 The tracheal system of an insect

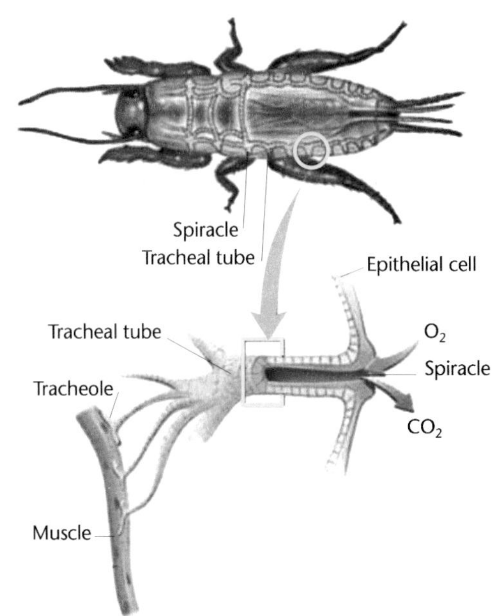

10 It has been estimated that oxygen is delivered to insect mitochondria about 200 000 times faster than oxygen is transferred by blood to the human mitochondria. Suggest an explanation for this.

how science works

The spiracle controversy

The spiracles in the cuticle behave like valves, opening and closing to allow or restrict the insect's gas exchange. Scientists have observed a peculiar rhythmic respiratory behaviour referred to as the **discontinuous gas-exchange cycle (DGC)**. In insects exhibiting DGC, the spiracles close for long periods (up to several hours or even days) and open occasionally for only a few minutes. This unusual respiratory pattern has been observed in many adult insects, as well as in resting butterfly and moth pupae. Two main hypotheses have been proposed to explain why some insects display DGC:

- to reduce water loss through the spiracles
- to adapt to living underground.

Scientists Hetz and Bradley have proposed a different hypothesis to explain DGC. Their experiments provide evidence that insects use DGC not to acquire but to avoid oxygen. Using the pupae of the moth *Attacus atlas* as a model system, these researchers varied the environmental oxygen concentrations from partial pressures of 5 to 50 kPa (the normal atmospheric oxygen partial pressure at sea level is about 21 kPa). They found that the concentration of oxygen levels in the resting pupae remained low, close to 4 kPa, across the whole range of partial pressures. It seems that the moth pupa limits the amount of oxygen taken in by keeping the spiracles closed for as long as possible, and opening them only to get rid of the accumulated carbon dioxide.

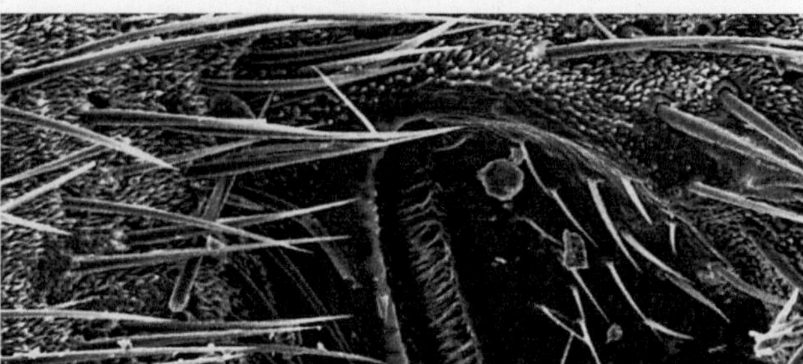

A scanning electron micrograph of an insect spiracle.

11 How might DCG help an insect living underground to get rid of carbon dioxide?

Fig. 2 Patterns of gas exchange in insects

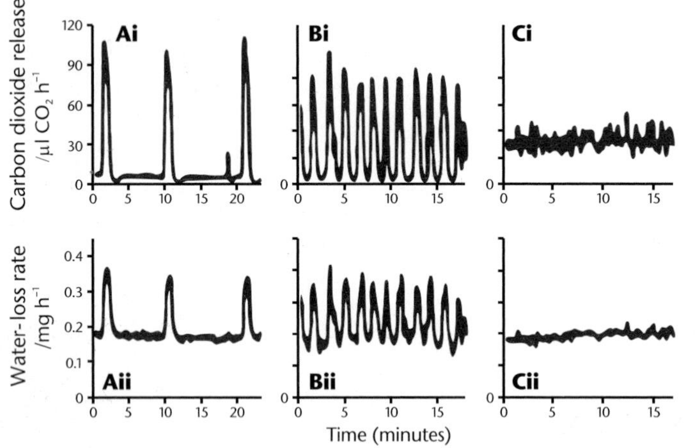

12 The graphs in Fig. 2 show three different patterns of gas exchange:

- DGC – graphs Ai and Aii
- cyclic gas exchange – graphs Bi and Bii
- continuous gas exchange – graphs Ci and Cii.

a What do the data show about the relationship between water-loss rate and carbon dioxide release in each of the three patterns of gas exchange?

b Do these data support the hypothesis that DCG reduces the rate of water loss by insects? Explain the reason for your answer.

key facts

- Insects do not use blood to transport respiratory gases.
- Gases enter and leave an insect's body through holes in the cuticle called spiracles.
- Gases diffuse to and from the exchange surfaces via tracheae and tracheoles.
- The exchange surface is the interface between the ends of the tracheoles and the insect's cells.

Fig. 3 The fish gill

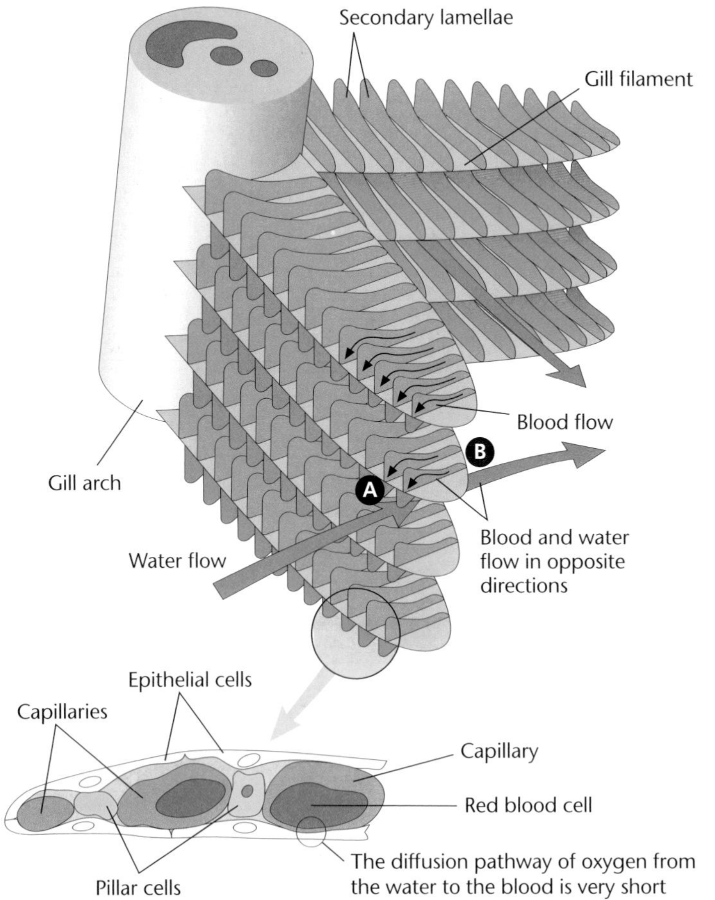

Gas exchange in fish

In fish, water is pumped over the specialised respiratory surface in the **gills**. Inside the gills, projections called secondary **lamellae** (Fig. 3) provide a large surface area. The diffusion path for oxygen is short, because the blood that flows within the lamellae is separated from the water outside them by a very thin layer of cells.

However, obtaining oxygen from water rather than air presents particular problems for aquatic animals. Oxygen is not very soluble in water and so water contains only one thirtieth as much oxygen per unit volume as air. Water also has a higher density than air, and is therefore harder to move over a respiratory surface during ventilation. In order to obtain maximum oxygen uptake from the water, fish have evolved the strategy of **countercurrent flow** (Fig. 4). The fish gill acts as a **countercurrent multiplier**.

Countercurrent multiplier

The water that flows over the secondary lamellae and the blood that flows through them travel in opposite directions. This is the countercurrent flow. It increases the efficiency with which oxygen can diffuse into the blood because:

- at point X in Fig. 4, the water that has just entered the fish's gills is saturated with oxygen, so there is a large diffusion gradient for oxygen between the water and the blood;
- at point Y in Fig. 4, although much of the oxygen has already diffused from the water into the blood, the blood here has just entered the gills. It therefore has a very low concentration of oxygen and so there is still a large diffusion gradient that favours the movement of oxygen from the water to the blood.

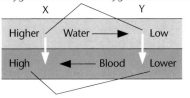

Fig. 4 Countercurrent flow

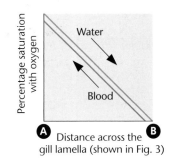

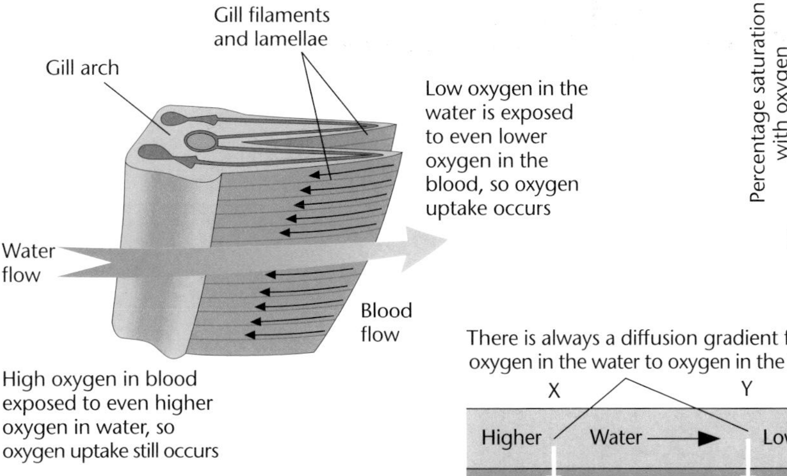

The countercurrent system is very effective: up to 80% of the oxygen dissolved in water is extracted by the gills. In comparison, our lungs can extract only a maximum of 25% of the oxygen from the air we breathe.

The diffusion of carbon dioxide from the blood into the water is also helped by the countercurrent flow. The countercurrent mechanism works only when the ventilation current moves in one direction only; notice that ventilation in fish is not tidal.

13 List three similarities and three differences between the methods used by mammals and fish to obtain oxygen.

key facts

- **Gill lamellae** comprise the respiratory surface in fish. The total surface area of the lamellae is enormous.
- The length of the diffusion path for oxygen is very short, because the layer of cells that separate the blood from the surrounding water is very thin.
- Water contains only one-thirtieth as much oxygen per unit volume as air, and water is more difficult to push over the respiratory surface.
- Fish use a **countercurrent system** to maximise the rate of gaseous exchange across the respiratory surface.
- A **countercurrent multiplier** ensures the oxygenating medium flows in the opposite direction to the blood. This means that a diffusion gradient exists across the whole length of the exchange membrane.

Gas exchange in plants

Plants do not move from place to place and so do not need anywhere near as much oxygen as active animals. But they do need surfaces for gas exchange because the processes of cell respiration and photosynthesis both involve an exchange of gases with the atmosphere.

The main gas exchange surface in plants is inside the leaves (Fig. 5). Root and stem cells obtain most of their oxygen as dissolved oxygen in the water that comes in through the roots.

The leaf has a large internal surface area in relation to its volume. Oxygen and carbon dioxide enter and leave a leaf mainly via microscopic pores called stomata (singular = stoma). There are huge numbers of stomata – several thousand per cm^2. Gases that enter the stomata diffuse through the intercellular spaces in the **mesophyll**. The surface of the mesophyll cells acts as the gas exchange surface. The cell walls are thin and permeable to gases. As in the lungs, water escapes, so the surface is moist. However, excessive loss of water into the atmosphere is avoided because the exchange surface is inside the leaf.

Each stoma is surrounded by two **guard cells** that can vary the width of the stoma, allowing more or less gas to enter or leave.

14 What are the main gases that the leaf exchanges with the atmosphere

a at noon on a bright summer's day?

b at midnight in summer?

key facts

- The main gas-exchange surface of a leaf is the outer surface of the mesophyll cells.
- Gases enter and leave a leaf mainly via holes called stomata.
- Each stoma is surrounded by two guard cells, which can open or close the stoma.
- A consequence of having stomata for gas exchange is that water vapour can diffuse through them, out of the leaf.

Fig. 5 Gas exchange in the leaf

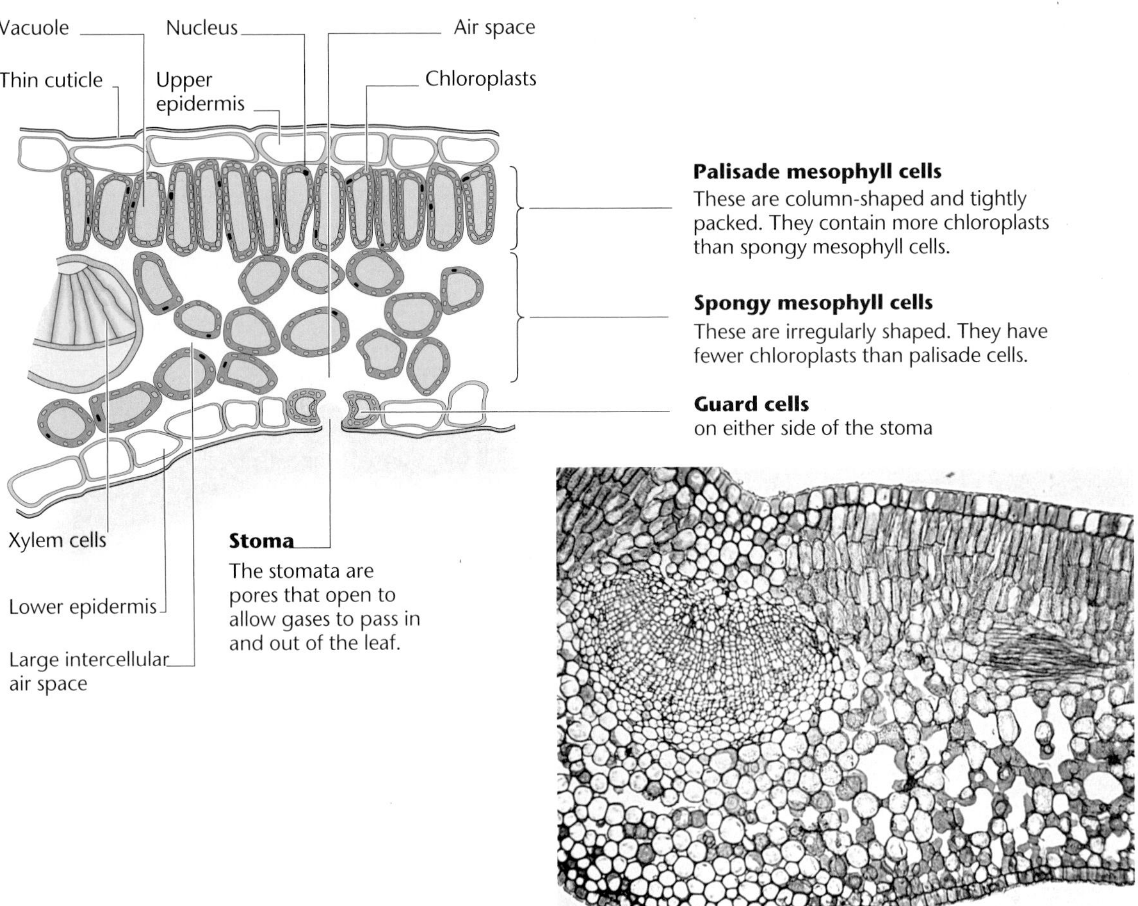
Vacuole
Nucleus
Air space
Thin cuticle
Upper epidermis
Chloroplasts
Palisade mesophyll cells
These are column-shaped and tightly packed. They contain more chloroplasts than spongy mesophyll cells.
Spongy mesophyll cells
These are irregularly shaped. They have fewer chloroplasts than palisade cells.
Guard cells
on either side of the stoma
Xylem cells
Stoma
The stomata are pores that open to allow gases to pass in and out of the leaf.
Lower epidermis
Large intercellular air space

A light micrograph of a section through a leaf. (approx. ×70)

12.4 The human blood system

Every organism moves substances around its body, including dissolved gases such as oxygen and carbon dioxide; nutrients such as glucose; waste products and hormones. Diffusion is efficient only over very short distances so most multicellular organisms use **mass transport**. This is the movement of relatively large amounts of material at relatively high speed. In most animals, mass transport forces liquid to flow along a system of tubes.

15 Which substance enters blood in the alveoli in the lungs?

16 Which substances enter blood in the small intestine?

17 Where do all these substances leave the blood again?

In the human mass transport system, materials are carried mainly by blood through a network of vessels called the **circulatory system** (Fig. 6).

The heart, a muscular pump, forces blood through a series of vessels called **arteries**, **capillaries** and **veins**. As blood flows through the body, substances are exchanged between the blood and the tissues. In this section we investigate how blood acts as a mass transport system.

Blood vessels

Fig. 7 shows the structure of the major blood vessels. Study the difference between arteries, veins and capillaries and refer to the diagrams frequently as you read the next section.

Arteries

The walls of the aorta and other large arteries contain a thick layer of elastic tissue.

This is stretched by the pressure produced by the left **ventricle** during **systole**. During **diastole** the elastic recoil of the walls of the aorta and large arteries squeeze on the blood, which maintains blood pressure at the right level. The expansion and recoil of the arteries helps to smooth the surges in blood flow that are caused by contraction of the left ventricle. Some signs of these surges can be felt as the pulse. This is particularly obvious at points where arteries pass over a bone near to the skin, such as at the wrist and the temple.

Arterioles

In the arterioles, the thickest layer of the wall consists of muscle fibres. This muscle enables the arterioles to act as control points for the blood system. By contracting, the muscle can restrict the flow of blood through a particular blood vessel.

As the arteries subdivide to form narrower vessels, the pressure of blood in the vessels decreases (Fig. 8). This is because the *total* cross-sectional area of all the smaller vessels is greater than that of all of the larger vessels.

Fig. 6 The circulatory system

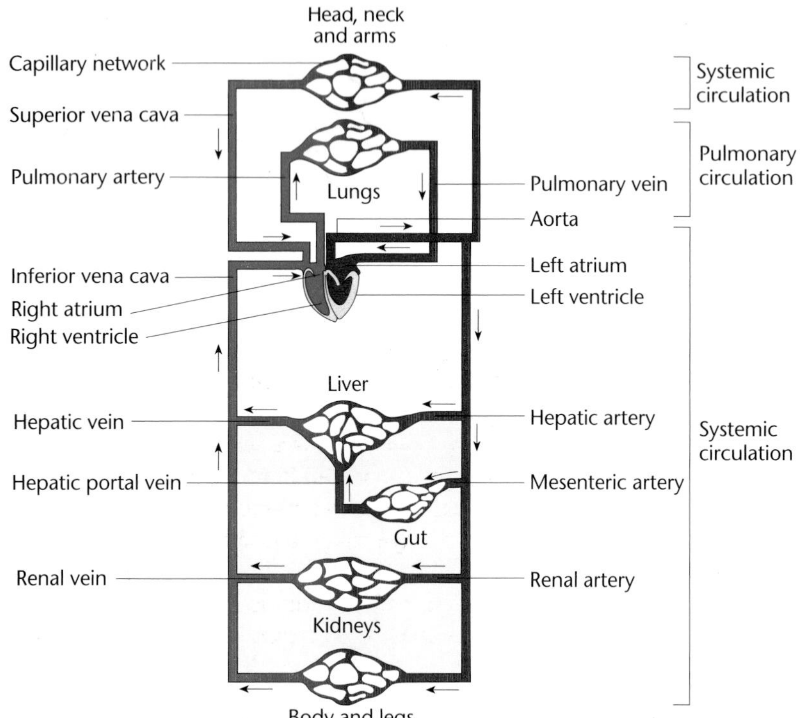

18 Suggest the effects of narrowing of the arterioles that supply the capillaries in the surface layer of the skin and that supply the capillaries in the villi of the small intestine.

Fig. 8 Arterioles

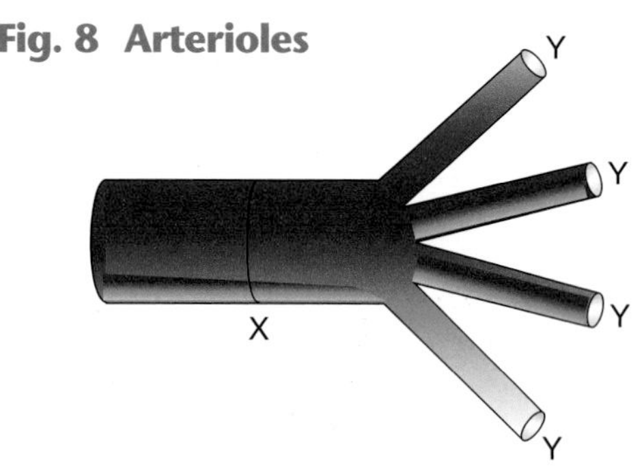

The total cross-sectional area at points Y is much greater than the cross-sectional area at point X

Fig. 7 Blood vessels

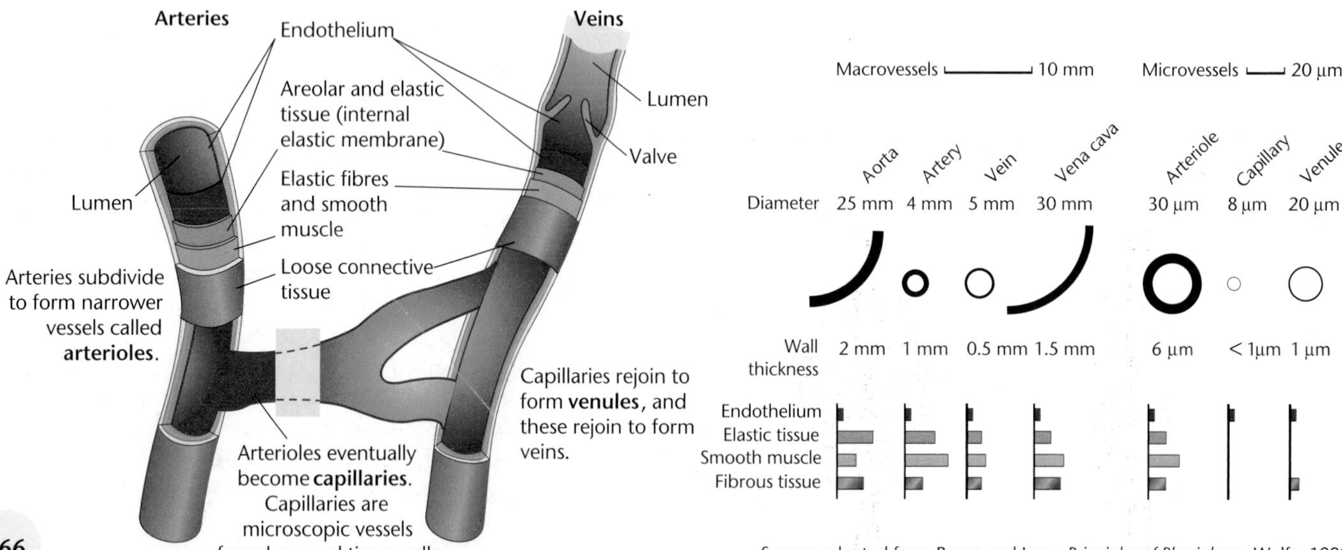

Source: adapted from Berne and Levy, *Principles of Physiology*, Wolfe, 1990

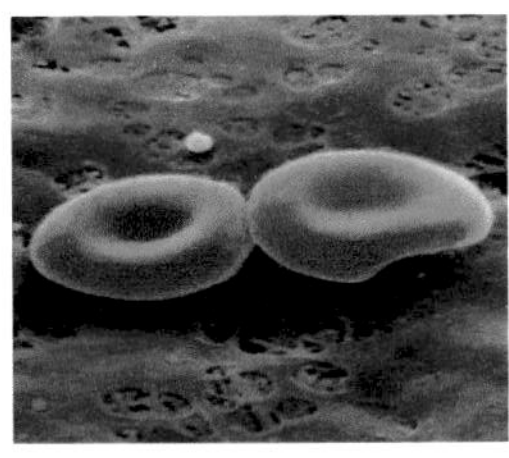

A false-colour scanning electron micrograph of red blood cells travelling through a capillary in the liver. The fenstrated endothlium of the capillary is characterised by a number of small holes through which nutrients can reach all the liver cells.

Capillaries

Capillary walls are made up from a single layer of **endothelial cells**. Capillaries have a diameter of 8 μm and a wall thickness of less than 1 μm. Blood flow and velocity are slowest through the capillaries. The velocity of the blood describes how fast it is moving, while the flow describes how much blood moves past a particular point. The flow depends on both the velocity and the size of the blood vessel. You can see how a large slow-moving river (low velocity) can shift much more water than a small fast-moving stream (high velocity) in the same time. Blood flow follows the same rules (Fig. 9). So, although blood flows through the capillaries in an organ at a low velocity, a large volume of blood flows through the many capillaries that serve that organ.

Exchange of materials such as oxygen, carbon dioxide and soluble food molecules occurs between the blood and the tissues at the capillaries. Some capillaries have small gaps in between adjacent endothelial cells. These gaps, called **fenestrations**, allow even faster rates of diffusion between the capillaries and the tissues.

19 Use information from Fig. 9 to explain why the velocity of blood decreases as the blood flows from arterioles into capillaries.

20 Explain why the velocity increases as the blood flows from the smaller veins into the larger veins.

21 Suggest two organs other than the liver where the fenestrations shown in the photograph might occur. Give reasons for your choices.

22 How does the size of capillaries and the rate of blood flow through the capillaries suit them for their function?

Fig. 9 Blood flow and velocity

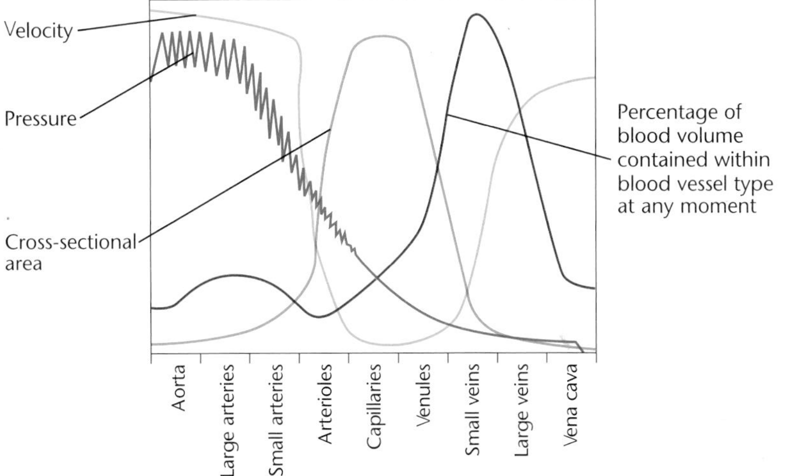

Fig. 10 The 'secondary heart' provided by the leg muscles

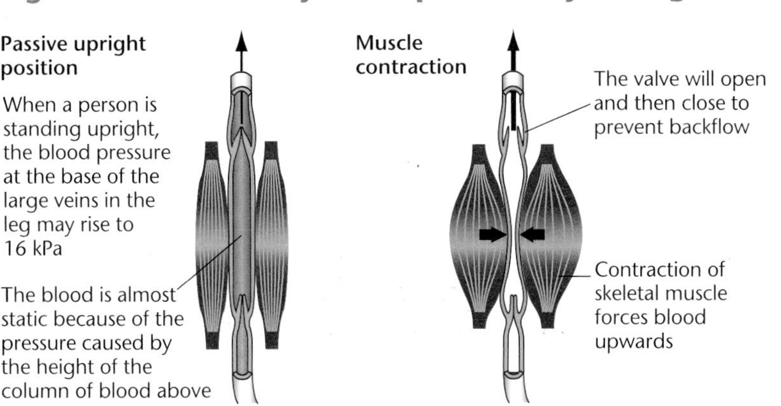

Passive upright position
When a person is standing upright, the blood pressure at the base of the large veins in the leg may rise to 16 kPa
The blood is almost static because of the pressure caused by the height of the column of blood above
Muscle contraction
The valve will open and then close to prevent backflow
Contraction of skeletal muscle forces blood upwards

Veins

Generally, veins have thinner walls than arteries and contain both elastic and muscle tissue. The pressure of blood in the veins is lower than in either arteries or capillaries, falling almost to zero in the largest vein, the vena cava, which carries deoxygenated blood back from the body into the heart. Unlike arteries, veins have valves to prevent backflow of blood.

Three main forces keep blood moving in the veins:

- the small residual pressure of blood coming from the capillaries
- the action of the leg muscles and valves in the veins
- the reduced pressure in the atria at **atrial diastole**.

Leg muscles act as a 'secondary heart' (Fig. 10): when they contract to move the legs, the contractions squeeze the veins in the leg and this pushes blood upwards, rather like squeezing toothpaste along a tube. Valves in the veins prevent the blood from moving downwards. This explains why people can faint if they stand still for a long time – particularly on a hot day. At atrial diastole, the muscles in the atrial walls relax and the atria increase in volume. This reduces the pressure and creates a suction force in the vena cava and other main veins that draws blood towards the heart.

23 Use the information from Fig. 7 to calculate the cross-sectional area of the space inside:

a an artery

b a capillary.

24 If the pressure inside the capillaries was exactly the same as the pressure in the artery supplying them, calculate the number of capillaries that would be supplied by the artery.

key facts

- Elastic tissue in the walls of arteries smoothes surges of blood pressure.
- Muscle layers in the walls of **arterioles** control the supply of blood to each organ.
- **Capillary** walls are one cell thick to enable the exchange of materials between the blood and the tissues.
- **Valves** in veins prevent backflow of blood, particularly in the limbs.
- Contraction of the skeletal muscles that surround veins, particularly in the legs, helps to return blood to the heart.

12.5 Blood as a transport medium

The main function of the blood system is to transport materials such as oxygen, food molecules and mineral ions to the cells of the body, and to take wastes such as carbon dioxide and urea away. In this section we look at the role that haemoglobin plays in transporting dissolved gases in the blood, and then go on to examine how other materials are exchanged between the blood and the tissues.

Haemoglobin

Haemoglobin transports oxygen around the body. It is this pigment molecule that gives red blood cells their colour. Haemoglobin is such a good oxygen carrier that it can be given on its own, instead of whole red blood cells, as a transfusion for patients who have lost a lot of blood.

The haemoglobin molecule is shown in Fig. 6 (page 61). It is a protein with a quaternary structure. The haem part of a haemoglobin molecule can combine temporarily with four oxygen molecules. In oxygen-rich situations such as in the capillaries of the lungs, haemoglobin and oxygen combine to form **oxyhaemoglobin**. In oxygen-poor situations, such as in the capillaries of exercising muscles, the oxyhaemoglobin **dissociates**: it splits up and releases the oxygen. A far greater mass of oxygen can be carried in the form of oxyhaemoglobin than can be carried in solution. The reaction between oxygen and haemoglobin is summarised by the equation:

$$\text{oxygen} + \text{haemoglobin} \rightleftharpoons \text{oxyhaemoglobin}$$

To understand why haemoglobin is such an efficient molecule for transporting oxygen we have to consider its **oxygen dissociation curve**. From Fig. 11 you can see that the percentage of haemoglobin molecules that combine with oxygen to form oxyhaemoglobin varies with the external partial pressure of oxygen (ppO_2). At the oxygen levels that occur in the lungs, haemoglobin readily absorbs oxygen to form oxyhaemoglobin. Part X on the graph shows this. Even if the oxygen level starts to fall, haemoglobin can still absorb oxygen – this means that the blood leaving the lungs is almost always completely saturated with oxygen. At point Y the graph starts to fall steeply as oxyhaemoglobin gives up its oxygen. This occurs in the tissues. The steepness of the graph at part Z means that a slight decrease in oxygen concentration in the tissues produces a large

Fig. 11 Haemoglobin oxygen dissociation curve

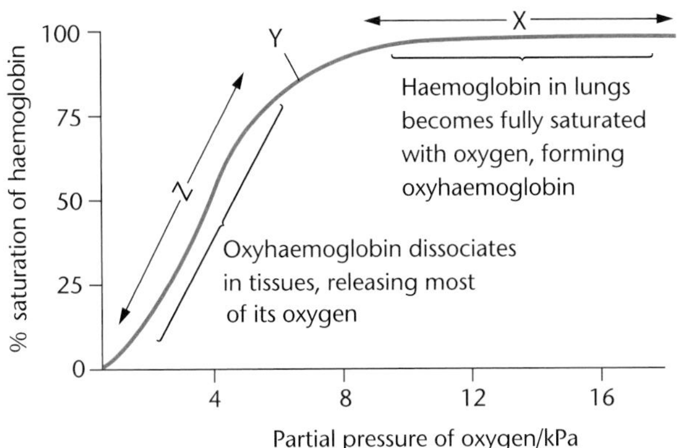

Fig. 12 The Bohr effect

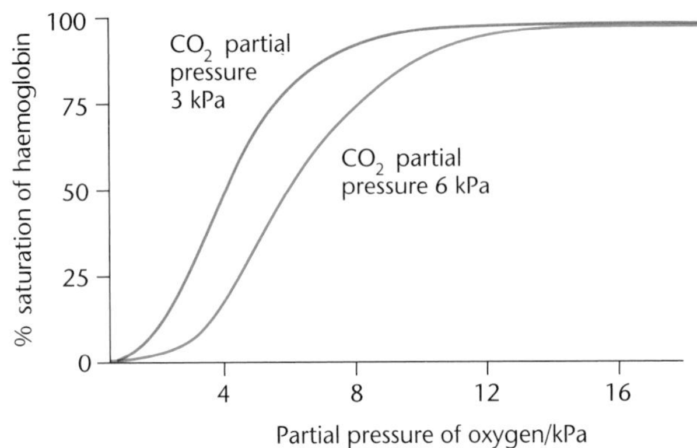

Fig. 13 *Tubifex* haemoglobin

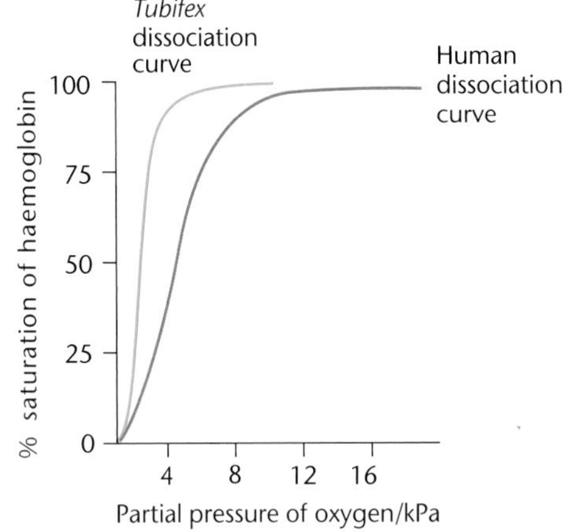

increase in the rate of oxyhaemoglobin breakdown. This delivers oxygen very effectively to the tissues that need it.

Haemoglobin has an 'S'-shaped oxygen dissociation curve: it becomes fully saturated with oxygen in the lungs but readily gives up this oxygen to respiring muscles. The tissue fluid in an active muscle has a high partial pressure of carbon dioxide due to the high rate of respiration in the muscle cells. The effect of this is to 'shift' the dissociation curve to the right (Fig. 12). This shift is known as the **Bohr effect**. This means that the oxyhaemoglobin gives up its oxygen even more readily in the prescence of carbon dioxide – exactly what is needed in actively respiring muscle.

Organisms that live in anaerobic conditions often have haemoglobin with a slightly different dissociation curve to that of humans. *Tubifex* worms live in mud at the bottom of lakes and rivers. The oxygen dissociation curve of *Tubifex* haemoglobin is shifted to the left compared with that of human haemoglobin (Fig. 13).

Look at Fig. 11.

25 hsw What percentage of haemoglobin is saturated at partial pressures of oxygen of:

a 8 kPa

b 12 kPa

c 16 kPa?

26 What is the lowest partial pressure of oxygen at which all the haemoglobin molecules combine with oxygen?

27 Blood moves from a partial pressure of oxygen of 12 kPa in the lungs to a value of 4 kPa in the muscles. What percentage of the oxygen is released?

28 What is the advantage to *Tubifex* in having haemoglobin with a dissociation curve of the shape shown in Fig. 12?

29 Llamas have haemoglobin adapted for life at high altitude. Copy Fig. 12, then draw on it the dissociation curve you would expect for the llama.

key facts

- Red cells contain **haemoglobin**. The principal function of haemoglobin is to carry oxygen.
- The reaction between haemoglobin and oxygen that forms **oxyhaemoglobin** is reversible.
- The oxygen dissociation curve of haemoglobin allows it to become fully saturated as it passes through the lungs and to release oxygen as it passes through the tissues.
- The oxygen dissociation curve is 'shifted' to the right at high carbon dioxide concentrations (e.g. in muscle) to allow more oxygen to be released.
- The oxygen dissociation curve can be shifted to the left in animals that live in situations where there is reduced oxygen availability, resulting in maximum uptake of oxygen in these conditions.

Exchange of other substances

Exchange of other materials occurs through the endothelial cells of the capillaries between the blood and the tissues. Different materials are exchanged using different mechanisms.

- Water is literally forced out of capillaries in the tissues by the pressure of the blood and it later re-enters the capillaries by **osmosis**.
- Sugars, mineral ions and waste products such as urea are exchanged between the capillaries and the tissues by **facilitated diffusion**.

Formation of tissue fluid

When blood travels through the tissues, the exchanges that occur between the blood and the cells determine the composition of the fluid that bathes the cells. This medium is called **tissue fluid**. Because not all of the contents of the blood pass into the tissue fluid, tissue fluid and blood are not the same. Red blood cells, platelets and plasma proteins remain inside the blood capillaries. The resulting tissue fluid is essentially blood plasma without the plasma proteins. Like blood plasma, its composition varies depending on its position in the body. Tissue fluid in the small intestine, for example, contains a high concentration of sugars in the couple of hours following a meal.

Tissue fluid also contains some types of white blood cells that squeeze out of the capillaries through the fenestrations. These white blood cells combat infection by ingesting bacteria or producing chemicals such as antibodies to combat the effects of bacteria and viruses.

How tissue fluid is formed

To understand how tissue fluid is formed, we first need to consider the composition of blood. This is shown in Fig. 14.

When blood flows into the capillaries, water and other materials are exchanged between the blood and the tissue fluid (Fig. 15). Two forces affect the exchange of water between the capillaries and the cells (Fig. 16).

Fig. 14 Composition of blood

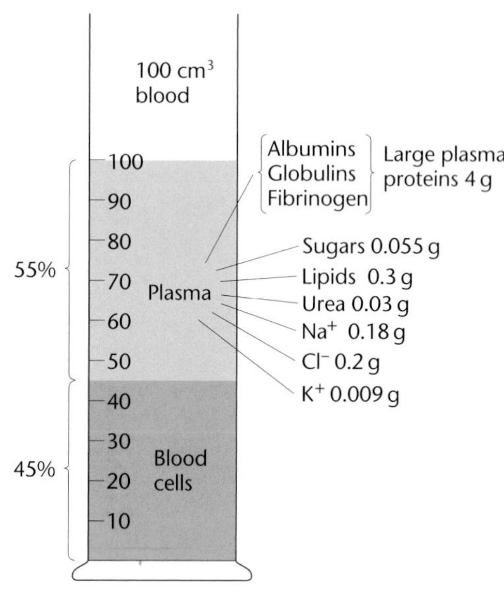

- **Hydrostatic pressure**. This is the pressure that is generated by the pumping force of the heart. It tends to push water out of the capillaries.
- **Water potential**. This is due mainly to the presence of large protein molecules in the plasma. These protein molecules attract large numbers of water molecules, so making the water potential inside the capillaries much more negative. This draws water into the vessels.

At the arterial end of a capillary, the hydrostatic pressure forcing water out is greater than the force due to water potential moving water back in, so there is a net movement of water into the tissues. The protein molecules in the plasma are generally too large to leave the capillaries so the water potential of the fluid that remains in the capillaries after water has moved out becomes more negative. This increases the force drawing water into the capillaries. The loss of water from the blood also reduces the hydrostatic pressure – again reducing the effect of

Fig. 15 Blood tissue fluid and lymph

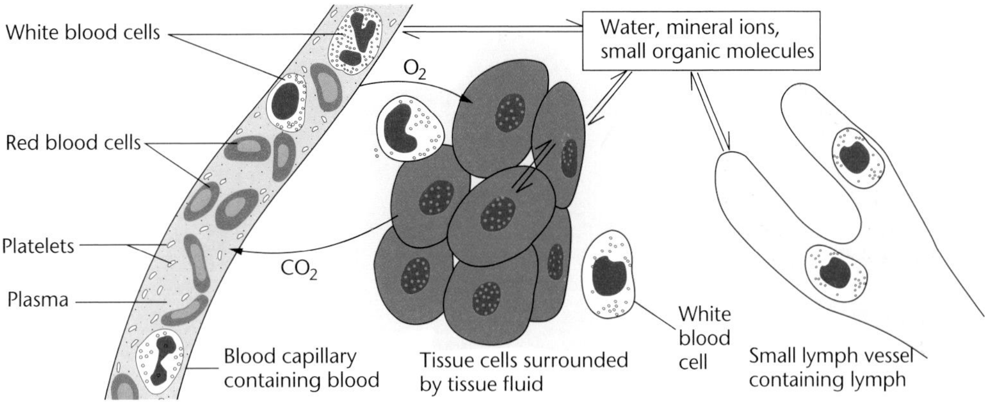

White blood cells
Red blood cells
Platelets
Plasma
Blood capillary containing blood
O_2
CO_2
Water, mineral ions, small organic molecules
Tissue cells surrounded by tissue fluid
White blood cell
Small lymph vessel containing lymph

Fig. 16 Comparing capillaries

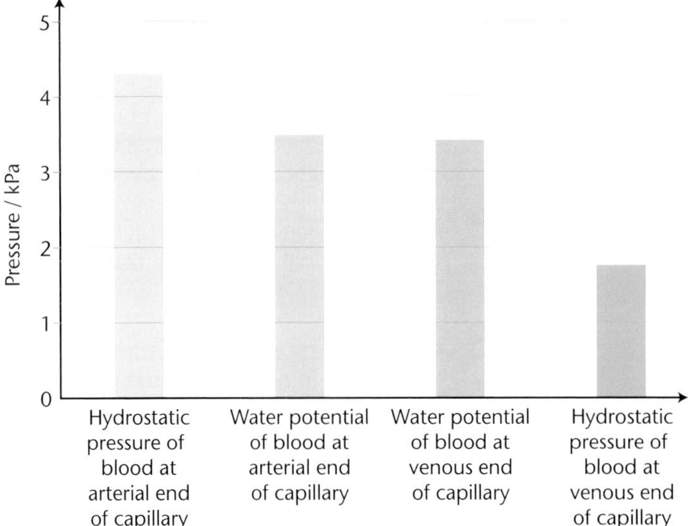

the force pushing water out of the capillaries. So, by the time blood reaches the venous end of the capillary, the effect of the water potential drawing water in exceeds the hydrostatic pressure pushing water out. This means that water re-enters the capillaries. Some tissue fluid drains into small capillary-like vessels called **lymph** capillaries, which drain into lymph vessels. Lymph vessels form a secondary drainage system that empties lymph back into the blood system.

Serum is a blood product used widely in medicine. It is particularly useful for replacing the blood plasma lost by patients with extensive burns. In these patients, the loss of the waterproof outer layer of skin means that tissue fluid, but not blood cells, is lost from their wounds very rapidly. To produce serum, whole blood is first centrifuged to remove blood cells, then treated to remove a blood protein called **fibrinogen**. The fibrinogen is an essential component in the clotting process; its removal prevents plasma clotting. The liquid produced is called serum. Serum is useful for burns patients, but patients who have lost a lot of whole blood need a whole-blood transfusion.

30 Construct a table to compare the composition of blood, plasma and tissue fluid.

31 Oxygen, carbon dioxide, glucose, amino acids, urea and proteins are all carried in the blood. Which substances pass from capillaries into tissues and from tissues into capillaries at the following sites: alveoli; intestinal villi; brain; leg muscles; liver; kidney?

32 Children suffering from protein deficiency often appear to be fat, but their abdomens are swollen because of the build-up of fluids in their tissues. Suggest why this fluid accumulation occurs.

key facts

- The main function of blood plasma is to transport materials around the body, in solution.
- Plasma is forced out of the capillaries by hydrostatic pressure to form **tissue fluid**. Tissue fluid does not contain plasma proteins, but may contain white blood cells.
- Tissue fluid returns to capillaries by osmosis.
- Much tissue fluid drains into lymph capillaries, which drain into lymph vessels that empty into the blood system.

how science works

Artificial blood

Scientists from the University of Sheffield are developing an artificial 'plastic blood', which could act as a substitute for real blood in emergency situations. The 'plastic blood' could have a huge impact on military applications.

Because the artificial blood is made from a plastic, it is light to carry and easy to store. Doctors could store the substitute as a thick paste in a blood bag and then dissolve it in water just before giving it to patients – meaning it's easier to transport than liquid blood.

Donated blood has a relatively short shelf-life of 35 days, after which it must be thrown away. It also needs refrigeration, whereas the 'plastic blood' will be storable for many more days and is stable at room temperature.

The artificial blood is made of plastic molecules that hold an iron atom at their core, just like haemoglobin, that can bind oxygen and could transport it around the body. The small plastic molecules join together in a tree-like branching structure, with a size and shape very similar to that of natural haemoglobin molecules. This creates the right environment for the iron to bind oxygen in the lungs and release it in the body.

Of course, doctors had the same hope back in 1868, when they first extracted haemoglobin, the oxygen-bearing protein in red blood cells. Haemoglobin failed as a blood replacement because it works only when intact and when assisted by a cofactor found in red blood cells. Stripped from its protective cell and its cofactor, haemoglobin is quickly snipped in two by enzymes, and the fragments can poison the kidneys.

While still in its development, the scientists hope this will make it particularly useful for military applications and being plastic, it's also affordable. The scientists are now seeking further funding to develop a final prototype that would be suitable for biological testing.

Dr Lance Twyman, from the Department of Chemistry at the University of Sheffield and who has been developing the artificial blood for the last five years, said: "We are very excited about the potential for this product and about the fact that this could save lives. Many people die from superficial wounds when they are trapped in an accident or are injured on the battlefield and can't get blood before they get to hospital. This product can be stored a lot more easily than blood, meaning large quantities could be carried easily by ambulances and the armed forces.

From an article published in *ScienceDaily*, May 14, 2007 and reproduced with permission from *ScienceDaily* and The University of Sheffield.

33 What are the advantages of using artificial blood rather than natural blood in the transfusion service?

34 What advantages has natural blood over artificial blood in the transfusion service?

35 Explain why extracted haemoglobin is not used in blood transfusions.

36

a Should people be paid to donate blood?

b Should blood donated freely be sold, and if so, where should the money go?

12.6 Transport in plants

When rains do fall, desert plants burst into flower, making the most of the water that is now available.

Much of the land that is desert today used to be fertile. The change to desert began about 5000 years ago, when a shift in the Earth's climate caused the rainfall in some regions to decrease. The land became drier and plants withered. The soil blew away, leaving large stretches of drifting sand. In some areas, humans made the problem worse by using poor farming techniques.

Once an area has become desert, the air above the land becomes very dry and no clouds form. In the burning heat of the day, temperatures can rise as high as 50 °C. However, as there are no clouds to provide any insulation, heat radiates out into space very rapidly at night and deserts can be freezing cold. Water from occasional rain collects underground and wherever this happens, plants grow. These plants need to conserve water to survive long dry spells. Plants that have adapted to arid conditions often look spectacular.

Fig. 17 Transpiration and the water potential gradient

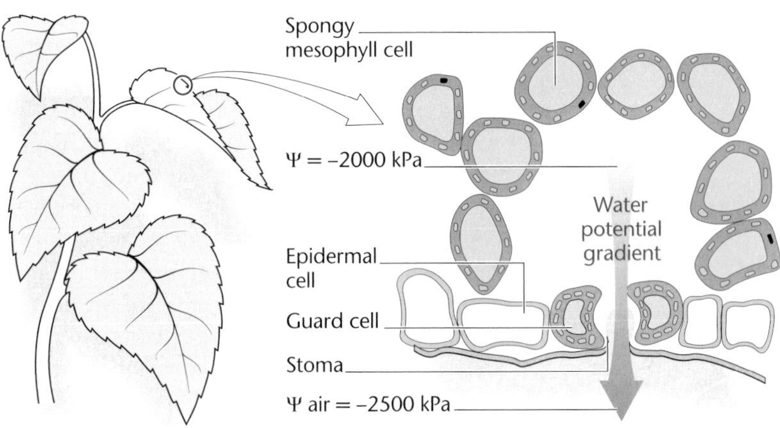

Reclaiming areas of desert is difficult; crops that are introduced to desert environments often fail. Scientists are increasingly coming to recognise that the solution to feeding people in arid lands is to grow plants that have become adapted to dry conditions over millions of years. Research into the survival methods of plants from arid areas could help turn desert into productive land. While science cannot solve the political and economic causes of famine, it can help to increase food production. In arid areas, this can mean the difference between life and death.

If you don't water a pot plant regularly, the soil in the pot will dry out and the leaves will droop. Plants are constantly absorbing water from the soil but, at the same time, they are also losing water vapour to the air. This creates a continual stream of water through the plant. Plants do not have complex circulatory systems with a heart to pump the fluid around the organism; movement of water and the substances dissolved in it is brought about by simple physical processes. In this chapter we will look at the way in which evaporation, osmosis and active transport can move materials round plants as small as a daisy and up trees taller than a cathedral. We shall also investigate some of the ingenious adaptations that plants have evolved for retaining water.

Transpiration

Photosynthesis is the primary function of leaves and for this they need carbon dioxide, water and sunlight. Leaves obtain carbon dioxide from the air through tiny pores. Each pore is called a stoma; the plural term is **stomata**. Two cells, called **guard cells**, surround each stoma; we find out more about their role in stomatal opening and closing later.

When the stomata are open, the hole that allows carbon dioxide to pass into the leaf also lets water molecules out. This is a disadvantage to the plant because it loses precious water; if the loss of water from the leaves is greater than the uptake of water by the roots, the plant wilts. For many plants, stomata are therefore a 'necessary evil'.

The loss of water vapour by evaporation from a plant to the air surrounding it, through the open stomata, is called **transpiration**. This loss of water vapour via the stomata is due to the **water potential (Ψ) gradient** between cells inside a leaf and the air outside (Fig. 17).

Factors that affect transpiration

Water molecules diffuse along a water potential gradient towards areas with a more negative water potential value. The greater the gradient, the faster the rate of movement. The air inside the leaf is always saturated with water vapour, so changes outside the leaf can alter the water potential gradient. Three factors increase this gradient and therefore the rate of transpiration:

- an increase in temperature
- a decrease in humidity
- an increase in wind speed.

An increase in temperature causes a higher rate of evaporation from the leaf surface by transferring more energy to the water molecules. The more energy a water molecule has, the faster it moves and the more likely it is to escape into the atmosphere outside the leaf and move away from the plant. Humidity is a measure of the number of water molecules in the air. Any decrease in humidity in the atmosphere around the leaf decreases the number of water molecules in the air and so increases the water potential gradient from the plant to the air. Warm air can hold more water vapour molecules than cold air; that is why the air in tropical forest feels so uncomfortable, and why dew forms as water condenses from cooling air. On a hot day the air in a tropical jungle may become fully saturated with water vapour, reducing the transpiration to zero, as there is no longer a water potential gradient between the plants and the air. When wind moves the air and water vapour away from around a leaf, this increases the water potential gradient from the plant to the air.

Air always has a value of water potential more negative than that of the leaf cells, except when it is fully saturated with water vapour. So water molecules will always diffuse out through the open stomata. The only way to stop the loss of water is to close the stomata. Most plants close their stomata during the night because they do not need to take in carbon dioxide for photosynthesis when it is dark. If a plant has lost too much water during the day, it gets the chance to replenish its supplies during the night.

37 Explain how an increase in temperature and wind speed and a decrease in humidity make the water potential of the air just outside the leaf more negative. Use the idea of the number of water molecules per unit volume of air in your answer.

Fig. 18 Stomatal opening and closing

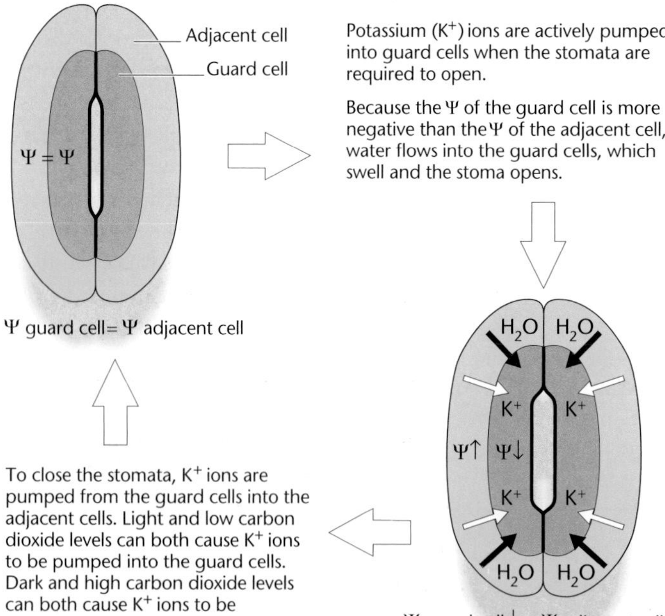

Adjacent cell
Guard cell
Ψ = Ψ
Ψ guard cell= Ψ adjacent cell
Potassium (K⁺) ions are actively pumped into guard cells when the stomata are required to open.
Because the Ψ of the guard cell is more negative than the Ψ of the adjacent cell, water flows into the guard cells, which swell and the stoma opens.
H₂O H₂O
K⁺ K⁺
Ψ↑ Ψ↓
K⁺ K⁺
H₂O H₂O
Ψ guard cell ↓ ≠ Ψ adjacent cell ↑
To close the stomata, K⁺ ions are pumped from the guard cells into the adjacent cells. Light and low carbon dioxide levels can both cause K⁺ ions to be pumped into the guard cells. Dark and high carbon dioxide levels can both cause K⁺ ions to be pumped out of the guard cells.

Stomatal opening and closing

Opening and closing of stomata is controlled by changes in **turgidity** of guard cells. But what is the mechanism that controls the turgidity of the guard cells? Any hypothesis for the mechanism of stomatal opening must account for the more negative water potential of the guard cells during the day, and also why some plants can keep their stomata closed during the day.

One hypothesis involves the movement of potassium ions (K^+) into and out of the guard cells. To open the stomata, adjacent cells pump K^+ ions into the guard cells, as shown in Fig. 18.

Light and carbon dioxide both affect the opening and closing of stomata. However, they can both be over-ridden by a plant hormone called **abscisic acid** (ABA). ABA is produced when the plant suffers water stress. This happens when a plant loses much more water through transpiration than it can take in through the roots and is common in plants growing in a desert environment. ABA causes a rapid pumping of K^+ ions from the guard cells into the adjacent cells. This explains how some plants can close their stomata at midday to conserve water. It also explains the complete closing of stomata when the soil is very dry.

how science works

Factors that affect stomatal opening

Graph A

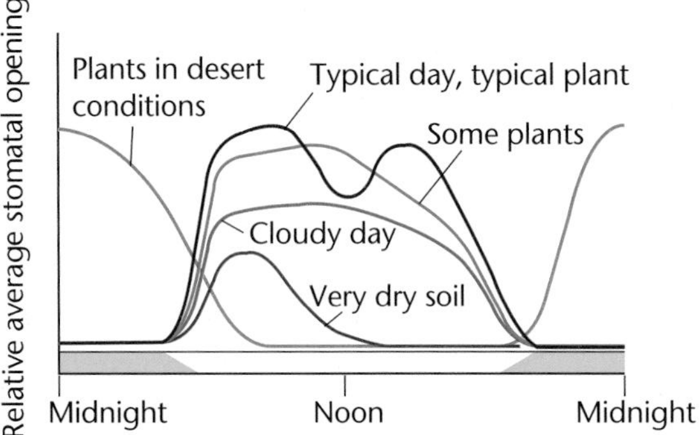

Source: Salisbury/Ross Plant Physiology 4e 1992 Brooks/Cole, a part of Cengage Learning, Inc. Reproduced by permission. www.cengage.com/permissions.

Graph A shows how external factors affect stomatal opening.

38 Study graph A carefully. Compared with a typical plant on a typical day, what effect does each of the following have on stomatal opening and closing?

a very dry soil

b a cloudy day

39 What are the advantages and disadvantages to a desert plant of opening its stomata during the night and closing them around midday?

Graph B

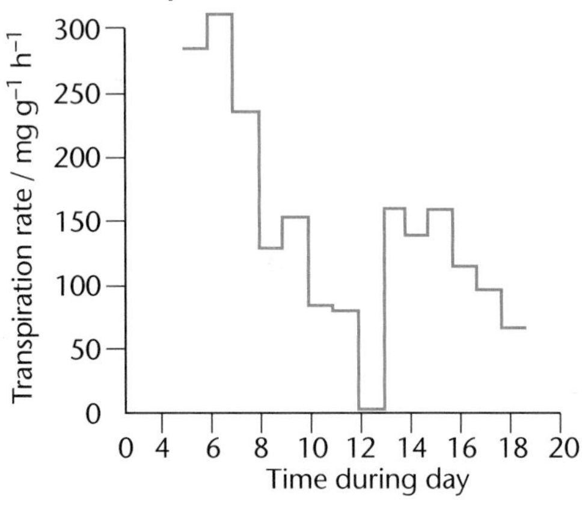

Some plants that are adapted to living in hot environments can close their stomata during the middle of the day, when the temperature is highest, and can open them at night. One such desert plant is the saltwort. The effect of its midday stomatal closing is dramatic: transpiration almost ceases, as Graph B shows.

40 What type of plant could keep its stomata open all day without suffering the effects of severe water loss?

Other plants keep their stomata open all day, even at noon (see the light-blue curve in Graph A). They only close their stomata at night.

key facts

- Leaves must allow the entry of carbon dioxide needed for photosynthesis. Most have stomata – pores in the leaf that allow gas exchange to occur.
- Open stomata allow water vapour to diffuse out of the leaf, a process called transpiration.
- Water vapour diffuses out through stomata down a water potential gradient between the leaf cells and the air outside.
- This water potential gradient, and so the rate of transpiration, is increased by an increase in air temperature, a decrease in humidity outside the leaf and an increase in air movements around the leaf.
- The stomata of most plants are open during the day and closed during the night. In some desert plants the stomata close during the day to conserve water and open at night to collect carbon dioxide.

A practice ISA on the measurement of the rate of water uptake by means of a simple potometer can be found at www.collinseducation.co.uk/advancedscienceaqa

12.7 Xerophytes

Plants that live in dry places usually have adaptations to survive long periods of drought. Such plants are called **xerophytes**.

Many desert trees, such as the quiver tree, drop their leaves to stop the whole tree from dying from desiccation. This certainly works, as it stops transpiration completely, but it is a drastic solution. A tree without leaves can no longer photosynthesise, so it must rely on the store of carbohydrates made during better times.

Cacti also have many adaptations that enable them to survive in deserts.

- The spines on the cactus trap a layer of air that is rich in water vapour next to the plant. This reduces the chances of wind moving the moist air layer away from around the plant. Many cacti have a dense covering of hairs that traps even more water vapour.
- The spines and stiff hairs on cacti also help to deter animals that try to eat the cactus to get at the stored water.
- The **epidermal cells** of cacti have a thick outer layer of wax that reduces water loss by evaporation.
- Many cacti store water in their stems; plants that do this are called **succulents**.

The epidermal cells of cactus stems have **chloroplasts**. These enable the stems to photosynthesise, since the spine-like leaves have lost this function. The thick waxy layer that prevents evaporation of water from cactus stems also prevents carbon dioxide from getting in, so the epidermis of a cactus stem has stomata. However, these stomata are closed during the day, so how does the plant obtain the carbon dioxide it needs for photosynthesis? Cacti have a special way of obtaining carbon dioxide for photosynthesis – **crassulacean acid metabolism (CAM)**. CAM plants open their stomata at night to absorb carbon dioxide. The epidermal cells combine this gas with an organic compound to form malic acid. This is stored in the vacuole. The large amounts of malic acid that are stored in the vacuoles at night make the plant taste sour. During the day the malic acid in the vacuole is broken down to release carbon dioxide, which diffuses to the chloroplasts to be used in photosynthesis:

Organic compound + $CO_2 \rightleftharpoons$ Malic acid

Quiver trees in Nambia can shed their leaves and store water in their branches to survive periods of extreme drought. They are called quiver trees because native people clean out the soft branches and used them as a quiver (a pouch) for their arrows.

41 Although cacti seem well adapted to life in deserts, their rate of growth is usually very slow. Suggest what factor limits their rate of growth.
Explain your answer.

The leaves of the teddybear cactus, found in Arizona, USA, no longer function as photosynthetic organs. Millions of years of adaptation have reduced them to spines.

how science works

Resourceful xerophytes

Xerophytes are all adapted to dry conditions, but the strategies and adaptations they use differ widely from plant to plant.

The sand dunes on the seashores in Britain have soil that often contains very little water, but some plants are very well adapted to living there. Perhaps the most successful is marram grass. When the soil is dry, hinge cells in the leaf make it roll into a cylinder. The air inside this cylinder quickly becomes saturated with water, resulting in a reduction in transpiration rate. When the soil is moist, the leaves uncurl.

A rather strange plant of the *Tortula* species, twisted moss, commonly called the 'all-screwed-up moss', also lives on sand dunes. It spends much of its life all screwed up, but when rain falls its leaves unwind.

Some xerophytic plants have different shapes of leaf in wet and dry seasons.

Collecting every drop of available water is vital to desert plants. Often the roots of these plants spread over a very wide area.

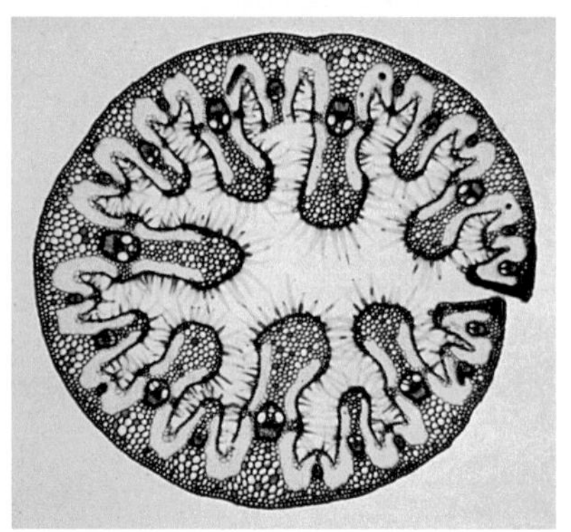

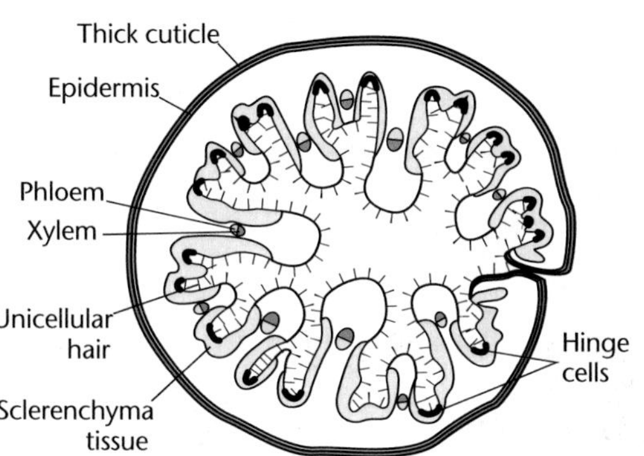

Micrograph of Marram grass (above left) and explanatory diagram (above right).

Cross-sectional structure of the winter and summer leaf of the grey sagebrush

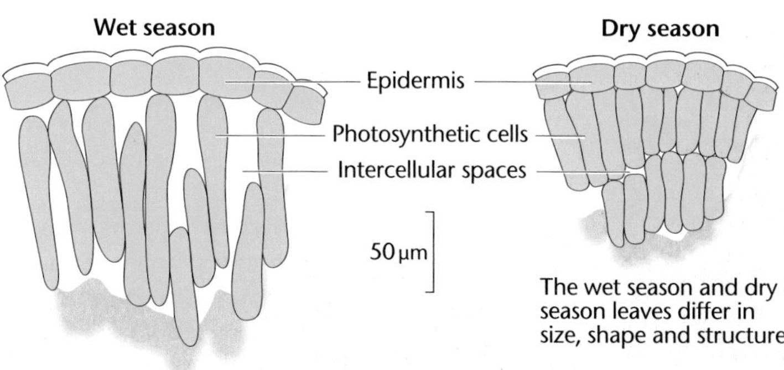

The wet season and dry season leaves differ in size, shape and structure

42 Suggest two adaptations that reduce the rate of transpiration even when the marram grass leaves are uncurled.

43 Explain how the behaviour of the all-screwed-up moss helps it to survive 8dry periods.

44 Explain the advantage to the sagebrush of having smaller leaves during the dry season.

45 Explain why the root systems shown below are xerophytic adaptations.

A schematic drawing of the root system of the bean caper

0
40
80
cm 120
160
200

320 280 240 200 160 120 80 40 0 40 80 120 160 200 240 280 320 360 400
cm

A schematic drawing of a **hydrotropic** root system, exploiting water pockets beneath stones

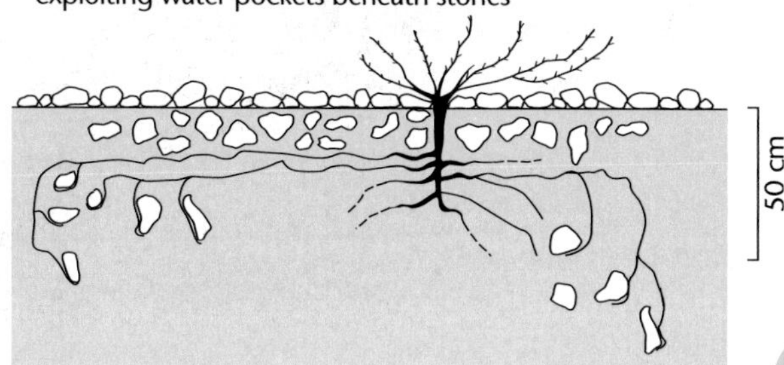

key facts

- Plants adapted to survive in dry desert conditions are called **xerophytes**.
- Some xerophytes reduce the area for transpiration by producing smaller leaves during dry periods, or by modifying the shape of the leaves.
- Many xerophytes sink their stomata into pits.
- Some desert plants use **crassulacean acid metabolism (CAM)** to collect carbon dioxide. CAM plants open their stomata only at night, store the carbon dioxide absorbed at night as malic acid, and then release it during the day for use in photosynthesis.
- Roots of desert plants often spread out over a wide area.

12.8 Movement of water up the plant

Fig. 19 Root structure

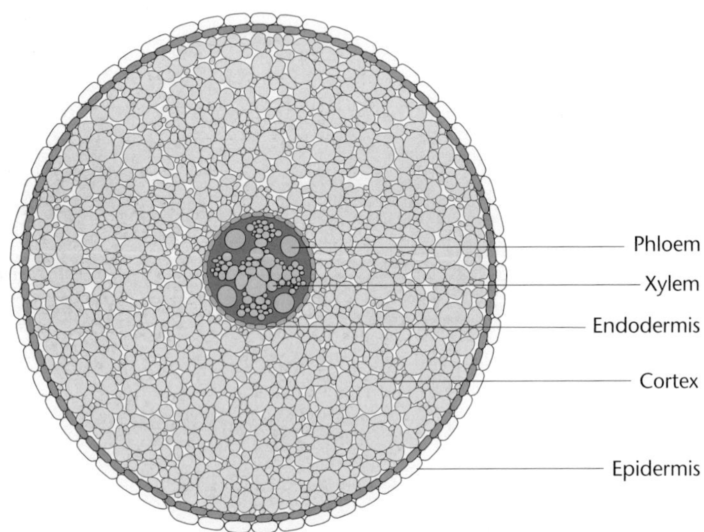

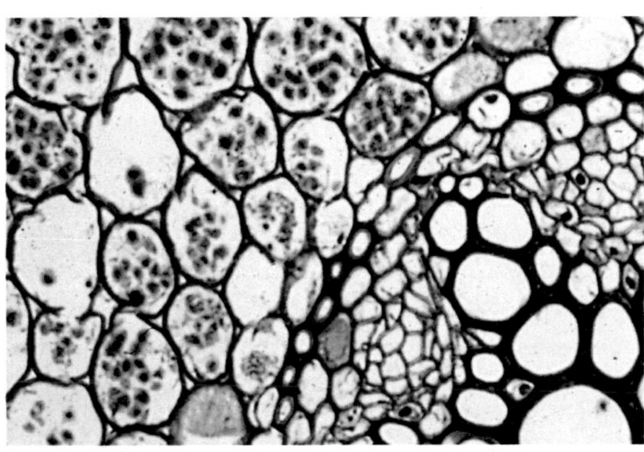

Diagram (top) and light micrograph (below), showing the cellular structure of a young root. Both are transverse sections through the root of a dicotyledenous plant.

How does water get into the roots and from there up to the leaves? The simple answer is that water enters a plant by osmosis through the root hairs. Whilst this does happen to a limited extent, the bulk of the water that travels through a plant from the soil to the air passes mainly through non-living parts of the plant. There is a continuous flow of water along a water potential gradient that extends from the air outside a leaf right down through the stem to the water in the soil. To see where these non-living parts are, we first have to consider the structure of the root.

Path of water through the plant root

Fig. 19 shows a drawing and photograph of a section through a young root. The main features of a young root are as follows.

- The single outer layer of cells is called the **epidermis**. These cells are where water and mineral ions enter the plant; the surface area of many of these cells is increased by projections called root hairs.
- The **cortex** is a layer of cells that are similar to mesophyll cells in a leaf but they do not have chloroplasts.
- The **endodermis** is a layer of cells each possessing a **Casparian strip**; this strip is impermeable to both water and mineral ions. It blocks the path to the centre of the root. All substances must pass through the cell surface membrane of the **endodermal cells** to get deeper into the root. In this way the endodermal cells control which substances pass from the roots to the rest of the plant.

Fig. 20 Symplast and apoplast

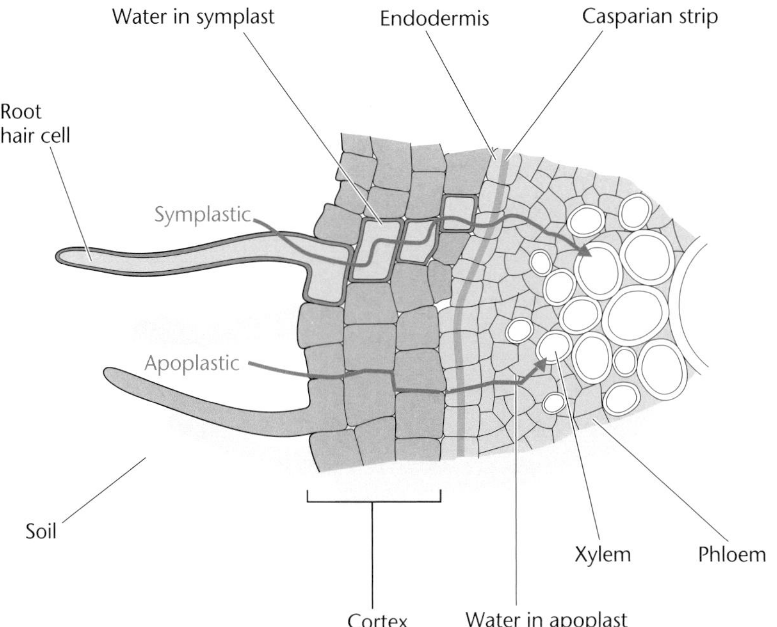

- A central core of vascular tissue consists of **xylem cells**, which transport water and mineral ions to the rest of the plant, and **phloem cells**, which transport organic materials.

Apoplast and symplast

Water passes into and through a root along two main routes: the **symplast** and the **apoplast** (Fig. 20). The symplast system involves the living contents of root cells including the membranes, cytoplasm and vacuoles. Minute strands of cytoplasm called **plasmodesmata** pass through the cell walls, connecting the symplasts of adjacent cells. Some water moves by osmosis along a continuous water potential gradient from the soil to the xylem cells in the centre of the root.

The apoplast is also a continuous pathway from the water in the soil to the endodermal cells but, in contrast to the symplast, the apoplast pathway comprises only non-living structures. The apoplast pathway includes the spaces between cells and the cell walls. Spaces between the cellulose fibres in the cell walls allow water to pass easily. When a plant is transpiring, a continuous stream of water is literally pulled through the apoplast of the root as far as the endodermal cells, then through the cytoplasm of the endodermal cells into the xylem cells. So, the apoplast and the symplast are separate continuous systems. Fig. 20 shows the movement of water through the apoplast and symplast systems in a root.

Xylem

Cells that make up the xylem vessels (Chapter 11 page 153) have no living contents at maturity. They are therefore entirely apoplast. Water is pulled through the apoplast on its way from the soil water to the air. The driving force is the gradient of water potential between air and soil water. Air has a very negative water potential (approximately –2500 kPa) whereas soil water has a value approaching zero.

Xylem cells join to form continuous tubes that stretch from the roots to the leaves. The tubes formed by the xylem cells function like the water pipes in your home – and, like your plumbing, they work only if there is nothing to block the flow of water. To move water along a stem requires a difference in water potential of 100 kPa for every 10 metres to overcome the forces between the water molecules and the walls of the xylem cells. Add to this a difference of 100 kPa needed to overcome the force of gravity for each 10 metres in height, and a difference of 200 kPa in water potential is required to move water 10 metres up a tree. Even pulling water this short distance up a tree needs a lot of energy. Californian redwood trees can reach heights of more than 110 metres – imagine the energy needed to pull water to that height.

46 If the difference in water potential between the air next to a leaf and the water in the soil is 2500 kPa, what is the theoretical maximum height for a tree?

As water is pulled out of a xylem vessel in a leaf, the column of water behind it is pulled upwards (see Fig. 21). As water evaporates from the leaves, the water columns in the xylem vessels are pulled upwards because the water molecules stick together. This is because there are forces, called hydrogen bonds, holding one water molecule to another. If you pull a cylinder of modelling clay at both ends, the middle of the cylinder will get thinner and will eventually break. What stops this happening to the columns of water in the xylem vessels?

Fig. 21 The cohesion–tension mechanism

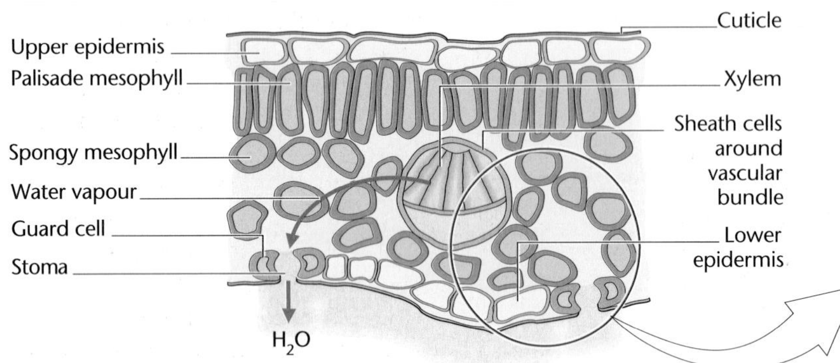

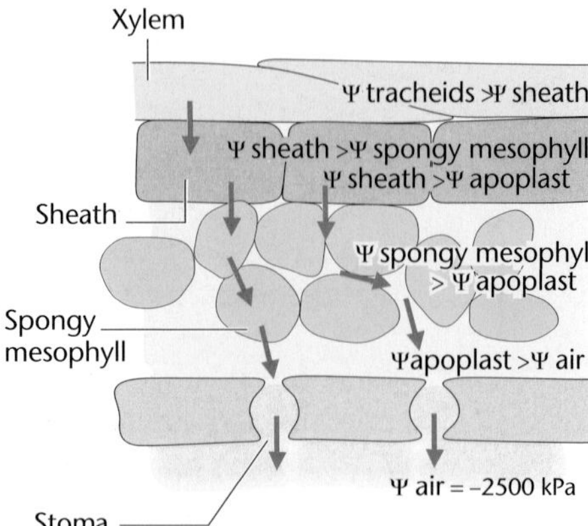

The process begins in the leaf. As water evaporates from the leaf cells into the air, a water potential gradient is set up between the water and the xylem.

The water molecules are strongly attracted to the walls of the xylem vessels. This attraction between the xylem walls and the water is so strong that as the water columns are put under tension by evaporation of water from the leaves, they become thinner and actually pull the walls of the xylem vessels inward. One function of the thickening in the xylem walls is to prevent the xylem vessels collapsing. However, the combined effect of the tension on all the xylem vessels in a tree trunk does produce a measurable narrowing of the trunk when the tree is transpiring (Fig. 22).

This mechanism for the movement of water up the xylem is known as the **cohesion–tension mechanism**, since there is a cohesive force between the water molecules themselves, and the water columns are under tension because of the forces resulting from evaporation and gravity. As a molecule of water evaporates from the leaf, the next molecule is pulled into the empty place next to it in the xylem cells, and the water molecule below that is pulled along after the moving water molecule, and so on.

As water leaves the xylem at the top, the xylem cells at the bottom of the column pull water from the cells in the root cortex. The water potential of the root cortex now has a value that is more negative than that of the root epidermis and root hairs. This means that water

Fig. 22 Variations in tree trunk circumference

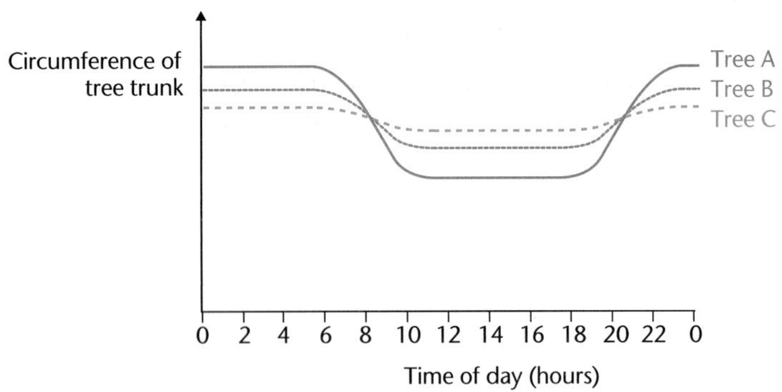

47 Which source of energy is used to pull water up a plant?

48 hsw The graph in Fig. 22 shows changes in the circumference of three different species of tree during a hot summer day. Which of the trees is best adapted for reducing the rate of transpiration? Explain the reason for your answer.

moves into the cortex cells. The value of water potential for the epidermis is now more negative than that of the soil water and so water moves from the soil into the root.

Water moves across most of the root mainly through the apoplast system. However, the endodermal cells, which form a ring of cells around the xylem and phloem, have a strip of **suberin**, a water-repellent substance, in their walls. This forms the **Casparian strip**, which prevents water passing through. The water and mineral ions have to pass through the symplast of endodermal cells before entering the apoplast of the xylem vessels.

Root pressure

Before leaves develop on a young plant or on a tree in spring, evaporation cannot be the driving force for the movement of water up the xylem. So how does water get to the developing leaves? If the stem of a young plant is cut just above soil level, liquid soon begins to exude from the cut end. This liquid has been forced up from the roots by a mechanism called root pressure.

When ions reach the endodermis they are moved from the apoplast across the cell membranes by active transport into the symplast. The ions are then moved into the apoplast of the xylem cells, again by active transport. There is now a higher concentration of ions in the xylem cells than in the soil solution. So, the water potential of the contents of the xylem cells is now more negative than that of the soil solution. Water therefore moves by osmosis from the soil solution into the xylem cells to push fluid up the stem. This creates root pressure. Root pressure is only important as a method of moving water into the leaves of young plants at the beginning of the growing season. Usually, it does not provide a large enough force for moving water to the top of tall plants, but root pressure can be significant in tropical rainforests where the atmosphere is permanently saturated.

49 Where in the root is the partially permeable membrane for the osmotic movement of water that brings about root pressure?

key facts

- Water moves through a plant, from the soil to the air, along a water potential gradient.
- The air has a much more negative water potential than the water in the soil.
- Water moves from the roots to the leaves mainly through the xylem vessels.
- The cohesion–tension mechanism describes how water is pulled up through the xylem to replace the water lost by evaporation from the leaves.
- The energy that drives the movement of water is heat from the Sun, which causes water in the apoplast of the leaf to evaporate.
- As water is pulled up the xylem, water is drawn across the root cortex to replace it.
- Water is drawn out of the soil solution because the root epidermal cells have a more negative water potential than the soil solution.
- In young plants, root pressure is the mechanism for moving water up to the developing leaves.

examination questions

1 The diagram shows structures involved in gaseous exchange in insects.

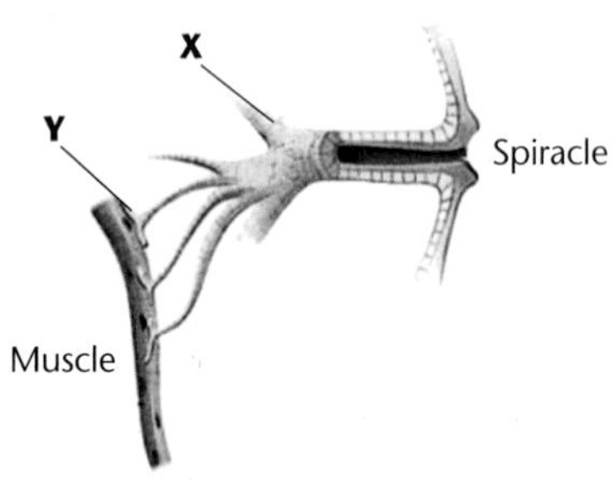

a Name the structures labelled **X** and **Y**. (2)

b Describe how oxygen reaches the muscle. (2)

c The spiracle has valves which can open or close. Explain the importance of these valves in the survival of terrestrial insects. (3)

Total 7

2 In fish, the flow of water over the gills and the flow of blood through the gills are in opposite directions (countercurrent flow) rather than in the same direction (parallel flow). The graph shows the effect of countercurrent flow on the oxygen saturation at different distances along a secondary gill lamella.

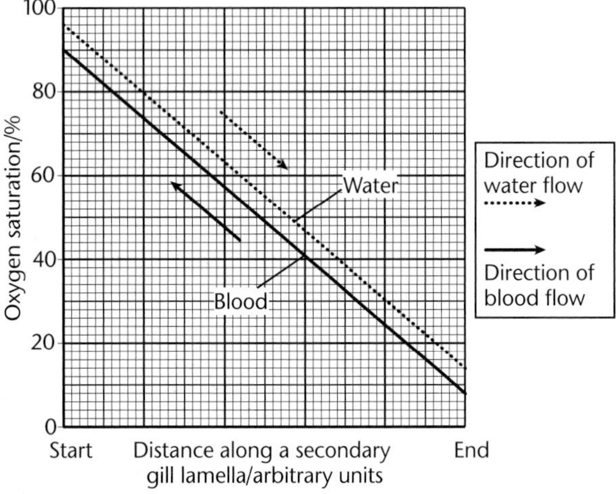

a What percentage of oxygen is removed from the water using this method of gas exchange? (1)

b Explain why countercurrent flow is more efficient for gas exchange than parallel flow. (2)

c The rate of diffusion across a membrane is related to a number of factors. These include surface area, membrane thickness and the difference in concentration on either side of the membrane. Use ticks to complete the table to show the effect of an increase in each factor on the rate of gas exchange

Factor	Increase	No effect	Decrease
Surface area			
Membrane thickness			
Difference in concentration			

(2)

Total 5

AQA, January 2002, Unit 1, Question 4

3

a Define:

i tissue

ii organ

iii system. (3)

b Single-celled organisms do not need a gas-exchange system to obtain oxygen. Explain why. (2)

c Explain why large animals need transport systems. (3)

d A walrus is a large animal that lives in the arctic.

An adult male has a length of approximately 4 m and a mass of approximately 1500 kg.
Use this data and information from the drawing to explain how a walrus is able to survive in very cold conditions. (3)

Total 11

4 The graph shows the oxygen haemoglobin dissociation curves for three species of fish.

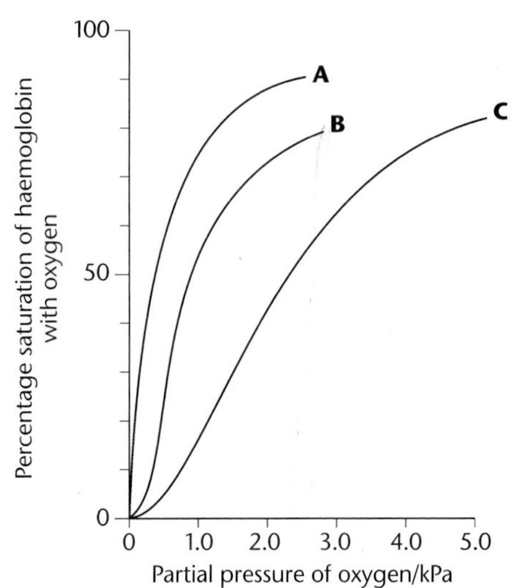

a Species A lives in water containing a low partial pressure of oxygen. Species **C** lives in water with a high partial pressure of oxygen. The oxygen haemoglobin dissociation curve for species A is to the left of the curve for species **C**. Explain the advantage to species A of having haemoglobin with a curve in this position. (3)

b Species **A** and **B** live in the same place but **B** is more active. Suggest an advantage to **B** of having an oxygen haemoglobin dissociation curve to the right of that for A. (2)

Total 5

AQA, June 2006, Unit 3, Question 7

5 The chart below shows the change in the speed of flow and pressure of blood from the start of the aorta into the capillaries.

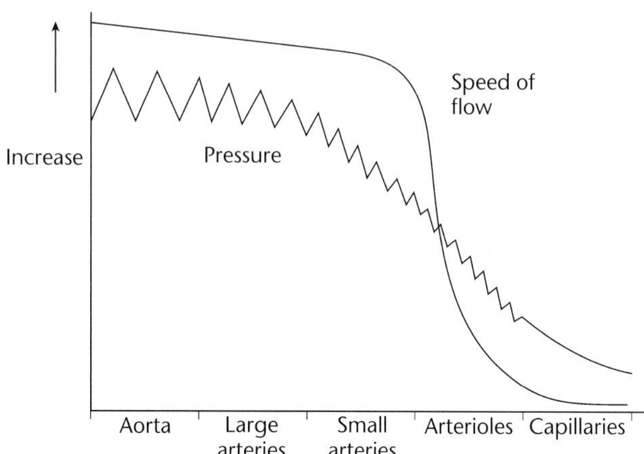

a Describe and explain the changes in the speed of flow of the blood shown in the chart. (2)

b Explain how the structure of the arteries reduces fluctuations in pressure. (2)

c Explain how the structure of capillaries is related to their function. (2)

d In one cardiac cycle, the volume of blood flowing out of the heart along the pulmonary artery is the same as the volume of blood returning along the pulmonary vein. Explain why the volumes are the same although the speed of flow in the artery is greater than in the vein. (1)

Total 7

AQA, June 2006, Unit 3, Question 6

6 The diagram shows vessels in a small piece of tissue from a mammal. The chart shows the hydrostatic pressure of the blood as it flows through the capillary.

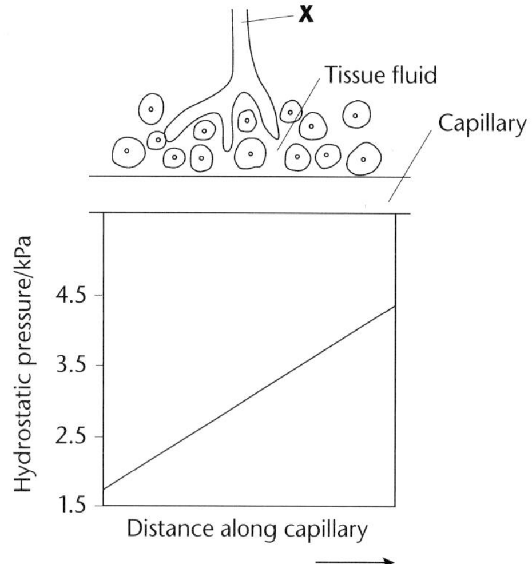

a Name the fluid contained in vessel **X**. (1)

b Draw an arrow on the capillary to show the direction of the flow of blood. Describe the evidence from the chart to support your answer. (1)

c Describe and explain how water is exchanged between the blood and tissue fluid as blood flows along the capillary. (4)

d Shrews are small mammals. Their tissues have a much higher respiration rate than human tissues. The graph below shows the position of the oxygen haemoglobin dissociation curves for a shrew and a human.

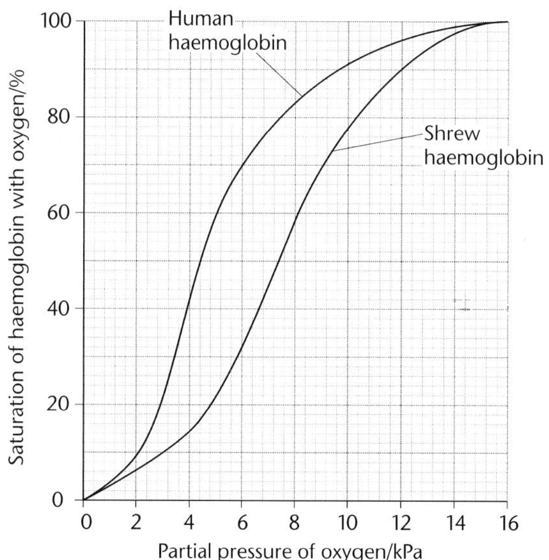

Explain the advantage to the shrew of the position of the curve being different from that of a human. (3)

Total 9

AQA, January 2006, Unit 3, Question 3

7 The volumes of water absorbed by the roots of a plant and lost by transpiration were measured over periods of 4 hours during one day. The bar chart below shows the results.

a i Describe the changes in the volumes of water absorbed and transpired between midnight and 1600. (2)

ii Explain these changes in the volumes. (2)

b Use your knowledge of the cohesion-tension theory to explain how water in the xylem in the roots moves up the stem. (4)

Total 8

AQA, June 2006, Unit 3, Question 3

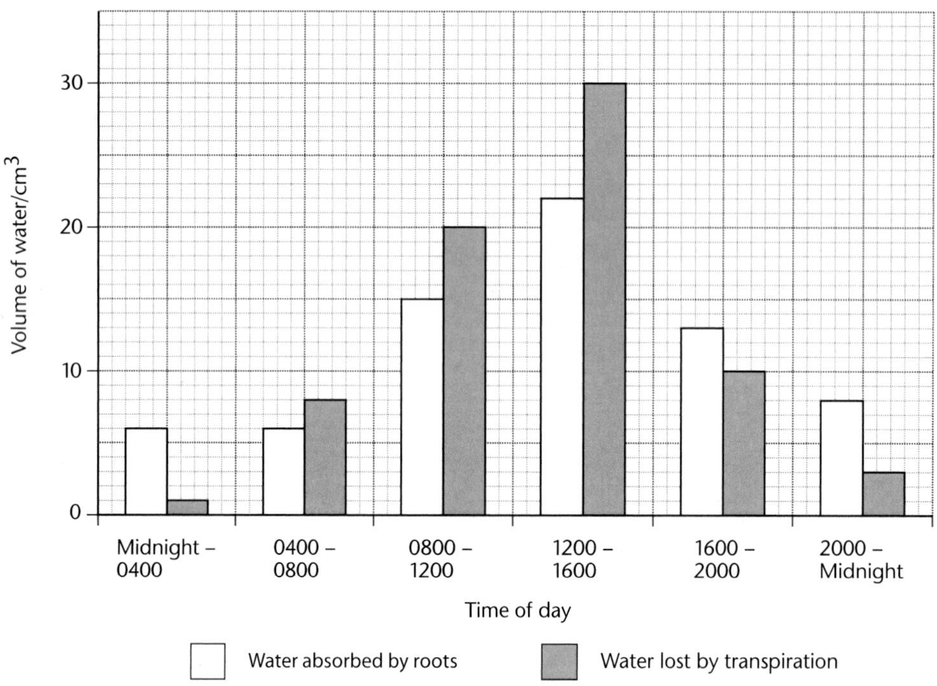

8 The diagram shows a section through a plant root.

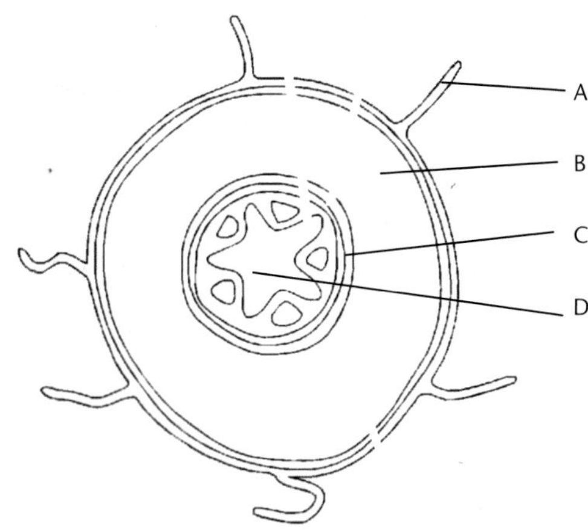

a Name the parts labelled **A–D.** (4)

b Explain how water is absorbed from the soil by structure **A**. (2)

c Through which system is most water transported across the tissue labelled **B**? (1)

d Explain the role of the tissue labelled **C**. (2)

Total 9

Greening the desert

Marula is a large deciduous tree found in Southern Africa. The tree is highly prized by local people for its fruits. The fruits are plum-sized with a sweet-sour flesh that can be eaten fresh or used to prepare juices and alcoholic drinks. Researchers introduced Marula trees into different sites in the Negev desert in Israel to evaluate its growth under different conditions, as shown in the table below. Thirty 1-year-old plants were planted at each of four sites.

The plants were drip fertigated every 1 or 2 days in summer and every 3–5 days in winter. In the fourth year the amount of water supplied was determined by the evaporation rate and varied from 17 m^{-3} per tree per year at Besor and Ramat to 25 m^{-3} at Qetura and Neot. The table shows the mean dimensions of the trees after 4 years.

The graph shows the development of new leaves in the fourth year. The percentage of the tree canopy with new leaves was estimated visually at the middle of each month on the following scale:

0 = no growth	2 = 20–60%
1 = less than 20%	3 = 60–100%

3.3 billion hectares of previously useful land are now classified as desert. If we can understand how plants manage their water economy we might be able to reclaim some of the land.

The above is an extract from research on introducing new crops into the desert (Adapted from A. Nerd and Y. Mizrahi 2003, pp. 496–9 in *New Crops* by Janick and Simon (eds), Wiley, New York).

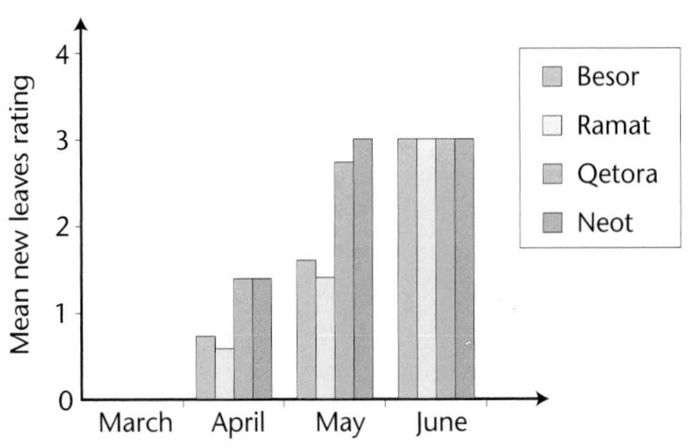

A1 The salinity (amount of mineral ions) of soil water is measured using electrical conductivity (EC).

a What is the unit of conductivity?

b Suggest why electrical conductivity is used to measure salinity rather than chemical methods.

c Suggest what is meant by 'fertigated'.

A2 Suggest why the Marula trees at Qetura and Neot were given much more water than those at Besor and Ramat.

A3

a What was the height after 4 years of the tallest tree at Qetura?

b What is the general relationship between the dimensions of the trees after 4 years and the conditions under which they were grown?

c Use data from the table to suggest why there was such a big difference in dimensions of trees grown at Qetura and Neot.

A4 The appearance of new leaves showed different patterns at Besor and Neot.

a Describe the patterns for the two sites.

b Suggest an explanation for the different patterns.

Site	Temperature extemes	Water supply	Height of tree/cm	Trunk circumference/cm
Besor	Moderate	Fresh EC 1 dS m^{-1}	533 + 60	50 + 2
Ramat	Sub-freezing in winter	Brackish EC 3.5 dS m^{-1}	290 + 25	54 + 3
Qetora	High summer temperatures, warm winters	Brackish EC 3.5 – 4.5 dS m^{-1} Ratio of Ca^{2+} to Na^{+} 1.2	620 + 30	58 + 2
Neot	High summer temperatures, warm winters	Brackish EC 3.5 – 4.5 dS m^{-1} Ratio of Ca^{2+} to Na^{+} 0.8	413 + 39	40 + 4

13 Classification

The giant squid was thought to be a mythical creature until fishermen caught one just off the coast of New Zealand in the 1980s.

Scientists have so far given a name to about 2 million species of living organisms. Over half of these are insects. In Great Britain alone there are over 3500 different species of beetle. In the rainforests of the world there may well be millions of still unnamed species; one estimate says that there could be 28 million species still to find.

You may wonder how it is possible to estimate the number of unknown and unnamed species. One way is to collect, say, a hundred different species of insect from a sample area of rainforest, or even from one species of rainforest tree. The insects are then checked against the databases of organisations such as the Natural History Museum in London to see whether they have been described and named previously. From the proportion of unnamed species in the sample it is possible to calculate the likely number of unknown species in rainforests. Although most of the new discoveries are small and apparently insignificant, there are still occasional discoveries of much larger organisms in remote situations. For example, two new species of mammal, the Giant Muntjac and the Vuquang Ox, were found in the forests of Southeast Asia in the 1990s.

Studies made in other habitats and of other invertebrates, plants and microorganisms, such as bacteria, reveal our staggering ignorance about the vast majority of inhabitants of the planet. For example, we know little about the deep oceans, which may well conceal many species of fish and large invertebrates, such as types of squid, previously unknown to science. Huge numbers of species are destined to disappear without humans having any knowledge of them. Too often we have come to regret the havoc caused by moving a species into a new environment, such as introducing rabbits to Australia. We know even less about the complex relationships between different species, and how the results of human activities, such as global warming, may disrupt these relationships.

In this chapter we explain how vast numbers of species have evolved and also explain the system that biologists use to give them names and describe how they are related to one another. Classifying this biodiversity is a first step in understanding our environment and is important in enabling humans to live in harmony with the natural world, as we shall see in later chapters.

13.1 What is a species?

A Great Dane and a Yorkshire terrier.

We can recognise a dog, a pig and a human as being different **species** because they look completely different. However, difference in appearance is not enough to distinguish one species from another. The two dogs in the photograph are far from being lookalikes, yet they belong to the same species. We describe them as different breeds. Their distinctive features are the result of **artificial selection**, which is a process very similar to **natural selection**. Dogs with specific features have been chosen as mating partners so that those features are retained in the resulting litter of puppies. For example, the Great Dane was bred for hunting large animals such as wild boar and deer stags, and mating pairs were chosen for their size and strength. The terrier was selected for its ability to follow small prey into underground burrows. Despite the great difference in size, the two dogs would mate happily with each other given half a chance. The puppies that would result from such a mating would be mongrels – dogs showing a mixture of the features of the two breeds.

In some instances, different species can look very similar, for example the chiffchaff and willow warbler. These are difficult to tell apart, even for experienced birdwatchers. They do, however, have quite different songs, so the birds themselves are not confused. Although they may live together in the same wood, they do not interbreed.

These two examples lead us to one of the most important features of a species: members of the same species can interbreed. There are examples, however, where members of different species mate and produce offspring. A mule is the result of a mating between a male donkey and a female horse. Donkeys and horses would not normally mate in the wild, but do so when they are put in the same field. Genetically, donkeys and horses are quite distinct. Wild horses have 66 chromosomes but donkeys have only 62. While the mule offspring are generally healthy, they have one important deficiency; they are infertile, and so are unable to breed. They are an example of a hybrid (i.e. the offspring of closely related species or breeds).

A chiffchaff.

A willow warbler.

1 Use your understanding of **meiosis** to explain why mules are sterile.

A mule – a cross between a male donkey and a female horse.

A donkey.

Defining a species

From the examples we have just seen, we could say that members of a species interbreed and produce fertile offspring. But it is difficult to define a species with precision, particularly if their mating habits are not well known. Visible features are not reliable – many species of insects and roundworms, for example, can only be distinguished by careful microscopic examination, and some microorganisms can only be distinguished by biochemical tests. In practice, we have to take all these aspects into account. A good working definition is therefore as follows.

A species is a group of organisms that:

- have similar physical, biochemical and behavioural features
- can interbreed to produce fertile offspring
- do not normally interbreed with any other group of organisms.

How can new species develop?

Imagine that you can observe a species of animal over many thousands or even millions of years. You would almost certainly notice significant changes. The animals may increase in size, alter their diet, improve their camouflage and so on. As a result of random variation and natural selection, they will adapt to the conditions in which they live. The unsuitable, or unfit, individuals will be weeded out, and with them the less successful alleles. In effect, the species will move with the times.

It is important to realise that this process is random and is not necessarily 'progress'. It is also possible for changes to be reversed. A larger animal may have an advantage when food is plentiful, but it might not do so well after a drought when food is in short supply. In this situation, the smaller animals would become more successful again. There is no destination, no perfect form, towards which a species is moving.

What if, instead of being able to observe the changes, you could obtain fossil remains from stages many years apart? Animals from different times might look so different that you would think they were different species. In fact, each fossil represents a stage in a continuous process. Often it is difficult to pinpoint a specific time.

This fossil ammonite shows the appearance of the species at one point in time. Ammonites were common in the oceans for well over 100 million years. They were rather like modern squids, but with coiled shells that were about 2 metres in diameter in the largest specimens. Although we recognise many different species from their fossils, the fossil record also shows that some types changed very little over long periods of time.

2 Suggest why some ammonite species may have changed very little over long periods.

Fig. 1 Ammonites

An evolutionary time sequence of ammonite genera

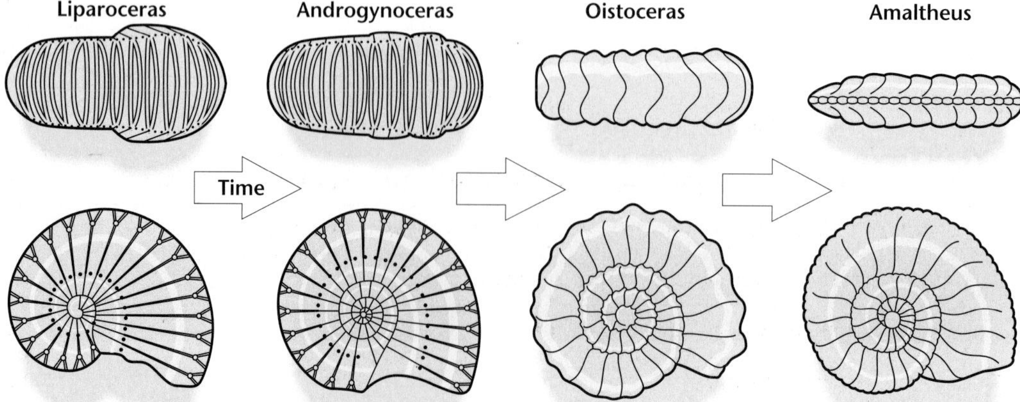

13.2 Classifying species

Classifying organisms involves more than giving them a name and then producing a catalogue. It helps biologists to understand the relationships between organisms, and to keep track of the changes that are occurring as a result of human pressures on habitats. The basic system of classification, which is called **taxonomy**, was devised before ideas of natural selection and speciation were developed.

Linnaeus and taxonomy

Carl Linnaeus, a Swedish biologist, devised the taxonomic system we use today in the eighteenth century. He classified living things into groups based on obvious similarities. The large groups, such as 'plants' and 'animals', were subdivided into smaller groups that showed even closer similarities. For example birds are clearly animals and not plants, but they are obviously similar in that they all have wings and feathers, features that distinguish them from other animals. Such a system in which large groups are split into smaller and smaller groups is called a **hierarchy**.

These are the basics of a taxonomic hierarchy.

- Organisms are classified into groups, which are further subdivided into yet more groups.
- Organisms are classified on the basis of similar or shared features. These features may be obvious, or they may be features determined by biochemical or genetic tests.
- There is no overlap beween the groups; an organism must be either a bird or a reptile, for example; it cannot fall halfway between the two.

Table 1 shows the groups, each called a **taxon**, that we use to classify living organisms. In order to describe a species precisely, biologists use both the **genus** and **species** name; this naming is called the **binomial system**. These names are international, understood by scientists in all countries in the world, and are often based on Latin or Greek. The binomial name for the seven-spot ladybird, for example, is *Coccinella septempunctata* (the Latin meaning 'bright red with seven spots').

Notice that a binomial name is always printed in italics, and that the genus name starts with a capital letter but the species name (second word) has a small letter.

The process of classification

Originally organisms were classified according to similarities in their appearance. We now know we cannot always rely on this. Species have arisen through natural selection operating over millions of years. Organisms have developed with many different adaptations. Some species will have arisen quite recently from a common ancestor, and they will share many features and genes. They can therefore be considered as close relatives. The common ancestor will itself have originated by a similar process. You can represent this process as a branching tree, with its main boughs and trunk going far back into geological time (Fig. 2 overleaf).

Table 1 Overview of how organisms are classified

Taxon	Description	Example
Kingdom	The largest group. Living organisms are divided into five kingdoms.	Animals
Phylum	Group of organisms that share a common body plan, such as having an external skeleton made of chitin and jointed limbs.	Arthropods
Class	A major group within a phylum, e.g. all the arthropods with 3 pairs of jointed legs.	Insects
Order	A subset of a class, with similar features, such as all the beetle-like insects.	Beetles
Family	A group containing organisms with very similar features.	Ladybirds
Genus	A clearly closely related group within a family.	*Coccinella*
Species	A specific type of organism.	Seven-spot ladybird

Fig. 2 Branching speciation tree

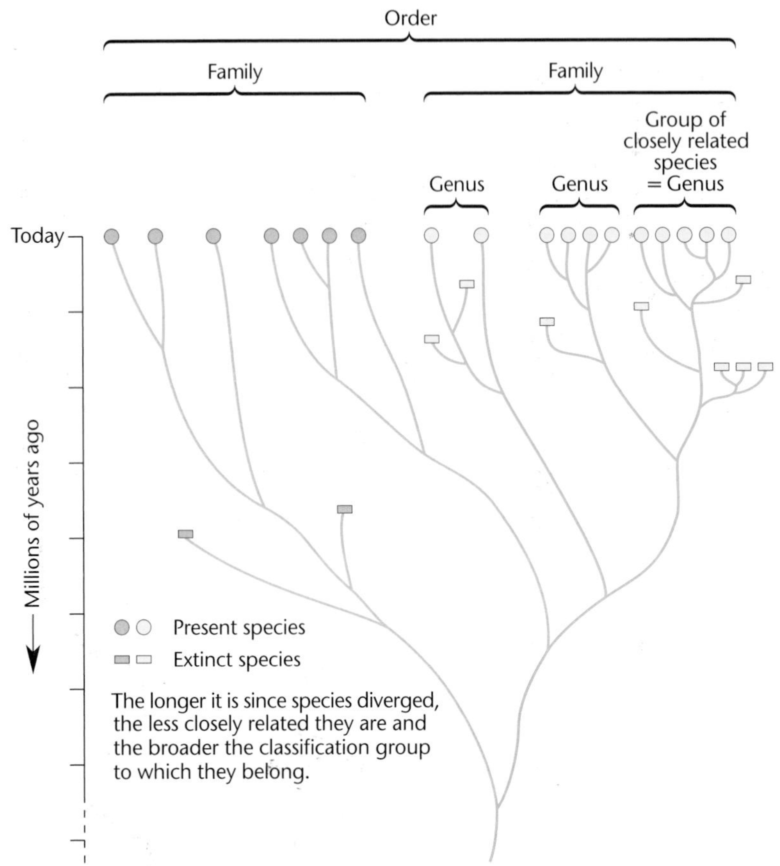

Modern biologists aim to reflect the ancestry of species when they classify organisms. This system of classification is said to be **phylogenetic** – it takes account of evolutionary history. To obtain evidence about evolutionary relationships, taxonomists study not only the anatomy of organisms and the fossil record, but also the structure of their proteins and DNA. New research means that detailed classification of some species is being revised as new information about how those species have evolved comes to light. However, it is never possible to be certain about the actual pathway of evolution, and scientists often disagree about how a particular species should be classified.

key facts

- A **species** is a group of organisms that generally are similar in appearance and that normally only interbreed with each other, producing fertile offspring.
- Species change over time as a result of **natural selection**.
- Evolutionary change over millions of years has resulted in huge numbers of different species.
- The classification of organisms is based on their presumed evolutionary history. This is called **phylogenetic classification**.
- Closely related species that diverged relatively recently in evolution are grouped together in one **genus**.
- Genera are grouped together into larger groups (**taxa**) called families. Increasingly large groupings comprise a **hierarchy**, with the largest being a kingdom.
- The order of the taxa from largest to smallest is: **kingdom, phylum, class, order, family, genus, species**.

13.2 Genetic comparisons

The famous anthropologist Dr Jane Goodall communicates with a baby chimpanzee.

Dr Jane Goodall is an expert on chimpanzees. Here she is communicating with a baby chimpanzee. But, how similar are we to our close relations, chimpanzees? Comparison of the **DNA** and proteins of humans and chimpanzees shows that our chemistry is very similar to theirs. The DNA base sequence that can be directly compared between chimpanzees and humans is almost 99% identical. At the protein level, 29% of genes code for the same amino sequences in chimpanzees and humans. In fact, the typical human protein has accumulated just one unique change since chimpanzees and humans diverged from a common ancestor about 6 million years ago.

To put this into perspective, the number of genetic differences between humans and chimpanzees is approximately 60 times less than that seen between humans and mice, and about 10 times less than between the mice and rats. But the number of genetic differences between a human and a chimp is about 10 times more than between any two humans. Similarities between the DNA of two species can be evaluated using the technique of DNA hybridisation.

DNA hybridisation

DNA hybridisation allows scientists to compare the total DNA of two organisms. Each DNA molecule is made of two strands. If the strands are heated, they will separate; as they cool, the nucleotides will bond back together again.

Fig. 3 DNA hybridisation

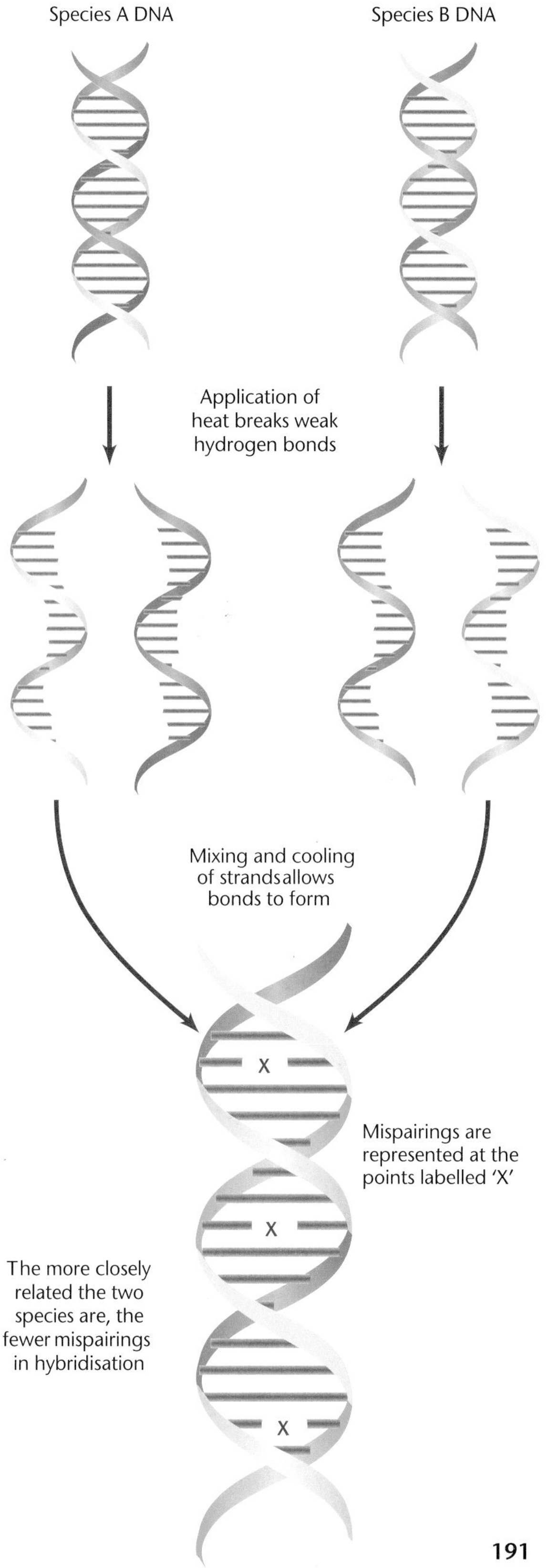

Species A DNA

Species B DNA

Application of heat breaks weak hydrogen bonds

Mixing and cooling of strandsallows bonds to form

X

Mispairings are represented at the points labelled 'X'

X

The more closely related the two species are, the fewer mispairings in hybridisation

X

To compare different species, scientists cut the DNA of the species into small segments, separate the strands, and mix the DNA together. When the two species' DNA bonds together – a process called hybridisation – the match between the two strands will not be perfect since there are genetic differences between the species. The more imperfect the match, the fewer bonds between the two strands. These DNA molecules can be broken into two chains with just a little heat. The closer the match, the higher the temperature needed to separate the strands again.

Fig. 4 shows a phylogenetic tree of the higher primates based on DNA hybridisation.

Fig. 4 Phylogenetic tree of primates based on DNA hybridisation

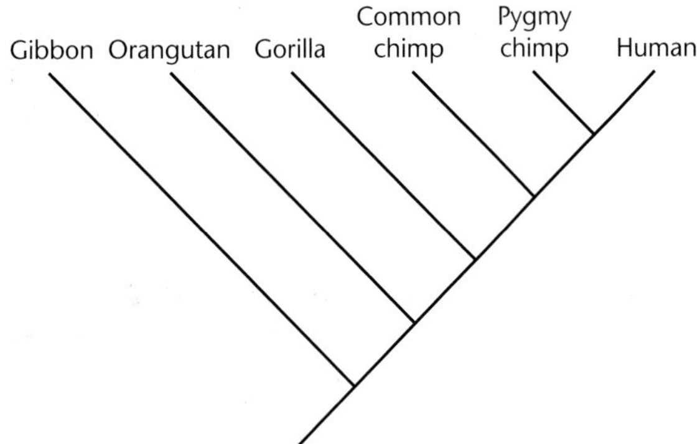

3 What type of bonds in the DNA are broken by heating?

4 What causes the two strands to join together again?

5 Did gorillas evolve from orangutans? Explain the reason for your answer.

DNA hybridisation technique

The main steps in the technique are as follows.

- The total DNA is extracted from the cells of each species and purified.
- The DNA is heated so that it becomes single strands (**ssDNA**).
- The ssDNA of species A is made radioactive.
- The radioactive ssDNA is then allowed to rehybridise with non-radioactive ssDNA of the same species (A). It is also allowed to hybridise with the ssDNA of species B. These rehybridised mixtures are called **DNA duplexes (dsDNA)**.
- After hybridisation is complete, the mixtures (A/A) and (A/B) are individually heated in small (2–3 °C) increments. At each higher temperature, a sample is analysed for the amount of radioactive strands (A) that have separated from the dsDNA.
- A graph showing the percentage of ssDNA at each temperature is drawn.
- The temperature at which 50% of the dsDNA have been denatured (**$T_{50}H$**) is determined.
- The lower the $T_{50}H$, the less hybridisation has occurred, and the more different the DNA sequences of A and B are.

A team led by scientists at The Royal Botanic Gardens, Kew, have recently devised a new classification of flowering plant families, based entirely on differences between DNA base sequences.

The scientists chose three genes found in all plants. They then chose 565 plant species to represent all the world's flowering plant families. For each plant, the DNA base sequences of the three genes were determined, and the sequences were compared using computer analysis. The result was a huge 'family tree' of plants, with branches showing how species have separated into natural groups. This new classification of plant families represents evolutionary relationships better than any other before it. Fig. 5 shows part of their classification.

6 Why is the ssDNA of species A made radioactive?

7 Explain why it is necessary to draw a graph of the results.

Fig. 5 A classification of flowering plants based on DNA base sequencing

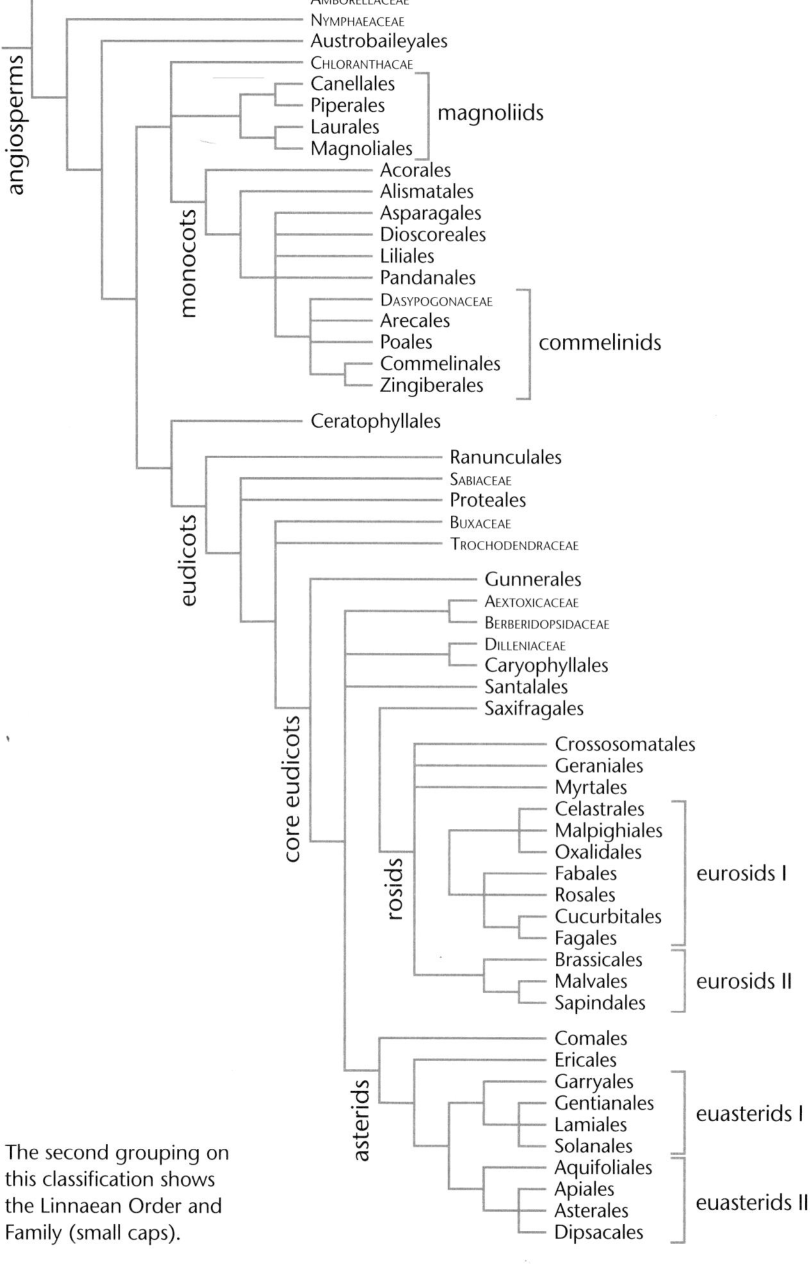

The second grouping on this classification shows the Linnaean Order and Family (small caps).

8 The magnolids, which include the magnolia tree, are said to be a primitive group of plants. Use information from the diagram to explain why.

Table 2 Number of differences in the amino acid sequence in the haemoglobin beta chain between the animal and a human

Animal	Number of differences
Gorilla	1
Gibbon	2
Rhesus monkey	8
Dog	15
Horse	25
Mouse	27
Grey kangaroo	38
Chicken	45
Frog	67

9 What is the relationship between the number of differences in amino acid sequence of the beta chain and closeness of kinship?

Amino acid sequencing

Haemoglobin is a protein found in a wide range of animal species. The beta chain of amino acids contains 146 amino acids in most species. Scientists have worked out the sequence of amino acids in the beta chains of many species. Table 2 shows some of their results.

Cytochrome c is a protein molecule found in the mitochondria of every eukaryotic cell, including all animals and plants. The amino acid sequences of many of these have been determined, and comparing them shows how closely species are related.

Human cytochrome c contains 104 amino acids; 37 of these have been found at equivalent positions in every cytochrome c that has been sequenced. Scientists assume that these molecules have descended from a precursor cytochrome in a primitive microbe that existed over 2 billion years ago.

Table 3 overleaf shows the N-terminal 22 amino acid residues of human cytochrome c. The corresponding sequences from five other organisms are aligned beneath. A dash indicates that the amino acid is the same one found at that position in the human molecule.

Table 3 Differences in the amino acid sequences of cytochrome c between humans and other species

Organism **Sequence of amino acid in terminal part of cytochrome c molecule**

Organism	1					6				10				14			17	18		20		
Human	Gly	Asp	Val	Glu	Lys	Gly	Lys	Lys	Ile	Phe	Ile	Met	Lys	Cys	Ser	Gln	Cys	His	Thr	Val	Glu	Lys
Pig	–	–	–	–	–	–	–	–	–	–	Val	Gln	–	–	Ala	–	–	–	–	–	–	–
Chicken	–	–	Ile	–	–	–	–	–	–	–	Val	Gln	–	–	–	–	–	–	–	–	–	–
Drosophila	–	–	–	–	–	–	–	–	Leu		Val	Gln	Arg		Ala	–	–	–	–	–	–	Ala
Wheat	–	Asn	Pro	Asp	Ala	–	Ala	–	–	–	Lys	Thr	–	–	Ala	–	–	–	–	–	Asp	Ala
Yeast	–	Ser	Ala	Lys	–	–	Ala	Thr	Leu	–	Lys	Thr	Arg	–	Glu	Leu	–	–	–	–	–	–

Table 4 Amino acid sequence in dogfish cytochrome c

Dogfish – – – – – – – – Val – Val Gln – – Ala – – – – – – Asn

10 Some columns in Table 3 are numbered. What do the numbered columns have in common?

11 Table 4 is data for dogfish cytochrome c. Between which two organisms in Table 3 should the data for dogfish be placed?

Immunological comparisons

You learned in Chapter 7 that **antigens** stimulate the production of **antibodies** by the body's immune system. A different antibody is produced for each antigen. Proteins are antigens. **Monoclonal antibodies** can be used to identify variations in protein structure. A variant of a protein will not bind to the monoclonal antibody stimulated by the original protein.

12 Why will the variant protein not bind to monoclo0nal antibodies produced in response to the original protein?

13 What causes a variation in the structure of a protein?

key facts

- Comparing the sequence of bases in DNA is used to elucidate relationships between organisms. The fewer the differences in base sequence, the more closely related the organisms.
- Similarities between the total DNA of organisms are determined by DNA hybridisation techniques.
- A new classification system for plants is based on data from the above techniques.
- Similarities and differences in the amino acid sequence of proteins such as haemoglobin and cytochrome c are used to elucidate relationships between organisms.
- Immunological comparisons use monoclonal antibodies to compare variations in proteins.

13.3 Courtship

Courtship is behaviour used to attract a mate. Courtship enables animals to:

- recognise their own species;
- approach each other closely without triggering aggression;
- choose a strong and healthy mate;
- form a pair bond and synchronise breeding behaviour.

Who makes the ideal mate?

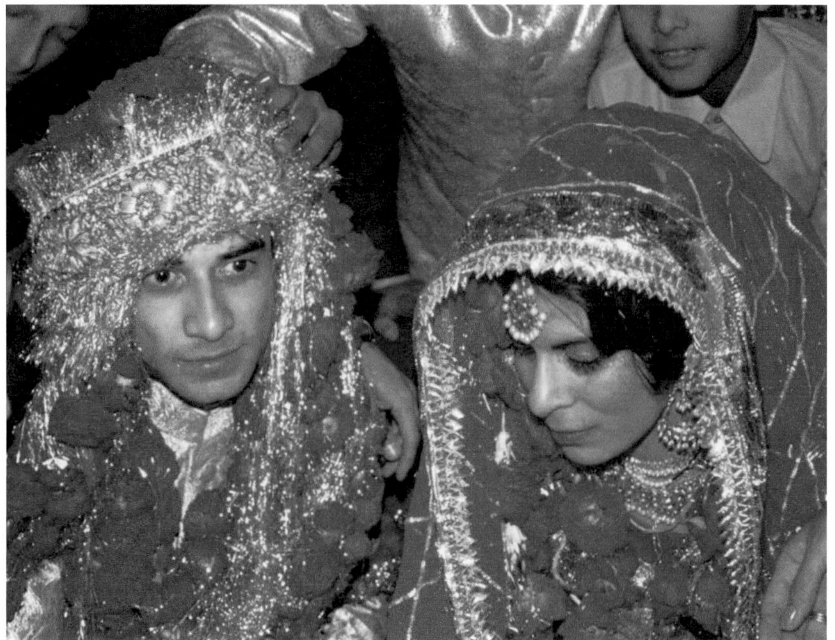

Species recognition

Every animal species has a different courtship display that helps it to recognise and attract a member of its own species of the opposite sex. Species recognition is especially important in closely related animals where a mistake could easily be made. Many kinds of ducks pair up on the wintering grounds where several similar species flock together. Each duck species has a different iridescent flash of colour in a bar on the wing, by which it can be recognised. Female ducks need to be camouflaged when sitting on their eggs, so their colour is dull and the wing bar is visible only in flight. Male ducks do not incubate the eggs and many have evolved spectacular plumage and courtship displays. However, male ducks on isolated islands tend to be dull in colour like the female. This is probably because only one species usually colonises each island, so species recognition is not a problem. Camouflage is then more important than having bright colours.

Insects such as butterflies also use brightly coloured patterns as recognition signals, while fireflies use characteristically timed sequences of flashes at night. Species recognition can also be by sound or smell: the calls of frogs and crickets differ between species, and many insects use chemical signals called pheromones for recognition.

The bird of paradise shows off its tail in courtship display.

Cuttlefish use rippling colour changes to attract a mate.

What qualities do we look for in a partner?

how science works

Stickleback courtship pattern

Male sticklebacks have a red patch during the breeding season. It is claimed that male sticklebacks will attack any red object regardless of its shape.

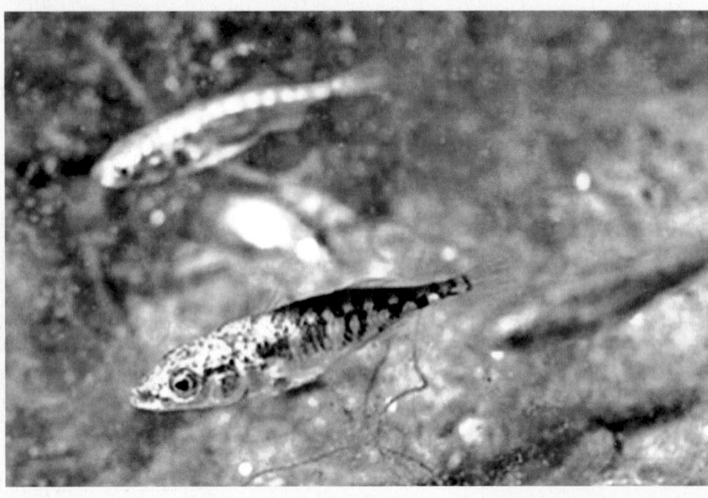

14 Explain how you could use models of sticklebacks to test this idea.

key facts

- Courtship enables animals to:
  - recognise their own species
  - approach each other closely without triggering aggression
  - choose a strong and healthy mate
  - form a pair bond and synchronise breeding behaviour.
- Every animal species has a different courtship display that helps it to recognise and attract a member of its own species of the opposite sex.

examination questions

1

a **i** What is meant by DNA hybridisation? (2)

ii Outline the steps in the procedure. (4)

b Scientists researching the classification of birds used the criteria shown in the table.

Group	Difference in temperature for 50% hybridisation in °C
Class	31–33
Order	20–22
Family	9–11

Explain what is shown by this data. (2)

c DNA hybridisation techniques have led to changes in the classification of birds. One example of this is the classification of pelicans. Until recently, pelicans were thought to be closely related to cormorants, boobies and gannets. DNA hybridisation shows the gannets are most closely related to shoebills.
Suggest why pelicans were thought to be closely related to cormorants, boobies and gannets until recently. (2)

Total 10

2 The table shows the percentage similarity in the sequence of amino acids in the beta haemoglobin chain of six different animals.

	% similarity in beta haemoglobin chains					
	Indian short nosed fruit bat	**California big-eared bat**	**White stork**	**Domestic pigeon**	**Lowland gorilla**	**Horse**
Indian short nosed fruit bat	100	89.7	68.5	69.9	85.6	84.9
California big-eared bat		100	66.4	67.1	84.2	82.9
White stork			100	86.3	68.5	65.8
Domestic pigeon				100	68.5	69.2
Lowland gorilla					100	82.9
Horse						100

Use evidence from the table to explain whether bats are most closely related to gorillas or to birds. (3)

Total 3

3 The diagram shows courtship behaviour in the stickleback.

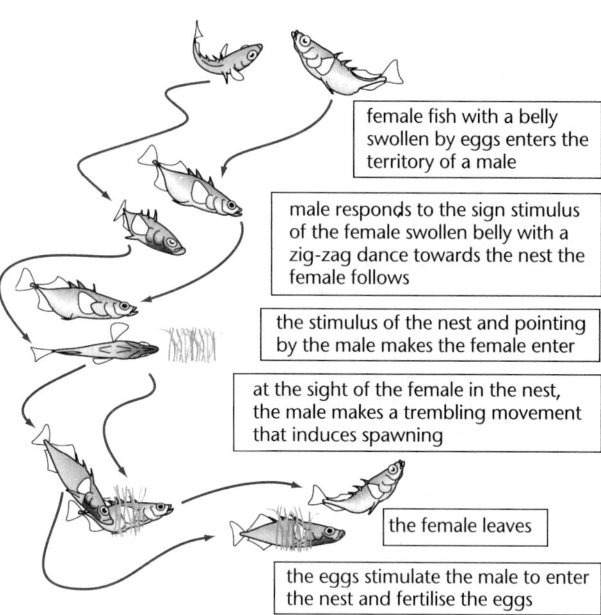

female fish with a belly swollen by eggs enters the territory of a male
male responds to the sign stimulus of the female swollen belly with a zig-zag dance towards the nest the female follows
the stimulus of the nest and pointing by the male makes the female enter
at the sight of the female in the nest, the male makes a trembling movement that induces spawning
the female leaves
the eggs stimulate the male to enter the nest and fertilise the eggs

Use information from the diagram to explain how courtship leads to successful reproduction in the stickleback. (6)

Total 6

4

a The table below shows the classification of *Fucus vesiculosus*. Enter the three missing groups in the table.

Kingdom	Protoctista
	Phaeophyta
Class	Phyophyceae
	Fucales
Family	Fucaceae
	Fucus
Species	*vesiculosus*

(2)

b Samples of DNA were removed from three species of *Fucus*. The DNA in each sample was separated into its two strands. This single-stranded DNA was then mixed with single-stranded DNA from another sample, either from the same species or from a different species. This allowed sections of DNA with complementary base sequences to join together to form 'new' double-stranded sections. The percentage of double-stranded DNA resulting is shown in the following table.

examination questions

Fucus species from which DNA was taken, separated into strands, and mixed together		Percentage of double-stranded DNA
F. vesiculosus	*F. vesiculosus*	99.8
F. vesiculosus	*F. serratus*	81.3
F. vesiculosus	*F. spiralis*	85.4
F. serratus	*F. spiralis*	94.6
F. spiralis	*F. spiralis*	99.9

Which **two** of these species seem to be the most closely related? Explain your answer. (3)

Total 5

AQA, January 2002, Unit 4, Question 6

5 Read the following passage.

Higher animals seem never to have evolved cellulases. Ruminants, such as cattle, deer and camels, house certain types of bacteria in a four-chambered stomach. Cellulose is digested by the prokaryotes, which possess cellulases. The rabbit, *Oryctolagus cuniculus*, has, like the horse, an enlarged caecum in which cellulose breakdown occurs, also brought about by resident bacteria. Cellulose is also digested in the intestine of the garden snail, *Helix pomatia*, although, in experiments, extracts of the digestive gland lack cellulase activity. A protoctist, *Trichonympha*, inhabits the intestine of wood-eating termites, and is responsible for their being able to digest cellulose in their diet of wood. Bacteria are also present here, but do not produce a cellulase. Instead they seem to be nitrogen-fixing, which may explain how termites are able to thrive on a diet so low in nitrogen-containing compounds.

a Give the names of **two** genera mentioned in the passage. (1)

b Organisms from three kingdoms are mentioned in the passage. Name each of these kingdoms. (2)

c A garden snail was mated with a snail from a different continent. Offspring were produced. What would you need to know about these offspring to be certain that both parents belonged to the same species? (1)

Total 4

AQA, June 2002, Unit 4, Question 2

6

a The mammals form a class called the Mammalia within the animal kingdom. The grey wolf is a species of mammal. The figure shows the groups within the Mammalia to which the wolf (labelled **W**) belongs.

i Label the figure to show the names of the groups. (2)

ii The lion, *Panthera leo*, belongs to another group in the Carnivora, called the Felidae. Add this information to the figure using the letter L to represent the lion species. (1)

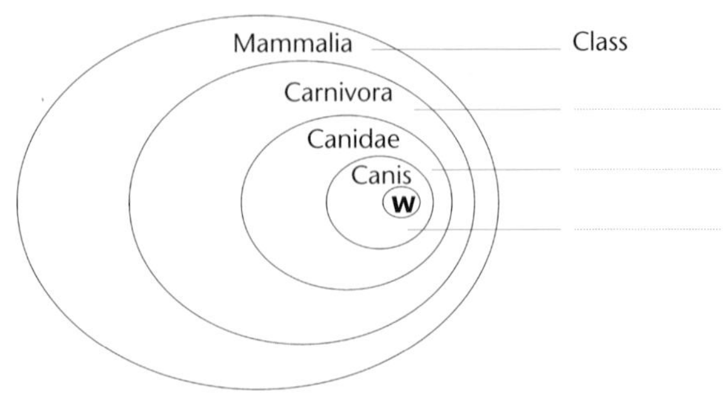

b The diagrams below show two systems of classification of mammals: Simple hierarchy (left), Phylogenetic system (right).

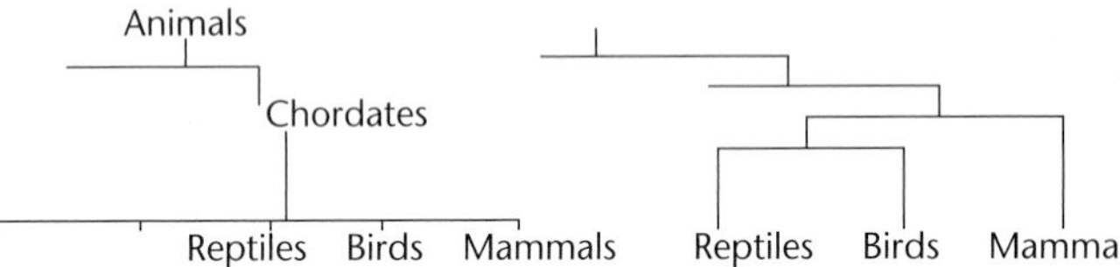

i What is meant by a hierarchy? (1)

ii By reference to the figures above, explain how a phylogenetic system differs from a simple hierarchy. (3)

Total 7

AQA, June 2004, Unit 4, Question 4

7

a The cheetah, *Acinonyx jubatus*, and other cat species belong to the family Felidae. Complete the table to show the classification of the cheetah.

Kingdom	Animalia
	Chordata
	Mammalia
	Carnivora
Family	Felidae
Genus	

(2)

b This system of classification is described as hierarchical. Explain what is meant by a hierarchical classification. (1)

c Despite differences in form, leopards, tigers and lions are classified as different species of the same genus. Cheetahs, although similar in form to leopards, are classified in a different genus.

i Describe **one** way by which different species may be distinguished. (1)

ii Suggest **two** other sources of evidence which scientists may have used to classify cheetahs and leopards in different genera. (2)

Total 6

AQA, June 2006, Unit 4, Question 3

The naked ape

Humans differ from other primates in that most parts of the body are hairless and they walk entirely on two legs, as well as having a much larger brain. Biologists have speculated for many years about how these features evolved, and several hypotheses have been put forward. This article describes one suggestion.

Men and women walk tall to stay cool. Researchers have shown that a two-legged gait allowed early humans to screen out most of the searing heat of our equatorial African homeland.

Instead of beating down on our backs, the Sun's rays would have fallen vertically on our heads – a far smaller area. We would have kept cooler and have needed less precious water, giving us a kick-start to biological supremacy.

In addition, cool humans would no longer have needed the thick pelts that shielded other animals on the grassland plains from the Sun. So we became naked apes.

Also, 'By walking on two feet, humans developed the animal world's most powerful cooling system, and that allowed us to acquire large brains', said Professor Peter Wheeler of John Moore's University in Liverpool.

To demonstrate this theory, Professor Wheeler made a model of *Australopithecus*, which he named Boris. This model could be bent to either an upright or a four-legged position while the movements of the Sun overhead were simulated. A camera recorded the area exposed to the rays of the Sun. 'We discovered that there was a 60% reduction in the heat received by Boris in his two-legged position compared with his quadruped posture', reported the Professor. He also calculated that far more of the body is raised away from the heat radiating from the hot ground, and that only just over half as much water would be lost in sweat.

Source: adapted from an article by Robin M McKie, copyright Guardian News & Media Ltd 1993.

An artist's impression of an *Australopithecus* mother and child.

A1 If Professor Wheeler's hypothesis is correct, how might natural selection have resulted in hominids losing most of their body hair and becoming 'naked'?

A2 Suggest why a larger brain would require a more efficient cooling system.

A3 Another hypothesis is that early hominids evolved around shores, lagoons or lakes, where they moved around in shallow water. It would be an advantage to be able to stand upright and keep the head above water. Support for this idea derives from pollen analysis of deposits from 2 to 3 million years ago, which indicates that most of Africa was forested at the time and that grassland plains were uncommon. Suggest how each of the following hominid features might support this 'aquatic origin' hypothesis.

a There is a much thicker layer of fat under the skin than in other primates.

b The teeth are smaller and less suited to chewing tough vegetation.

c Newborn babies can float and swim.

d Humans produce large quantities of dilute urine.

e Humans can deliberately stop breathing for short periods, whereas in other primates breathing is not under voluntary control.

A4 Suggest some counter-arguments to the hypothesis suggested in question A3.

14 Antibiotics

It was once thought that the discovery of antibiotics would mean an end to infectious diseases. But bacteria are fighting back – many strains are now resistant to most antibiotics.

HOSPITAL SUPERBUG KILLED MY WIFE

"All I want to do is prevent the same thing happening to other people."

Patricia Arthur, 73, who went into St Helier hospital in Carshalton, Surrey, had a benign obstruction to her bowel. She underwent a small 10-minute operation to clear it and remained in hospital to recover.

But Dr Arthur, of New Malden, Surrey, said his wife started to feel unwell a few hours after returning home.

"I couldn't get her back into hospital the next morning because they said her bed was taken, but they said she could go in as a private patient, so I paid £2,000 for that."

Dr Arthur had requested a blood test for MRSA infection during his wife's first stay in hospital after becoming aware that other patients on the ward were infected. It confirmed his wife had MRSA.

He said patients with MRSA had been on his wife's ward during her first stay.

"In St Helier, they had a principle where MRSA patients were put on special wards, but the wards were all full. After she was re-admitted, she deteriorated over the next three days, and died of septicaemia."

Her cause of death was registered as MRSA septicaemia (blood poisoning).

Instead of flowers for his wife's funeral, Dr Arthur asked for contributions to the hospital hygiene campaign he is setting up.

He said: "I've never been interested in litigation. I can't do anything about what's happened. All I want to do is prevent the same thing happening to other people. I want cleanliness to be the most important thing in any doctor's training, and in the retraining of older doctors. It should be absolutely supreme for all staff."

hsw 14.1 Antibiotics

Alexander Fleming discovered penicillin.

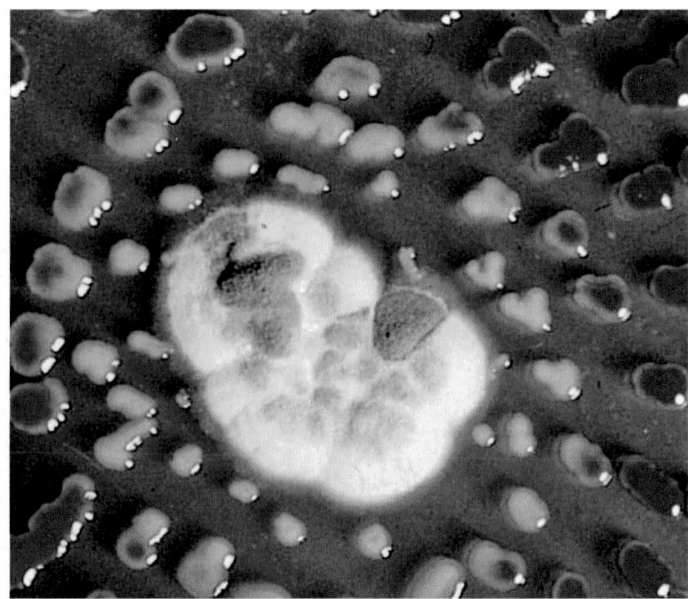

A substance released from the *Penicillium* fungus at the centre of the Petri dish kills bacterial colonies (grey) close to the fungus, but colonies further away (red) are still viable.

One day in 1928, Alexander Fleming came back from a 2-week holiday and noticed mould growing in a Petri dish containing a bacterial culture on a plate he had left lying on his desk. No bacteria grew around the mould. From this observation it occurred to him that the mould might be producing some substance that was killing the bacteria. He embarked on the series of experiments that led to the discovery of the first antibiotic for medical use, penicillin.

This is one of the most famous examples of a chance observation leading to a major scientific breakthrough. Fleming's brilliance was to realise the possible significance of his observation and to propose an explanation. A suggested explanation that can be tested by experiment is called a hypothesis. Fleming's hypothesis was that the mould secreted a substance that killed the bacteria in the dish. He set up experiments to determine whether the secretions really did kill bacteria, for example by taking samples of agar medium from around growing mould and testing their effect on bacterial cultures. This led to many more experiments with different substances extracted from the secretions, different types of bacteria and different concentrations. Once the antibacterial activity of the substance produced by the penicillin mould had been established, it was a long time before the penicillin could be used as a drug. Alexander Fleming had neither the expertise nor the interest to purify the substance and work on it further, and it was then up to other scientists, notably Howard Florey and Ernst Boris Chain, to carry the research to its next stage. Within 15 years of Fleming's initial observation, a vast pharmaceutical industry had grown up and antibiotics found widespread use against many previously incurable infections.

This little girl was seriously ill with a massive infection and would have died without antibiotics. She was one of the first people to receive penicillin. The second photograph was taken 6 weeks later.

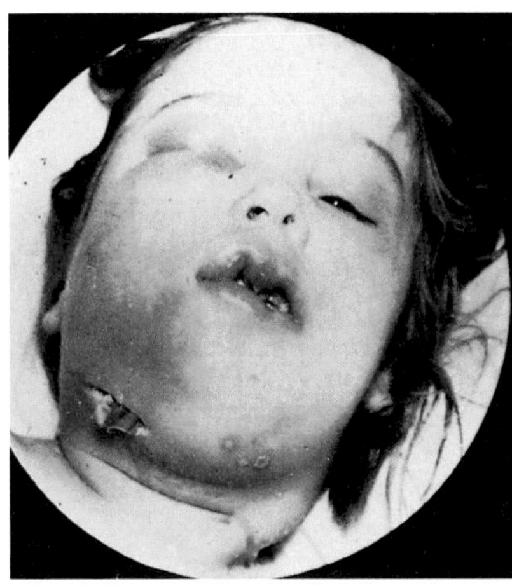

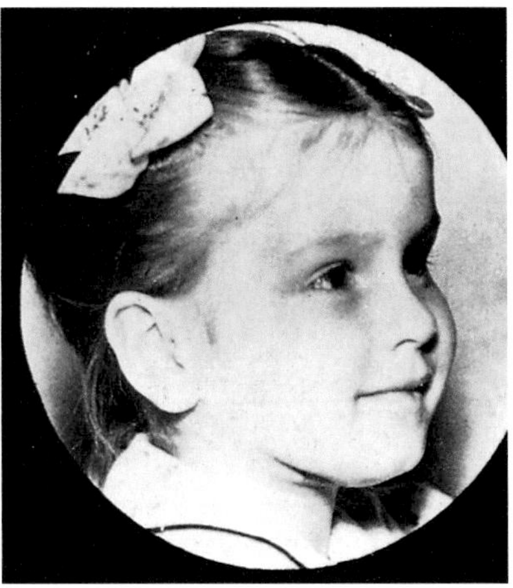

Many such infections that were once killers are now preventable by vaccine and treatable by **chemotherapy**. Chemotherapy is the use of chemical agents, including **antibiotics**, that are toxic to selected microorganisms but do not harm human cells. There are now about 50 antibiotics in common use – about 1% of these have been successfully isolated from their microbial source. Some are effective against only a few types of bacterium and are called **narrow-spectrum antibiotics**. Others kill or inhibit a wide range of bacteria and are called **broad-(wide) spectrum antibiotics**. Some antibiotics have been chemically altered to make them more effective – these are termed semi-synthetic antibiotics. Antibiotic effectiveness can be determined by disc diffusion tests.

Penicillin, ampicillin and the cephalosporins prevent bacterial growth by inhibiting cell wall synthesis. One function of the cell wall in bacteria is to prevent the cells bursting by taking in excess water via osmosis. Without the cell wall the bacterial cells undergo **osmotic lysis**.

Streptomycin, gentamicin and the tetracyclines bind across bacterial **ribosomes**, inhibiting protein synthesis. Bacterial cells are **prokaryotic** and the ribosomes are different from those in human cells, which are eukaryotic, so protein synthesis in the host cells is not inhibited.

Nucleic acid in prokaryotic cells is similar to nucleic acid in eukaryotic cells so antibiotics that inhibit the synthesis of nucleic acid are not as specific as some others. However, ciprofloxacin interferes with the replication and transcription of prokaryotic DNA, and rifampicin blocks the synthesis of prokaryotic RNA by binding to RNA polymerase. Polymyxin B binds to the cell membrane and alters its structure making it more permeable, leading to cell death.

Inhibition of prokaryotic cell metabolism is brought about by a number of antibiotics that act as **antimetabolites**. Antimetabolites inhibit key reactions in cells and result in a shortage of essential metabolites. For example, sulfonamides inhibit folic acid production in prokaryotic cells. As humans do not metabolise folic acid but must obtain it from food, sulfonamides have no detrimental effect on human metabolism.

It is more difficult to treat fungal infections than bacterial infections. This is because fungi are eukaryotic organisms and their cell processes are similar to those of human cells. This makes most antifungal agents toxic for humans, but

1 Why are humans not affected by the actions of:

a antibiotics?

b sulfonamides?

there are a few exceptions. Polyoxin D inhibits chitin synthase – an enzyme essential for the synthesis of fungal cell walls, which human cells do not have. Another useful antifungal agent is nystatin, which damages the fungal cell membrane so the cell contents leak out. It is used to treat skin infections.

Selecting a particular antibiotic or antifungal agent to treat a specific disease means considering

- how the drug works
- its side-effects
- possible reactions with other drugs that the patient might be taking
- whether the patient is pregnant, and so on.

2 Why are bacterial infections easier to treat than fungal infections?

14.2 Resistance to antibiotics

Fig. 1 Mechanisms of bacterial resistance to antibiotics

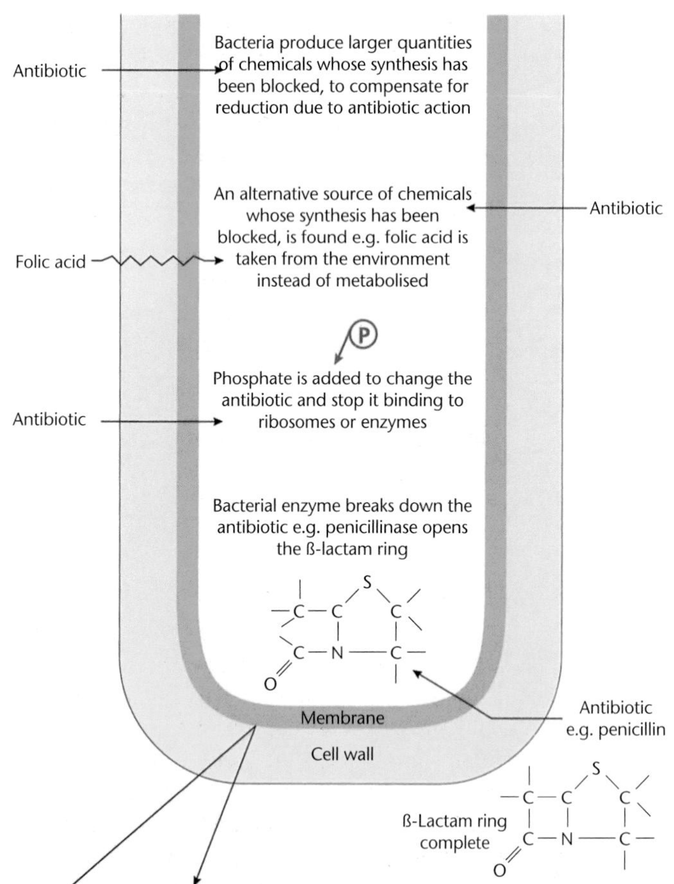

Antibiotic
Bacteria produce larger quantities of chemicals whose synthesis has been blocked, to compensate for reduction due to antibiotic action
An alternative source of chemicals whose synthesis has been blocked, is found e.g. folic acid is taken from the environment instead of metabolised
Antibiotic
Folic acid
P
Phosphate is added to change the antibiotic and stop it binding to ribosomes or enzymes
Antibiotic
Bacterial enzyme breaks down the antibiotic e.g. penicillinase opens the ß-lactam ring
Membrane
Cell wall
Antibiotic e.g. penicillin
ß-Lactam ring complete
Antibiotic is prevented from entering the cell e.g. by the outer envelope of Gram-negative bacteria or by a mutation causing decreased permeability of the cell membrane in *Neisseria meningitidis*

For as long as antibiotics have been used to treat bacterial infections, the bacteria have been fighting back. Not all bacteria are susceptible to all antibiotics. For example, it is pointless to use penicillin to treat an infection caused by bacteria that have a capsule, as the capsule prevents penicillin from entering the cell.

There are a number of different mechanisms by which bacteria are able to resist the action of antibiotics (Fig. 1).

Transmission of antibiotic resistance

Resistant strains have also developed among bacterial species that were generally susceptible to particular antibiotics. It is important to remember that antibiotics do not cause resistant bacteria. An antibiotic-resistant organism develops and is able to grow and reproduce successfully despite the presence of the antibiotic. In other words, the resistant bacterium is *selected*. Bacteria become antibiotic-resistant when they obtain the genes for drug resistance. There are two ways by which this can happen:

- spontaneous **mutation**
- transfer of genes for resistance from other bacteria.

Mutations in the bacterial chromosome do not occur very often but when they do, they can make the bacterium resistant to an antibiotic.

The most common mutations change the binding site for an antibiotic (for example, the ribosomes) so that the antibiotic cannot bind. These changes in the DNA can be passed on when the bacterial cell divides (**vertical gene transmission**) and also by conjugation (**horizontal gene transmission**). Many resistant mutants do not survive the normal defence mechanisms of the host, but some do. In a host that is being treated with an antibiotic, these surviving mutants are at a selective advantage because non-mutated bacteria are inhibited or killed by the antibiotic. The mutants can then grow and reproduce with much less competition for nutrients.

Besides the bacterial chromosome, bacteria may also contain genetic information in small circular pieces of DNA called **plasmids**. If a gene for antibiotic resistance is among the genetic information in a plasmid, it is called an **R plasmid**. Once a bacterial cell contains an R plasmid, the resistance gene on the plasmid can be transferred to other cells by **conjugation** (Fig. 2). Plasmids can cross species boundaries.

Fig. 2 Conjugation in bacteria with a plasmid

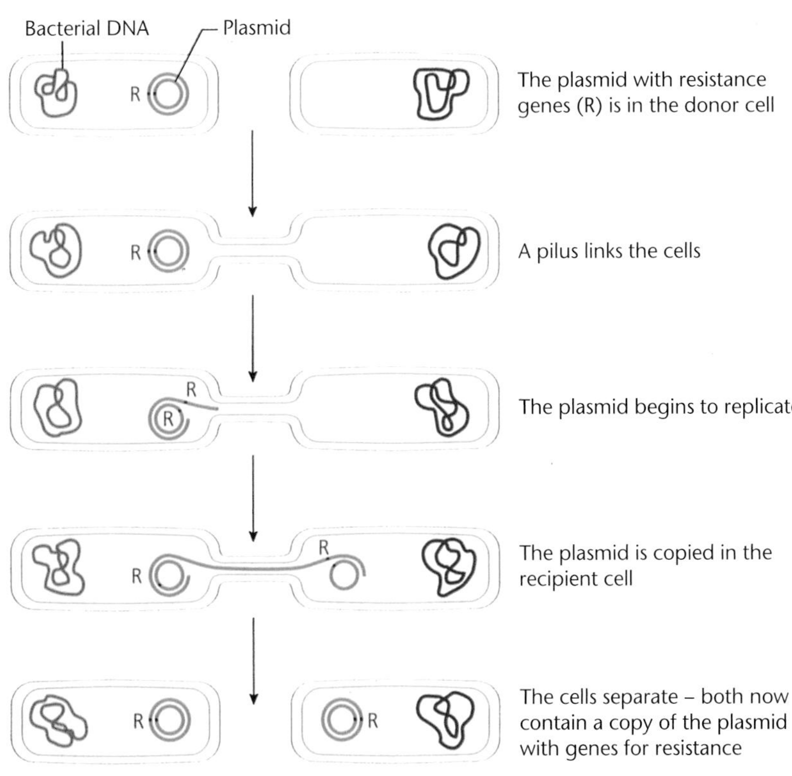
Bacterial DNA
Plasmid
R
The plasmid with resistance genes (R) is in the donor cell
R
A pilus links the cells
R
R
The plasmid begins to replicate
R
R
The plasmid is copied in the recipient cell
R
R
The cells separate – both now contain a copy of the plasmid with genes for resistance

> **3** Why are resistance genes in plasmids more of a problem to medical staff than resistance genes on the bacterial chromosome?

Resistant infections and super-infections

Many diseases that were once thought to be under control are now making a reappearance. This is because the pathogenic bacteria have developed genetic changes that allow them to survive antibiotic attack. The widespread and sometimes indiscriminate use of antibiotics for 40–50 years has provided an environment for the growth of resistant bacteria.

Hospitals are a focus for bacteria. People going in for treatment often have infectious diseases, and take the bacteria into hospital with them. Sometimes these bacteria include resistant strains that are at a selective advantage in hospitals because the generally high concentration of antibiotics allows the resistant strains to multiply as the rest fall victim to the effects of the drugs. Today, a major reservoir of genetically altered resistant strains is found in hospitals. This reservoir of resistant bacteria carries a pool of resistance genes in bacterial plasmids. A single plasmid can contain genes for resistance to more than one antibiotic. This means that resistance to several antibiotics can spread rapidly through the bacterial population. Pathogens with multiple resistance have caused difficult-to-treat infections called super-infections (Fig. 3 overleaf).

> **4** How does the indiscriminate use of an antibiotic lead to an increase in bacteria resistant to this antibiotic?

Where resistance seems likely, doctors now use combinations of two or more antibiotics to reduce the possibility of infectious bacteria surviving. Before antibiotics, tuberculosis (TB) was a serious disease for which there was no cure. Then antibiotics brought it under control and patients recovered. But the bacteria that cause TB have now developed resistance and TB has re-emerged as a serious infection. You are advised to read again the passage on resistant TB on page 78 in Chapter 5.

Fig. 3 How bacteria can become resistant to more than one antibiotic

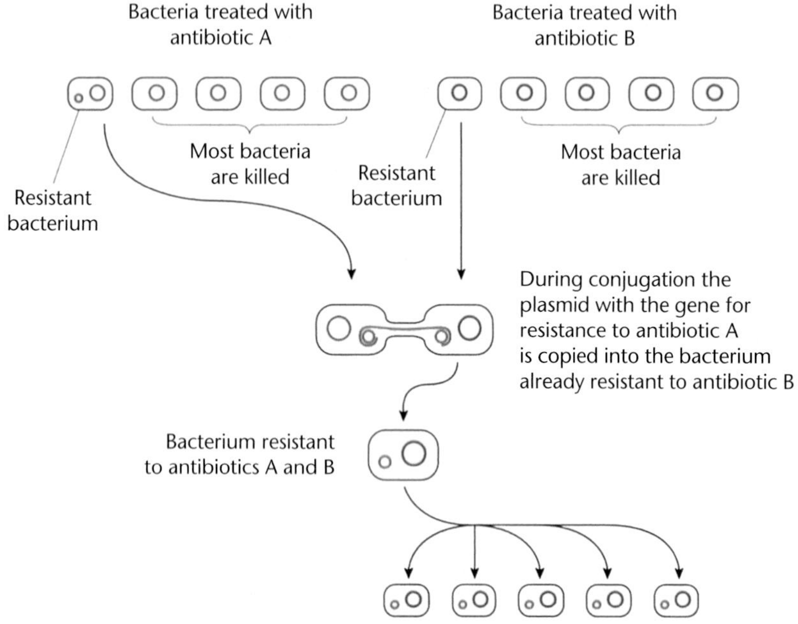
Bacteria treated with antibiotic A
Bacteria treated with antibiotic B
Most bacteria are killed
Resistant bacterium
Resistant bacterium
Most bacteria are killed
During conjugation the plasmid with the gene for resistance to antibiotic A is copied into the bacterium already resistant to antibiotic B
Bacterium resistant to antibiotics A and B
Resistant bacterium multiplies to form a new strain resistant to antibiotics A and B

MRSA

Staphylococci are a family of common bacteria. Many people naturally carry these bacteria in their throats, and they can cause a mild infection in a healthy patient. MRSA stands for methicillin-resistant *Staphylococcus aureus*, but is shorthand for any strain of *Staphylococcus* bacteria that is resistant to one or more conventional antibiotics. Antibiotics are not completely powerless against MRSA, but patients may require a much higher dose over a much longer period, or the use of an alternative antibiotic to which the bug has less resistance.

MRSA infections can cause a broad range of symptoms depending on the part of the body that is infected. These may be surgical wounds, burns, skin and blood. Infection often results in redness, swelling and tenderness at the site of infection. Some people carry MRSA without having any symptoms.

Why does MRSA exist?

The advice from doctors who give you antibiotics is always to finish the entire course – advice that many of us ignore. When you don't finish the course, there's a chance that you'll kill most of the bacteria, but not all of them – and the ones that survive are of course likely to be those that are most resistant to antibiotics. Over time, most of the *Staphylococcus* strains will carry resistance genes, and further mutations may add to their survival ability. Strains that manage to carry two or three resistance genes will have extraordinary powers of resistance to antibiotics.

Hospitals seem to be hotbeds for MRSA because so many different strains are being thrown together with so many doses of antibiotics, vastly accelerating this natural selection process.

Why is it so dangerous?

It is a fact of life that patients are at higher risk than normal of picking up a staphylococcus infection on hospital wards. This is for two reasons. Firstly, the people in hospitals tend to be older, sicker and weaker than the general population, making them more vulnerable to the infection.

Secondly, hospitals provide the perfect environment for the transmission of all manner of infections. Many people are in a relatively small area, and nurses and doctors risk carrying bacteria from patient to patient. Staphylococcal infections can be dangerous in weakened patients, particularly if they can't be cleared up quickly with antibiotics.

What can we do about it now?

One of the main reasons behind this swift evolution of bacteria into 'superbugs' is the overuse of antibiotics, in both human and veterinary medicine. Until recently, patients visiting their doctor with a viral infection might demand, and be given, a prescription for antibiotics – despite the fact that antibiotics have no effect on viruses. All those patients were doing was strengthening the communities of bacteria in their bodies. Doctors have now been told to reduce the prescribing of antibiotics.

Hygiene is another tried and tested way to protect the most vulnerable patients from the most dangerous strains. Doctors and nurses *must* wash their hands before treating any patient, so that they minimise the risk of transferring bacteria from patient to patient.

Government ministers are trying to improve overall standards of hygiene, perhaps by reintroducing the ward matron, with responsibility for cleanliness. New bedside phones for patients are being introduced that include speed dial buttons to alert staff to the need to deal with a hygiene problem. Whether a dirty ward rather than a dirty hand is a reservoir for *Staphylococci* is a matter of debate. But patients with MRSA are also increasingly being treated in isolation where possible. In the long run, many experts suggest that it may take a breakthrough akin to the discovery of penicillin before humans can regain a temporary upper hand over the bacteria again.

5 What is MRSA?

6 Why are hospitals likely to be a source of antibiotic-resistant bacteria?

7 How can the spread of MRSA in hospitals be reduced?

hsw

how science works

The graph and the passage give information about deaths involving *Staphylococcus aureus*.

Some of the recent increases in mentions of MRSA on death certificates may reflect improved levels of reporting, possibly brought about by the continued high public profile of the disease.

Most of the deaths involving *S. aureus* or MRSA were in the older age groups. In 2005, MRSA was involved in the deaths of 85 men and 702 women per million population aged 85 years and older. In the under 45-age group there were 1.1 and 0.8 deaths per million population for males and females, respectively.

8 Describe the trend in deaths involving *Staphylococcus aureus* between 1993 and 2005.

9 Describe how the proportion of deaths involving resistant forms of *S. aureus* changed between 1993 and 2005.

10 Are the data for resistant forms of *S. aureus* likely to be accurate for the earlier years in the data set? Explain the reasons for your answer.

11 Suggest reasons for the effect of age on the risk of death from MRSA.

Fig. 4 Number of death certificates mentioning *Staphylococcus aureus* infection

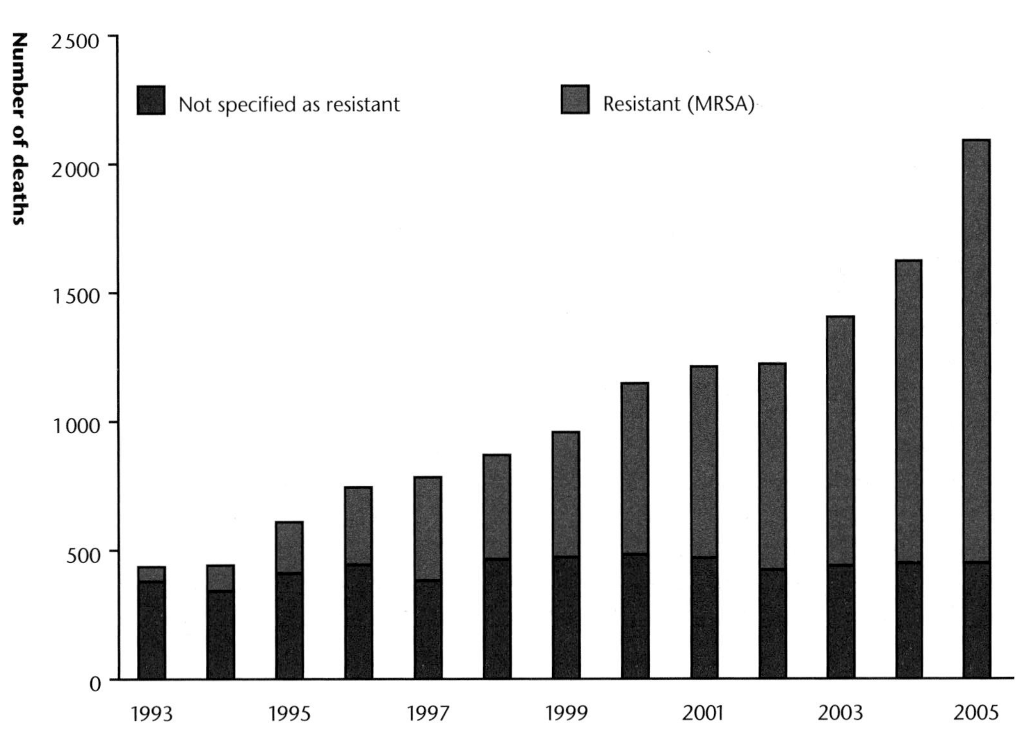

Fighting resistance

Although the search for new drugs is ongoing, merely looking for or designing new antibiotics is not enough to combat resistance. As fast as we develop a new antibiotic, resistance follows on through the combination of mutations and cross-species transfer of resistant genes. So, what other approaches to combat resistance are possible? Plasmid-based research is looking for ways to:

- eliminate R plasmids from bacteria
- prevent the expression of the resistance genes carried by the plasmids.

Ascorbic acid (vitamin C) seems to be able to remove the resistance without killing the bacteria. It is important not to kill the bacteria because doing so would encourage the natural selection of bacteria capable of resisting the ascorbic acid treatment. Other researchers are exploring how to stop the spread of plasmids between bacteria.

Another approach is to reconsider the way in which antibiotics are used. If we were more sparing in our use of them, they would remain effective for longer in cases where they are really needed. International cooperation is needed so that reservoirs of resistant strains do not build up in countries that do not control the use of antibiotics.

12 What ways of fighting bacterial resistance are being investigated?

key facts

- Penicillin was the first antibiotic. It was discovered in 1928 and was first used to treat soldiers in the early 1940s.
- Different antibiotics act on microbial cells in different ways: cell wall synthesis is inhibited by penicillin, resulting in osmotic lysis.
- Not all bacteria are susceptible to antibiotics. Some bacteria have always been resistant to some antibiotics; some bacteria develop resistance to some antibiotics
- Resistance to antibiotics develops when the bacteria obtain genes for resistance. This can happen by spontaneous mutation or by gaining the genes from other bacteria.
- The genes for resistance can exist on the bacterial chromosome or on R plasmids.
- The indiscriminate use of antibiotics for 40–50 years has provided an environment for the growth of antibiotic-resistant bacteria.
- MRSA stands for methicillin-resistant *Staphylococcus aureus*.
- MRSA is spread mainly by poor hygiene practices in hospitals.
- Resistance genes can be passed on by vertical transmission when a bacterium divides, or by horizontal transmission when bacteria conjugate. Plasmids can cross species boundaries.
- The reservoir of resistant bacteria found in hospitals carries resistance genes in bacterial plasmids. A single plasmid may contain genes for resistance to more than one antibiotic. Pathogens with multiple resistance cause super-infections.
- The fight against bacterial resistance includes: searching for new antibiotics; eliminating R plasmids from bacteria; preventing the expression of the resistance genes on R plasmids; stopping the spread of plasmids; reducing the prescribing of antibiotics.

examination questions

1

a Explain one way in which antibiotics kill bacteria. (3)

b Antibiotics do not kill viruses.

i Why are viruses not affected by antibiotics? (1)

ii Explain why doctors sometimes prescribe antibiotics to a patient suffering from a viral infection. (2)

c Explain the dangers of over-prescription of antibiotics. (4)

Total 10

2 Read the passage.

Scientists investigated the occurrence of MRSA skin infection in an American football team. These infections caused skin abscesses.

The scientists collected information about players' field positions, demographic characteristics, health care exposures, antibiotic use, close contact with other persons with skin infections, skin-abrasion management, hygiene practices, and use of saunas, whirlpools, and training and therapy equipment.

They also evaluated antibiotic use among the players by reviewing the team pharmacy log and calculating the average number of antibiotic prescriptions per player per year.

The scientists found that skin abrasions occurred frequently among players. Approximately two to three turf burns per week were acquired from sliding on the field during competition or practice. Players reported that abrasions were more frequent and severe when competition took place on artificial turf than when it took place on natural grass. Trainers, who provided wound care, did not have regular access to hand hygiene, and alcohol-based hand-hygiene products were not available near areas where wound care or physical therapy was provided. Towels were frequently shared on the field during practice and games, with as many as three players using the same towel. Players often did not shower before using communal whirlpools. At the training facility, weight training and therapy equipment was not routinely cleaned. According to the team pharmacy log for the 2002 football season, maintained at the training facility, a team player on average received 2.6 antibiotic prescriptions per year. This rate was greater than 10 times the rate among persons of the same age and sex in the general population (0.5 prescriptions per year). In their survey responses, approximately 60 percent of players indicated they had taken or received antibiotics during the 2003 football season.

The investigation revealed that a cluster of skin abscesses among professional football players was caused by an emerging MRSA clone.

a What was the dependent variable in this investigation? (1)

b Give three ways in which this MRSA strain might have been transmitted to players. (3)

c Suggest an explanation for the origin of this strain of MRSA. (3)

d Suggest how an 'emerging MRSA clone' is produced. (2)

Total 9

3

a What is an antibiotic? (1)

b What is meant by mutation? (2)

c Explain what is meant by

i vertical gene transmission (2)

ii horizontal gene transmission. (2)

Total 7

4 The graph shows the proportion of bacteria resistant to the antibiotic vancomycin in samples taken from patients in an intensive care ward.

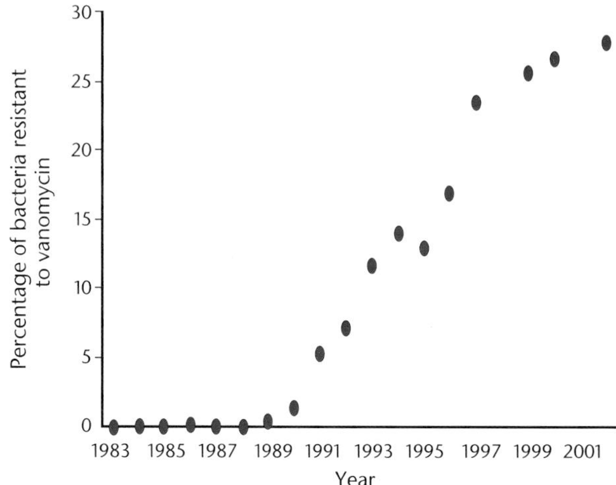

a Describe the trend shown by the data. (3)

b Suggest an explanation for this trend. (3)

c Due to the rise in the proportion of resistant bacteria it was decided to trial the use of a different antibiotic in the intensive care ward.

Discuss the ethical issues involved in a trial such as this. (4)

Total 10

examination questions

5 Penicillin was first used to treat infections in the 1940s.

a Describe how penicillin prevents the growth of bacteria. (1)

b The bacterium *Staphylococcus aureus* is a common cause of life-threatening infections. By the 1960s it had already become resistant to the antibiotic penicillin.

i Describe **one** mechanism of resistance to penicillin. (1)

ii Explain how *S. aureus* evolved resistance to penicillin. (3)

c If a patient fails to respond to treatment with penicillin, another antibiotic, called vancomycin, may be used. In 2002 a patient was found to be infected with a strain of *S. aureus* containing a gene that made it resistant to vancomycin. The same gene was found to be very common in bacteria of the species *Enterococcus faecalis* from the gut of the patient. Suggest how some of the *S. aureus* bacteria came to contain the vancomycin-resistance gene. (2)

Total 7

AQA, June 2005, Unit 7, Question 4

6 The diagram shows the structure of a bacterium and the different sites of action of two antibiotics.

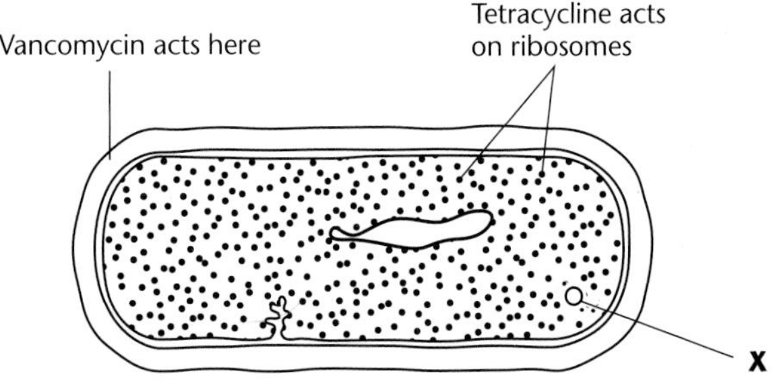

a Vancomycin is a *narrow* spectrum antibiotic. What is meant by a narrow spectrum antibiotic? (1)

b Using information in the diagram explain

i why vancomycin does **not** affect mammalian cells (1)

ii how tetracycline prevents bacterial growth. (1)

c Explain how the structure labelled **X** gives a bacterium resistance to an antibiotic. (2)

Total 5

AQA, January 2004, Unit 7, Question 2

7

a Describe **one** way in which antibiotics prevent the growth of bacteria. (1)

b Filter paper discs soaked in two types of antibiotic were placed on a lawn of bacteria growing in a Petri dish. The concentration of antibiotic dissolved in each disc is shown.

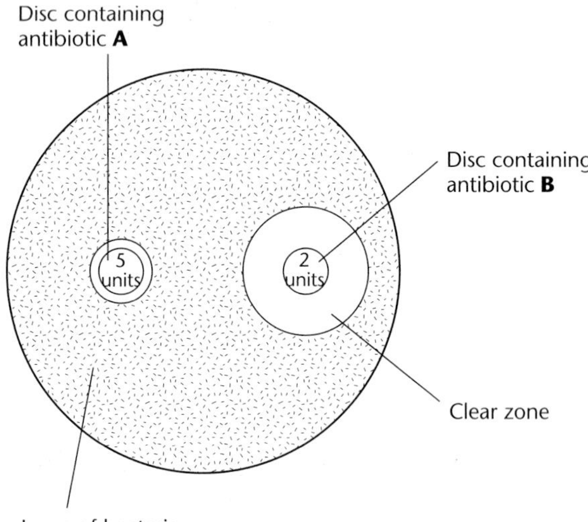

i Explain why clear zones formed around both filter paper discs. (1)

ii How many times more effective is antibiotic **B** than antibiotic **A?** Explain how you arrived at your answer. (2)

Total 4

AQA, June 2003, Unit 7, Question 5

Antibiotics and the common cold?

Trying to stop parents asking for antibiotics.

Most children will have between three and eight colds per year. Over half of patients seen by doctors for the common cold are given a prescription for antibiotics.

Many studies have investigated the efficacy of using antibiotics to treat the common cold in children.

The results of some of these studies are shown in the table below. Symptomatic treatment includes 'self-treatment' using medicines such as paracetamol, and decongestants.

A1 Explain the need for campaigns to reduce the number of prescriptions for antibiotics.

A2 Which study gave the most reliable data? Give the reason for your answer.

A3 Which studies gave the least valid data? Give the reason for your answer.

A4 What conclusions can be drawn from these studies?

A5 Evaluate the ethics involved in these studies.

Study	Number of children in study	Comparison groups	Outcome
A	2177	Penicillin or symptomatic treatment	Required return outpatient visit(s): penicillin group 26%; symptomatic-treatment group 20%
B	217	Antibiotic or placebo	Rate of all infectious complications: antibiotic group 15%; placebo group 15%
C	845	Antibiotics or symptomatic treatment	Rate of all infectious complications: antibiotic group 14%; symptomatic-treatment group 9%
D	781	Antibiotic or symptomatic treatment	Rate of complications (e.g. acute ear infection): antibiotic group 3.5%; symptomatic-treatment group 2.6%
E	261	Penicillin or placebo	Not improved or complications: antibiotic group 5%; placebo group 5%
F	212	Doxycycline or placebo	Runny nose at day 5: doxycycline group 14%; placebo group 30%
G	197	Amoxicillin, co-trimoxazole or placebo	Runny nose at day 8: amoxicillin 6%, cotrimoxazole 4%, placebo 15%. Normal activity at day 8: amoxicillin 89%, cotrimoxazole 95%, placebo 97%

A placebo is a tablet or medicine that looks like the active drug but does not contain any active ingredient. It is used in comparisons to determine how many patients would get better anyway without active treatment.

15 Biodiversity

Food production versus conservation

Modern supermarkets offer a massive range of foods. People have become used to being able to buy most types of high-quality fresh vegetables, fruit and meats all year round. Although food may seem expensive, the proportion of income that most people spend on food is much lower than it was 30 or 40 years ago. This extension of choice and quality has come about partly as a result of improved efficiency in agricultural practices and partly because modern transport means that products can be moved rapidly around the world.

A typical supermarket trolley with mixture of foods, mostly mass-produced and produced by intensive farming and processing.

A selection of organically grown vegetables, which are often relatively expensive, but which may be better for us and the environment.

Some of the improved efficiency in agriculture in the UK has been at the expense of wildlife and the environment. Politicians therefore face a dilemma. Most people, and that means most voters, want continued access to cheap food. On the other hand, there is increasingly vociferous campaigning for more organic farming and for conservation of the environment.

Populations of many previously common birds have declined steeply in recent years. The house sparrow was one of our most common birds, but it is a seedeater that depended for much of its food in winter on the seeds left behind after harvesting crops. This source has been greatly reduced by the practice of ploughing in autumn soon after harvest and planting winter wheat, which improves yields because it has a head start and grows more rapidly than spring-sown wheat.

Skylarks and lapwings both nest on the ground in long grass, and in many parts of the country their populations have plummeted because of the ploughing of grassland. Also, farmers often cut grass early to make silage, rather than leaving it until later in the summer and then harvesting the hay. This often destroys nests before the young have hatched.

Changes such as these in bird populations are often the most visible signs of much greater changes taking place in the ecosystems of the countryside.

Skylark (top), lapwing (below left) and house sparrow (below right).

15.1 How humans have shaped the countryside

Typical unspoilt British countryside, as most people perceive it, is an attractive patchwork of fields, growing a variety of crops and separated by hedges. But things have been changing over the past few decades – pasture fields have been ploughed, and the mixture of native grasses and wild flowers that used to thrive has been replaced by a single fast-growing species of grass. The demise of wild flower has also been hastened by regular use of fertilisers, which enhance crop growth but suppress recolonisation by wild flower species. Modern farming methods, which often remove hedges and give enormous areas of land over to a single crop, are bad news for animals too. More efficient harvesting machines mean that fewer seeds get left behind for birds and small mammals to feed on in winter, and autumn sowing of cereal crops brings earlier crops and earlier harvests, making it difficult for these animals to complete their breeding cycle in the fields. Birds like the skylark have disappeared from many areas.

Farmland in the Vale of Clwyd, North Wales, shows the typical patchwork apperance of countryside in Britain.

There has also been increased pressure on the variety of **habitat** as a result of building development, roads, drainage schemes, forestry and other human activities. The bar chart in Fig. 1 shows the habitat losses in various parts of just one English county. As you can see, there have been significant losses from a wide range of different types of habitat. In some areas, a high proportion of the habitat that was present only 20 years ago has now vanished.

Fig. 1 Loss of habitat in the English countryside between 1979 and 1997

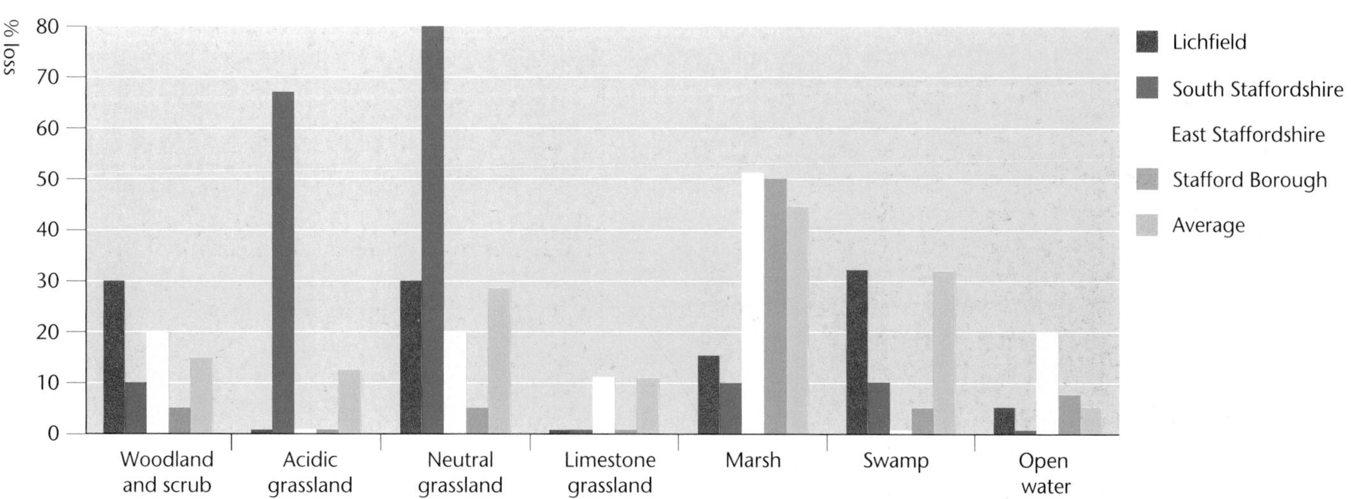

How can we balance such damaging effects on the environment with the need for a reliable and inexpensive food supply, increased demand for housing, areas for recreation and so on? In this chapter we look at some examples of the impact of farming on the environment, and consider how pressures for food production can be balanced against the demand for conservation.

Hedges and farming

A farmer's first priority is to make the best possible use of the land, either for growing crops or raising animals. Traditionally, farmers produced food for themselves and for the local community by growing a variety of different crops and keeping a range of different animals. In many parts of the world, subsistence farmers are still doing this. For this type of mixed farming, it is an advantage to divide the land into small fields to allow animals to be separated from crops, and for crops and stock to be rotated between fields. Hedges and walls are a commonplace feature of the landscape.

Today, there is increasing pressure on farmers to specialise and grow only a single crop, or maybe two or three, and to concentrate on one form of animal husbandry, such as dairy cattle or pig breeding. This is more economical because better use can be made of equipment, and organisations such as supermarkets have fewer farmers to negotiate purchases with. Also, transport is now much easier, so produce can easily be moved to more distant markets. Consequently, in many parts of the country, large areas are used for single crops, such as wheat or oil-seed rape. This is called **monoculture**. In these areas most of the hedges, walls and fences have been removed. Removing field boundaries offers several advantages to the farmer.

Wheat growing in a huge single field in the Vale of Pickering, North Yorkshire.

- The space previously taken up by hedges can be used for growing the crop, which increases the yield per unit area.
- There is no need to cut and maintain hedges, which reduces labour costs.
- It is easier to manoeuvre large machines, such as combine harvesters, which speeds up sowing seeds, harvesting and spraying.
- It removes the shading effect of tall hedges.
- It removes places that might harbour pests.

However, there are also disadvantages in taking away hedges, both to the farmer and to the environment as a whole.

- When not covered with growing crop, the soil is much more exposed to wind erosion, and, if sloping, to erosion by rainwater. In drier parts of the country there can be considerable loss of the more fertile topsoil.
- There is reduced flexibility of land use on farms that keep some animals, since fences have to be erected to keep them separate. There is also less shelter for the animals.
- There are far fewer habitats for wildlife. For instance, birds have little cover and few nesting sites and, with fewer plant species and insects, less food is available.
- Pressure to improve yields by cutting down competition with weeds leads to intensive use of **herbicides**. This further reduces the variety of food plants available for insects such as butterflies and for seed-eating birds and small mammals.
- **Predators** that normally keep insect pests under control lose their habitat, and crops are more exposed to large-scale infestations by pests. As a result, farmers apply large quantities of **pesticide**, which tends to reduce natural populations even further.
- Growing the same crop on land for several years reduces the availability of mineral ions at the soil depth of the crop roots. With no animals on the farm, **organic manure** is not easily available, so the amount of humus in the soil declines. This affects the soil structure, for example by reducing its water-holding capacity, and lowers the bacterial and earthworm content. The soil dries out easily and is therefore more susceptible to erosion. Monoculture farmers usually use large amounts of **artificial fertilisers** to counteract the loss of nutrients, but this does little to restore soil structure.

- Heavy machinery can compact soil and reduce the amount of air spaces, which ultimately slows down root growth.

Overall, therefore, monoculture can increase yields and offer significant economies in the short term; in the long term, however, productivity is liable to decrease. In the mid-west of the USA, where monocultural farming has been carried out on a huge scale for many years, the land in some areas has become so degraded that it is now an almost infertile desert.

A pine tree killed by the arid conditions of the encroaching desert in Colorado, USA.

hsw

how science works

The effect of hedges on crop yield

Figure 2 shows the effect of a 2-metre high hedge on crop yield and on some abiotic factors at various distances from the hedge.

Fig. 2 The effect of a high hedge on crop yield

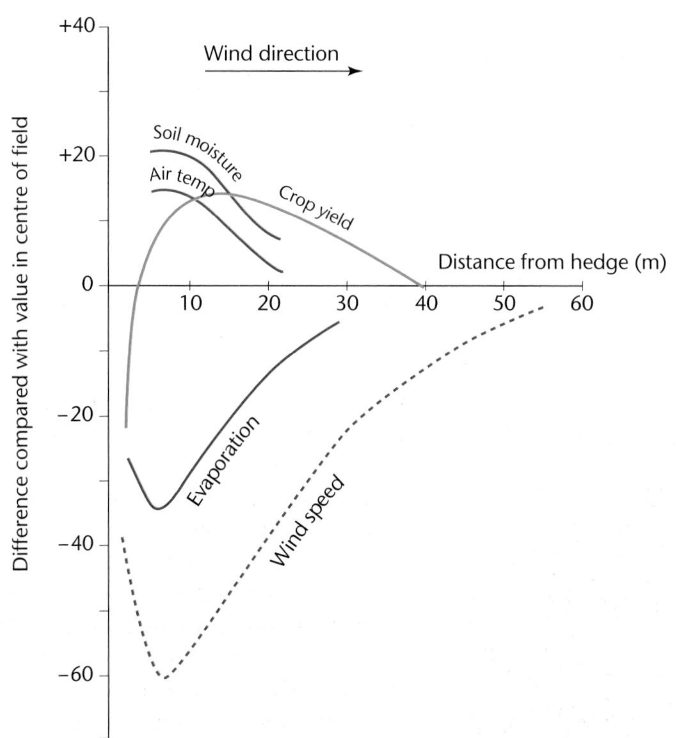

1 Explain the effects of the hedge on soil moisture and air temperature.

2 How far from the hedge is the greatest increase in crop yield?

3 Suggest why the yield close to the hedge is decreased.

4 What is the evidence from the graph that the net effect is a gain in yield in the area within 40 metres of the hedge?

5 Suggest why this gain in yield may not be found in practice.

Deforestation

International efforts are being made to conserve some of the world's increasingly endangered tropical rainforests. Tropical rainforests currently cover about 6% of the Earth's surface, yet they are thought to contain around half of the total number of species on Earth. Almost 75% of all known arthropods are found only in tropical rainforests. No one is sure why **biodiversity** in these forests is so high, but it is probably due at least in part to the ideal conditions for plant growth that are found there. High rainfall and high temperatures provide an environment in which many different plant species have evolved, and provide a wide range of niches for animals that interact with them.

Tropical rainforests are under considerable threat: it is estimated that at least 1% of these forests are being lost each year. Commercial logging companies cut down large areas of trees each year to sell as timber. Forest is also cut down and burnt to produce new land for growing crops. This happens both on a small scale – for example, in slash-and-burn systems of agriculture, where families burn an area of forest and then grow crops on the land for a few years before moving on – and on a very large one, where forest is destroyed so that huge stands of crops such as oil palms and coffee plantations can be planted. Fires started locally may run out of control, especially in years when rainfall has been low. People also cut down trees to use as fuel.

Tropical rainforest is by no means the only type of habitat under grave threat, but the loss is particularly worrying because of the exceptionally high biodiversity. When forest is lost, so is the habitat for large numbers of species. Many of these species may not have been discovered yet, especially those that live high in the canopy about which relatively little is known.

In many tropical countries, some areas of rainforest are protected, or at least are undamaged so far. However, these areas are often relatively small, and may be separated from each other by unforested areas. This situation is known as **fragmentation**, and it reduces biodiversity. Small areas of forest cannot support the high biodiversity of large areas. Isolated populations of a particular species living in a forest fragment are much more likely to die out than if they could freely intermix with other populations. And some species, particularly top carnivores, need very large areas in which to hunt.

Plants are affected by fragmentation as much as animals are. A study carried out in Brazil looked at more than 64 000 large trees over a period of 20 years. Some of them grew in continuous forest, while others grew in forest fragments ranging from 1 to 100 hectares in area. During this period, 6348 of the trees that were growing within 301 metres of the edge of the forest died; of those growing further than 300 metres from the edge, only 3523 died. Figure 3 shows the percentage difference in mortality of trees growing on the edges of patches of rainforest compared with trees growing well inside the forest.

These findings, and the results of many other research programmes looking at the effects of deforestation and fragmentation on species diversity, provide a strong argument for making sure that any rainforest reserves are as large as possible. It also helps for reserves in different parts of a country to be linked by corridors of protected forest so that animals can travel between them. For example, the charity Rainforest Concern is buying land to create a rainforest corridor between two large reserves in Ecuador. This corridor, the Choco-Andean Corridor, will have buffer zones on either side of it, in which the forest will not be completely protected, but where sustainable forms of income generation for the local population will be developed. These people, like most others who live near to rainforests, fully appreciate the value of the rainforest in both material and aesthetic terms. Some will be employed as rangers and guards.

An aerial view of Brazilian rainforest (bright green) with areas cleared for crops (yellow and brown).

Fig. 3 Effect of land fragmentation on tree mortality

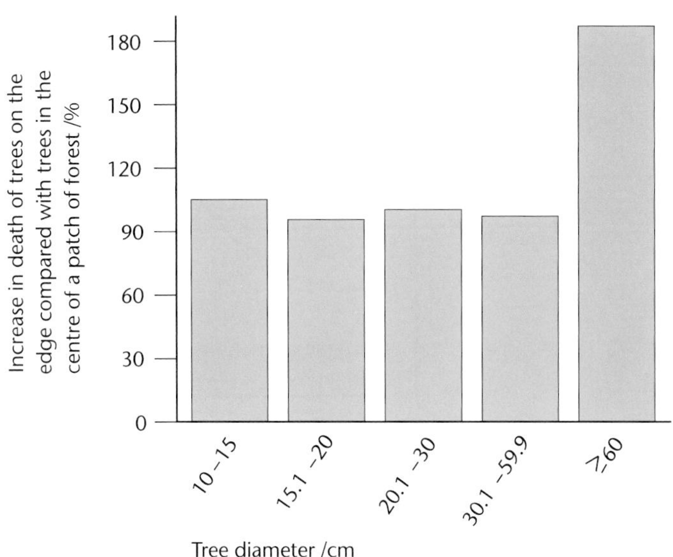

6 Give three reasons why rainforests are being cut down.

7 Explain why the loss of rainforests is a serious threat to biodiversity.

8 Suggest why a population of animals living in a small isolated fragment of rainforest is likely to have less genetic diversity than if it were in contact with other areas of the rainforest.

9 Describe the results shown in the bar chart, and suggest explanations for them.

key facts

- There is economic pressure on farmers growing crops to:
  - concentrate on a small number of crops and grow these as large areas of monoculture
  - remove hedgerows and field boundaries to make maximum use of land area
  - drain marshy areas and remove unprofitable pockets of woodland
  - use large quantities of chemical fertilisers to obtain maximum yield
  - use pesticides to deal with the increased damage from insects, plant diseases and weeds resulting from growing large areas of the same crop on the same land for many successive years.
- Tropical rainforests are rapidly being destroyed:
  - for timber
  - to provide space for growing cash crops, e.g. palm oil
  - for fuel.
- Deforestation of tropical rainforests is a serious threat to the high degree of biodiversity there.
- Fragmentation of forest areas reduces biodiversity.

15.2 Bottleneck and founder effects

An elephant seal.

Bottleneck effect

Genetic diversity in organisms may be influenced in many different ways. Hunting animals can result in a **population bottleneck**. Population bottlenecks occur when a population's size is reduced for at least one generation. Bottlenecks can severely reduce a population's genetic variation, even if the bottleneck doesn't last for many generations. The photograph shows an elephant seal. There are two major populations of elephant seals, which live in the northern and southern hemispheres. The northern elephant seals were hunted almost to extinction. By the end of the nineteenth century there were only 20 of them left. Their population has since recovered to over 30 000, but their genes still carry the marks of this bottleneck: they have much less genetic variation than the population of southern elephant seals that was not intensely hunted.

Figure 4 shows the effect of a bottleneck on genetic diversity.

Other examples include the American bison whose number fell to a few hundred, and cheetahs, which are so closely related to each other that skin grafts between animals are not rejected.

Founder effect

A **founder effect** occurs when a few members of the original population start a new colony.

Cockleburr plants produce hundreds of little football-shaped burs, about 2.5 cm long and

A cocklebur plant.

Fig. 4 The bottleneck effect

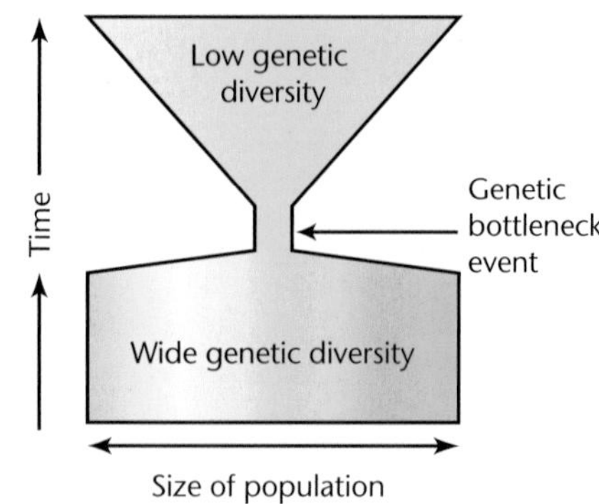

covered with stiff hooked spines. These easily hook on to the coats of animals, which may move a considerable distance before the bur drops off. The two seeds in the bur germinate, producing two plants, which in turn will produce a new population of cocklebur plants. This population will have a very much-reduced genetic variability.

There is some disagreement between botanists as to how many varieties of common cocklebur (*Xanthium strumarium*) exist, and precisely where their native (indigenous) habitat is. Several named varieties are listed in the botanical literature, including var. *canadense* and var. *glabratum*; however, some authorities believe that *Xanthium strumarium* is one cosmopolitan species with many highly variable populations around the world. These populations are different because of the founder effect.

The Afrikaner population of Dutch settlers in South Africa is descended mainly from a few colonists. Today, the Afrikaner population has an unusually high frequency of the gene that causes Huntington's disease, because the original Dutch colonists carried the gene for it with unusually high frequency.

10 Suggest why the southern group of elephant seals was not hunted so intensely.

11 Suggest why skin grafts between cheetahs are not rejected.

The Millennium Seed Bank

Scientists at the Royal Botanic Gardens at Kew have been collecting and storing seeds since 1974. The Millennium Seed Bank is a new and ambitious project, which opened at Wakehurst Place in Sussex in 2000, providing ideal facilities for storing seeds from many more species. Its aims are:

- to collect and conserve representative samples of the 1400 native British plants that set seed
- to collect and conserve seeds from 10% of the world's flora (seeds from more than 24 000 species), concentrating on plants that grow in dry places in the tropics and sub-tropics.

The Seed Bank opened just in time to play a major role in saving the St Helena boxwood from extinction. St Helena is a small island in the South Atlantic Ocean, thousands of miles from the nearest mainland in southern Africa. Like many isolated islands, it has a number of species that have evolved there, and are found nowhere else in the world. The St Helena boxwood, *Mellissia begoniifolia*, is one such species. First described in 1875, when it was already very rare, the St Helena boxwood was thought to be extinct by the mid 20th century. In 1999, a single remaining plant was found. It looked as though it was dying.

The last plant on St Helena has now died and *Mellissia begonifolia* is truly extinct in the wild. Hopes rest on the 53 plants like this one growing in controlled conditions.

Scientists from Kew and St Helena worked hard to try to save this species. They took cuttings from the plant, but none of them rooted. None of the seeds lying in the soil around the plant germinated. It looked as though, having been rediscovered at the last minute, the boxwood was about to become *genuinely* extinct.

Eventually, about 400 seeds from the dying plant and the soil around it were collected and shared out between several different organisations, in the hope that someone would manage to get some to germinate. And this has resulted in success. By 2001, there were five plants on St Helena, two at the Royal Horticultural Society at Wisley in Surrey, and 46 at Kew. Of those 46 plants, 38 have been grown from seed and eight propagated from cuttings.

Studies of the genome of the individual *Mellissia* plants have been made at Kew and it has been found, not surprisingly, that there is virtually no genetic variation between the plants. This is not good news for *Mellissia*, because without genetic variation the plant will not be able to adapt to changes in its environment. This lack of genetic variation probably reflects the death of almost all of the individuals of this species some time ago – all the seedlings that are now growing have inherited their genes from a single parent plant.

Nevertheless, there is still hope that *Mellissia* can eventually be returned to the wild. The original seedlings grown at Kew flowered in September 2000, and hand pollination resulted in the production of a number of fruits. The first 35 ripe fruits were harvested in January 2001, each containing 20–30 seeds. These have all been sent to the Millennium Seed Bank, where they will be kept safely in store, in the hope of eventual repatriation to St Helena.

Thanks to Dr Wolfgang Stuppy and the Royal Botanic Gardens, Kew.

12

a Why is there little genetic variation between the *Mellissia* plants?

b What is the disadvantage of this lack of variation?

13 Suggest what needs to be done before attempts are made to return plants of the St Helena boxwood to the wild on the island.

15.3 Selective breeding

Selective breeding – crops

For thousands of years farmers and other plant breeders have used **selective breeding** to create crops that have many desirable characteristics and few undesirable ones. Some of the characteristics that are bred for are: higher yield, better taste, richer colour, and increased tolerance to heat, cold or drought.

In selective breeding of plants, those that exhibit the most desirable characteristics are used to generate seed. The future crops that are produced from this seed are in some way better than the original plant.

Rice is the world's most vital food crop. The grain accounts for 80% of the total calories consumed by 2.7 billion Asians – half the world's population. We urgently need more to keep up with the world's rapid population growth. Selective breeding is helping this.

Producing better rice by selective breeding.

A high-yield, semi-dwarf variety of rice called IR–8 was produced by crossing a high-yield rice called PETA with a variety called DGWG, which had short stiff stalks. It was hailed as 'Miracle Rice' and helped to spark what is now known as the Green Revolution. IR–8 gave double the yield of previous rice varieties when grown in irrigated conditions, had greater resistance to diseases and insects and was more responsive to fertilisers. It helped to avert famine for a generation.

14 Suggest the advantage to IR–8 of having short stiff stalks.

Scientists are currently working on a new plant type to address some of the shortfalls of the IR–8 variety; it has optimistically been dubbed 'Super Rice'. Work started on this highly productive rice in the 1990s.

Super Rice has fewer but stronger stems and there are many more seeds on each rice flower. In the IR–8 plant, half of the plant's weight is grain and half is straw, whereas the new Super Rice plant is 60% grain and 40% straw. So more energy goes into grain production, increasing yield potential by about 20%. The Super Rice also has a vigorous root system, and scientists are working on improving its resistance to disease and insects.

However, developing a new rice plant to save half the world from starvation is not all smooth sailing. The scientists have encountered many problems in the development of Super Rice. For instance, most current high-yielding rice varieties produce around 100 grains per panicle. The prototype Super Rice, on the other hand, produced 250 to 300 grains per panicle, which was too many. The plant simply couldn't supply enough carbohydrates and nutrients to fill the grains. The breeders overcame the problem by reducing the number of grains back to 200, which still makes Super Rice twice as productive as older plant types.

15 What are the advantages and disadvantages of 'Super Rice' over IR–8?

how science works

Rice production in Brazil

The graphs show changes in the area of land cultivated for rice in Brazil and the amount of rice produced.

16 Describe the trends for area of rice harvested and rice production.

17 Explain the relationship between the two sets of data.

Fig. 5 Area of rice cultivation and rice production in Brazil between 1985 and 2007

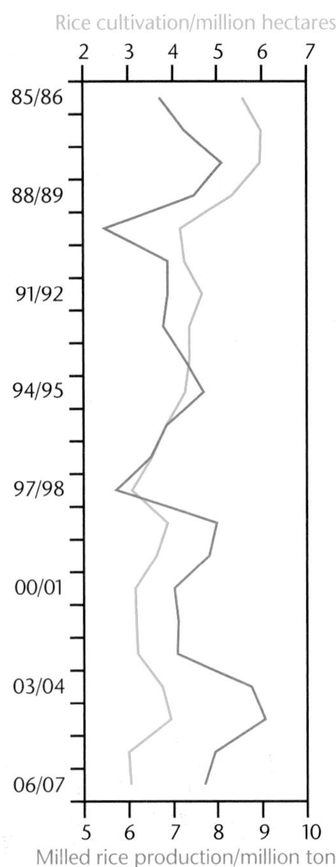

Battery hens.

Selective breeding – animals

The chickens in the photo on the left are egg-laying hens. They have been selectively bred to lay lots of eggs, but they grow at a normal rate. Most are still kept in battery cages, though this system is to be banned in 2012. The chickens in the photo overleaf are broiler chickens. They have been bred for meat. They grow twice as quickly and are usually slaughtered at 6 weeks of age. Most chickens reared for meat are raised intensively in large sheds.

All of these chickens have the same common ancestor. They are descended from the jungle fowl, which can still be found in the wild in the forests of India and South-East Asia. Chickens were first domesticated at least 3400 years ago.

Selectively bred chickens.

Farmers have been selectively breeding chickens for thousands of years. The basic method is quite simple. If you breed from the hen that lays the most eggs, the chances are that her daughters will also be good layers. This is because the number of eggs that a hen lays is partly controlled by the genes that a hen inherits from her parents.

Chickens raised for meat have been selectively bred using the same principle. A wild jungle fowl might lay 20–30 eggs in a year. Today's hens each lay over 300 eggs a year on average.

Many people are concerned about the welfare of selectively bred domesticated animals. These are some of their concerns.

- Battery hens now lay so many eggs that many suffer bone disease from lack of calcium.
- Broiler chickens are bred to grow so fast that they reach slaughter weight in just 6 weeks, (twice as fast as 30 years ago). Many suffer chronic pain from resulting leg disorders, and many die of heart failure.
- Turkeys are bred to be so large and fleshy that they cannot mate naturally and have to be artificially inseminated.
- Dairy cows now produce 10 times as much milk as their calves would have drunk – around 35 litres of milk every day. Many are prone to health problems such as lameness.
- Beef cattle, e.g. the Belgian Blue, are bred for increased muscle, with the result that some calves are so large that they have to be delivered by caesarean section.
- Sows are bred for larger litters. The sows grow so large that there is a very real risk of them crushing their piglets.
- Fattening pigs grow so fast that their legs are unable to keep up with the rest of the body, resulting in painful arthritic joints.

18 What characteristics would a breeder select for broiler chickens?

19 Discuss who is to blame for the above:

- breeders
- farmers
- scientists
- consumers who buy the products.

how science works

What makes chickens grow faster?

A group of students in Australia decided to investigate the commonly held view that the larger size and faster growth of the broiler chickens we eat today are due to the use of hormones.

The students obtained 15 egg-laying chickens and 15 broiler chickens as day-old chicks from a commercial supplier, and then raised them. All chickens were fed the same standard chicken feed product, made mostly from cereal grains and protein sources, obtained from a local feed supplier. The chickens were checked every day and weighed regularly for a period of 6 weeks. At the end of the 6-week period, the average weight of the chickens bred for egg laying was 592 g while that of the chickens bred for meat was more than four times larger, at 2388 g.

The photograph and the graph (Fig. 6) show the students' results.

Fig. 6 The students' results

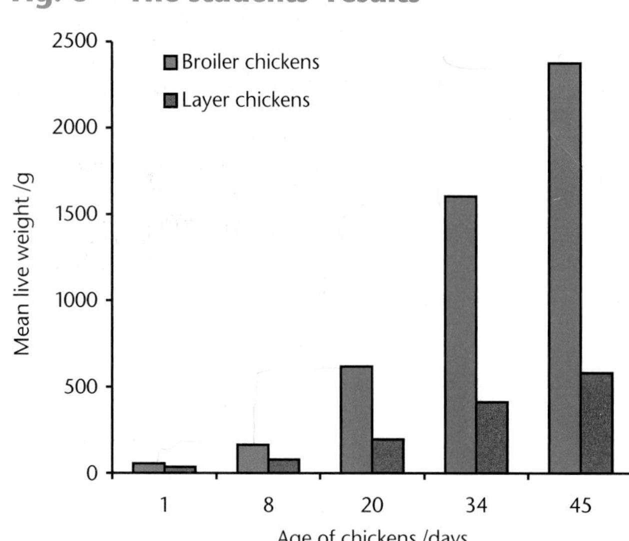

Broiler and egg-laying chicken at 38 days.

20 In the students' investigation:

a name the independent variable

b name the dependent variable

c give two control variables.

21 How could the students have made their results more reliable?

22 Use the graph to calculate the rate of growth of each batch of chickens between day 1 and day 45.

23 What conclusion can be drawn from the students' investigation?

key facts

- Genetic diversity in organisms is influenced by:
  - the bottleneck effect when the population of an organism is greatly reduced for at least one generation
  - the founder effect when a new colony is founded by a few individuals
  - selective breeding of domesticated animals and crops.
- Reduction in genetic diversity may affect the ability of an organism to evolve successfully in changing conditions.

15.4 Measuring biodiversity

Biologists studying an ecosystem often want to know the range of different species that live there. The more species there are, the stronger the argument for conserving the habitat.

A very simple statistic is the number of different species that live in a particular habitat. This is called the species richness. However, this does not tell you anything about the relative sizes of the populations of each species. A measure of species diversity takes into account not only the number of different species in a habitat, but also the relative sizes of the populations, and how well the species are spread through the habitat. A meadow with 20 plant species, in which nearly all the plants belong to six species and the other 14 are very rare and all are clumped in one corner, will have the same species richness as another meadow in which all 20 species are fairly common and are found all over the meadow. However, the first meadow has a smaller species diversity than the second.

Although it is easy to get a general idea of what species diversity means, there are no universally accepted definitions of it. There are many formulae for calculating species diversity. One commonly used example is Simpson's **Diversity Index**. The formula for calculating this is:

$$D = \frac{N(N-1)}{\Sigma n(n-1)}$$

where

D = Simpson's Diversity Index

N = the total number of individuals recorded in the sample

Σ = the sum of

n = the number of individuals of each species.

The larger the value for Simpson's Diversity Index, the greater the species diversity in the habitat.

Table 1 shows data taken from sampling a regularly mowed but rather weedy lawn.

Table 1 Species found in a regularly mown but weedy lawn

Species	n	n–1	n(n–1)
Grass species 1	80	79	6320
Grass species 2	45	44	1980
Clover	9	8	72
Black medick	22	21	462
Daisy	13	12	156
Dandelion	3	2	6
Germander speedwell	10	9	90
Self-heal	14	13	182
Moss	36	35	1260
Total	(N) = 232		Total n(n–1) = 10 528

The values for n are the number of individuals of that species.

So $D = \frac{232 \times 231}{10\,528}$

$= 5.09$

24 A student collected the following data from a grass field grazed by sheep.

Grass	185
Thistles	28
Stinging nettles	35
Moss	2

a Arrange these data in a table that will allow you to calculate Simpson's Diversity Index.

b Calculate the index, showing your working.

c How does the species diversity of the field compare with that of the lawn? Suggest reasons for the differences between them.

d How reliable do you consider this comparison to be?

key facts

- A measure of species diversity takes into account:
  - the number of different species in a habitat
  - the relative sizes of the populations
  - how well the species are spread through the habitat.
- The formula for calculating a diversity index is:

$$D = \frac{N(N-1)}{\Sigma n(n-1)}$$

where

D = Simpson's Diversity Index

N = the total number of individuals recorded in the sample

Σ = the sum of

n = the number of individuals of each species.

examination questions

1

a Read the passage
There is very little genetic difference among all living humans on earth. This is due to a series of bottlenecks in human evolutionary history. Geneticists studying many different parts of the human genome have concluded that the past effective population size (that is, the number of reproducing females) averaged only 10,000 individuals over the last one million years, and was as low as 5,000 around 70,000 years ago. Compare this to the approximately one billion reproducing females alive today, and it becomes clear just how narrow these bottlenecks were.

i Explain what is meant 'a series of bottlenecks in human evolutionary history'. (3)

ii Explain how this has affected the human genotype. (3)

b The Amish population in Pennsylvania was founded by about 200 German immigrants. It is a closed population.
One form of dwarfism, Ellis-van Creveld syndrome, involves short stature, polydactyly (extra fingers or toes), abnormalities of the nails and teeth, and, in about half of individuals, a hole between the two upper chambers of the heart.
This syndrome is much more common in the Amish population than in the rest of Pennsylvania. Explain why. (3)

Total 9

2 The table below shows the fastest time for a one mile horse race in three different years.

Age of horses in years	Fastest time for a 1 mile race in minutes and seconds		
	2002	**1990**	**1980**
2	1:53	1:55	1:57
3	1:51	1:52	1:55
4	1:50	1:53	1:54

a Summarise the trends shown in the data. (3)

b Explain how selective breeding could account for the changes in performance. (3)

c Discuss the ethics of using selective breeding in this way.

Total 6

3

a Explain what is meant by monoculture. (1)

b **i** Where monoculture takes place on a large scale, farmers often remove hedges.
Explain **two** benefits to the farmer of removing hedges. (2)

ii Usually, the older a hedge the more species of shrub it contains. Explain why removal of hedges that are several hundred years old affects more animal species than the removal of young hedges. (2)

c Monoculture often involves the use of large amounts of pesticides. Some of these pesticides are toxic to species that are not pests. These animals may be killed immediately when the pesticide is applied. Explain **one** other way by which the use of pesticides can lead to the death of animals that are not pests. (2)

Total 7

AQA, June 2006, Unit 5, Question 2

4 A hedgerow is a line of shrubs and trees bordering a field, together with the herbaceous plants at their base. In the last 50 years farmers have removed many hedgerows.

a Explain **two** advantages for a farmer of removing hedgerows. (2)

b In recent years some hedgerows have been replanted. Ground beetles, which are unable to fly, are predators of crop pests. The beetles overwinter in the shelter of grasses at the base of the hedgerow. In some large fields, a permanent strip of grass is left as shown in the diagram below.

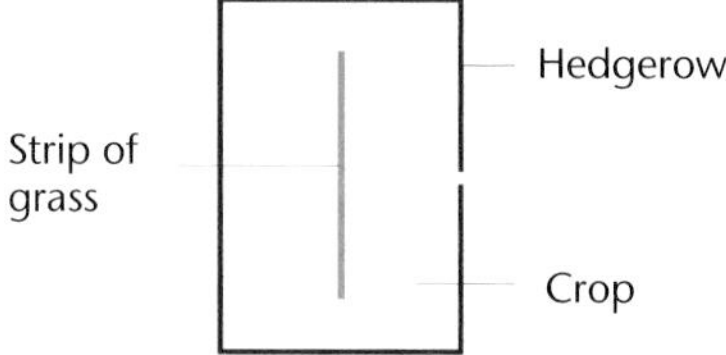

Suggest and explain the advantage of leaving the strip of grass in the middle of the field. (2)

c Apart from providing a habitat for predators of crop pests, give **two** biological benefits of replanting hedgerows. (2)

d A species of insect that feeds on crop plants has individuals that may be either black or green. Colour in this insect is a genetically determined characteristic. The colour of these insect pests in two fields of the same crop was recorded. Ground beetles were present in one field but not in the other. The proportion of black insect pests in the field with beetles was much smaller than in the field without beetles. Explain how the presence of the beetles may have caused this difference. (4)

Total 10

AQA, June 2005, Unit 5, Question 4

5 The numbers of species of woody plants in samples from 227 hedges of different ages were counted. The results are shown in the graph below. The size of the solid circles shows the number of hedges in each category.

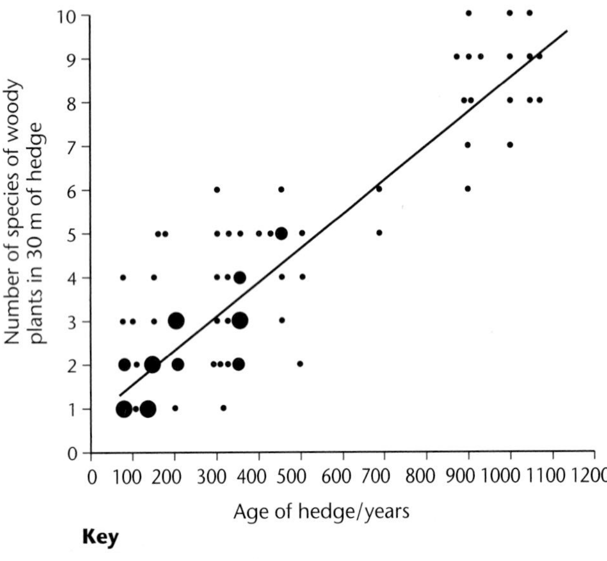

Key

• 1–4 hedges ● 5–14 hedges ⬤ 15–30 hedges

a i Suggest the age range of the hedges which are likely to support the most complex food webs. Explain your answer. (3)

ii Explain how the complex food webs maintained by these hedges may be of benefit to farmers. (2)

b i Calculate the **maximum** percentage of hedges which are 1000 years old or over in this sample. Show your working. (2)

ii Many hedges have been removed from arable land in recent years. Explain **two** advantages to farmers of removing hedges. (2)

Total 9

AQA, June 2003, Unit 5, Question 4

6 The table shows the numbers of adult butterflies in two areas of the same tropical forest. In the logged area some trees had been cut down for timber. In the virgin forest no trees had been cut down. The two areas were the same size.

	Logged forest		**Virgin forest**	
Butterfly species	**Number**	**$n(n-1)$**	**Number**	**$n(n-1)$**
Eurema tiluba	72	5112	19	342
Cirrochroa emalea	43	1806	132	17 292
Partenos sylvia	58	3306	14	182
Neopithecops zalmora	6	30	79	6162
Jamides para	37	1332	38	1406
Total	216	11 586	282	25 384

a Describe a method for finding the number of one of the species of butterfly in the virgin forest. (2)

b The index of diversity of a forest can be calculated using the equation $d = \frac{N(N-1)}{\Sigma n(n-1)}$.

Calculate the index of diversity for the virgin forest. Show your working. (2)

c What does the table show about the effects of logging on the butterfly populations? (2)

Total 6

AQA, June 2006, Unit 6, Question 3

7 The Solomon Islands are situated in the Pacific Ocean. The nearest large land mass is Australia, which is about 1500 km away. The biggest islands are mountainous, with large areas of tropical forest and a wide range of habitats. Some islands have a very high species diversity, and many species are endemic, that is they occur only in the Solomon Islands.

The table shows the total number of species on the islands in four vertebrate classes and the percentage which are endemic.

Vertebrate class	**Total number of species**	**Endemic species/%**
Mammals	53	36
Birds	223	20
Reptiles	61	16
Amphibians	17	53

a How many reptile species are endemic? (1)

b Suggest an explanation for the high proportion of endemic species on the Solomon Islands. (3)

c Many of the endemic species of plants and animals are threatened with extinction as a result of deforestation and the introduction of non-native species such as cats and rats. Some of the populations of endemic species are now very small.

Describe ways in which the endemic species could be conserved and suggest reasons for protecting them from extinction. (6)

Total 10

AQA, January 2005, Unit 6, Question 7

Species conservation – Somerset Levels and Moors Natural Area

Fig. 7 Location of the Somerset Levels and Moors Natural Area

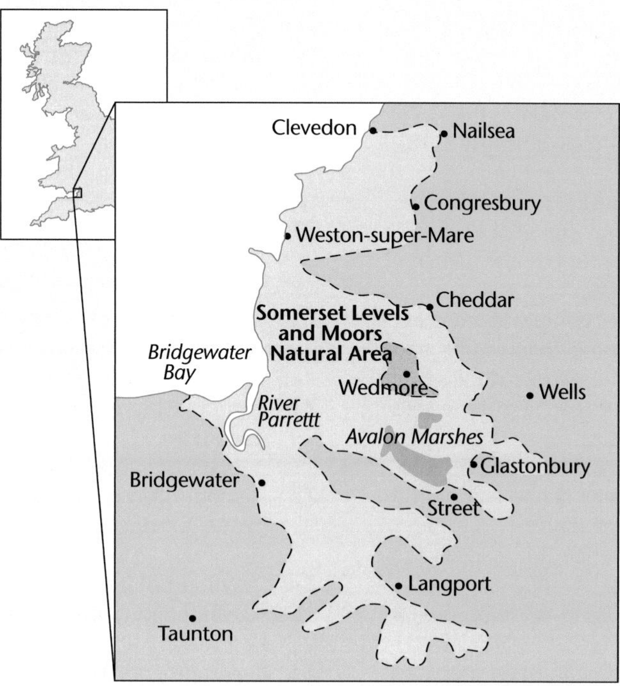

The Somerset Levels and Moors Natural Area is a large area of low-lying land in the south-west of England. The boundary of the Natural Area is defined by the 10-m contour line. The area contains the most extensive area of lowland wet grassland and natural floodplain in England, and is of international importance. Eight major rivers flow through it, eventually emptying their waters into the Severn Estuary.

Eight thousand years ago, this area was an inlet of the sea. As the sea receded, the land became salt marsh and bog. The remains of dead plants did not decay completely in this wet anaerobic soil, but turned into peat. People have lived and farmed in this marshy area for thousands of years. A network of rivers plus constructed drains and ditches helps to drain the land in winter, and supply water in summer. In 1987, the Ministry of Agriculture, Fisheries and Foods designated a large area of the Somerset Levels as an environmentally sensitive area (ESA).

Farmers are now given the opportunity to join a scheme in which, in return for managing their land in ways that benefit the environment, they are compensated financially for the loss in productivity. The scheme was revised and renewed in 1997, and currently more than 65% of the grassland is protected from cultivation. The main management objectives are now to ensure that the water levels are not lowered any further and, in some areas, are raised to what they would have been many years ago. New dams, sluices and culverts have been constructed to help with the control of water levels.

Fig. 8 Breeding territories of bird populations in the Somerset Levels

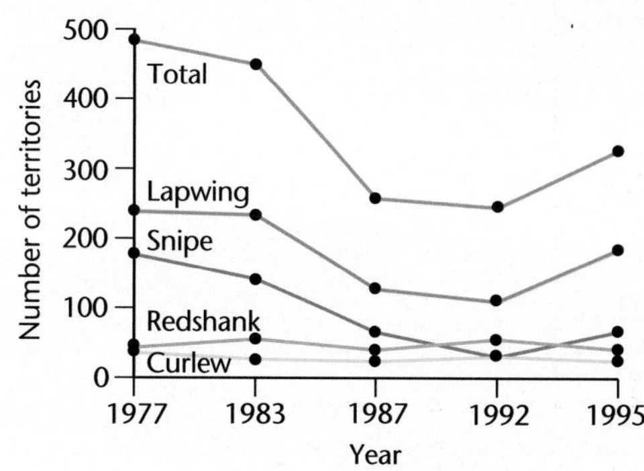

Water-level controls in the Somerset Levels are having a beneficial effect on wildlife. For example, winter flooding benefits waders, and provides suitable areas in which they can nest in spring. The graph shows changes in the number of breeding territories of five species of wading bird on eleven sites in the area. (Note that the area was designated an ESA in 1987.)

Thanks to Stephen Parker and the English Nature Somerset Team.

A1 What group of birds was particularly threatened by drainage of the Somerset Levels?

A2 Use the information in the passage and the graph to comment on the effects that designation of the Levels as an ESA in 1987 had on biodiversity.

A3 Suggest why farmers in the scheme are not allowed to spray pesticides.

A4 Grass is normally cut for hay once a year, in June. Grass for silage is normally cut much earlier in the year than this, and usually second or even third cuts are made later on. Suggest why farmers in the ESA scheme can only cut grass for hay, and not for silage.

Answers to in-text questions

Chapter 1

1 Bacteria have organelles (such as cell wall, cell membrane, ribosomes).

2 Skin, reproductive organs

3 Digestive system, breathing system

4 For: *Y. pestis* traces found in remains of victims.
Against: Two-thirds of people died from plague in Iceland, where there were no rats.
Biology of rats and fleas together with modern epidemiology show that this method of spread was most unlikely.

5 A microorganism that causes disease

6 Infection is the presence of a pathogen in the host's body. Disease occurs when the host shows signs and symptoms caused by the pathogen.

7 Damage cell structure; disrupt cell function; release toxins

8 Exotoxins are released by bacteria as they grow; endotoxins are released from bacteria when they die.

9 Proteins released by macrophages affect the body's temperature-regulating mechanism.

10 This would require infecting a healthy human with the disease, which is unethical.

11 By identifying the pathogen in blood, urine or other samples from the patient. The coincidence of disease symptoms and the presence of the pathogen indicates a link.

12 Effective treatment removes most of the pathogens from the water.

13 Increased breathing rate is caused by (i) decrease in oxygen concentration in the air breathed in; (ii) increase in carbon dioxide concentration of the air breathed in.

14a The heat is transferred to the blood, which in turn stimulates the temperature-control centre in the brain.
14b Germinating sunflower seeds give off a substance that inhibits the germination of seeds from other plants.
14c The cat's eyes contain more cells that are sensitive to low light intensities; the narrow pupil prevents light reaching these cells during the day.

15
- What the patients had eaten on the previous three days
- Where and when the food had been bought
- The conditions in the shop or restaurant
- How the food had been prepared and stored

16
- Age profile of the population
- Level of public hygiene
- Population density
- High-risk behaviour

17 Older people have higher level of acquired immunity since they have come into contact with a larger number of pathogens.

18
- Development of modern medical practices including use of antibiotics
- Higher standards of public health provision

19
- Old age results in a decline in the efficiency and effectiveness of the immune system.
- Old people in care homes live in a relatively small, closed environment, in close proximity to each other.
- They all eat the same food prepared in the same kitchen.

20 A diet low in saturated fat reduces risk of atheroma since these fats cause an increase in cholesterol levels. Reducing stress, not smoking and lowering salt intake all reduce blood pressure, which is a significant factor in atheroma formation. Regular exercise makes obesity less likely – obesity is a major factor for atheroma.

21 The blood clot might break away and be carried by the blood to a coronary artery. Blockage of a coronary artery reduces the supply of glucose and oxygen to heart muscle.

22 A mobile blood clot is called an embolism.

23 Uncontrolled cell division in a tissue, producing a tumour

24 Anything that increases the likelihood of getting the disease

25 Environmental factors

26 Cardiovascular and breathing systems

27 Stroke volume = 5.25/45
= 0.117 dm^3

28 The average adult has a lower stroke volume than an athlete since regular exercise increases the volume of the heart chambers.

29a The athlete has a slightly higher cardiac output at rest, but a much higher cardiac output during exercise.
29b Cardiac reserve of the average adult = 21–5 = 16 $dm^3\ min^{-1}$
Cardiac reserve of athlete
= 30 – 5.25 = 24.75 $dm^3\ min^{-1}$.
29c Regular exercise increases the cardiac reserve as the cardiac muscle becomes larger and stronger, and the peripheral resistance to blood flow is reduced because of the development of peripheral circulation in the muscles.

30a Greater cardiac output increases blood flow to the active tissues.
30b Increased capillary networks in the muscles increase gaseous exchange.
30c Improved capillary networks in the lungs, increased lung volume and better ventilation improve gaseous exchange in the lungs.

A1 Smoking, blood pressure, plasma cholesterol, age, sex.

A2 Male smoker age 60+ with high blood pressure and high plasma cholesterol

A3 Risk = 20–40%. Therefore between 100 and 200 men
If non-smokers, risk = 10–20% therefore 50–100 men

A4 None of the factors has a direct cause–effect relationship with heart disease, only an association. Therefore only statistical links exist.

A5 • Risk factors will change over 10 years.
- Smoking is not quantified.
- Other risk factors are not considered.

Chapter 2

1a Vaccination will reduce the incidence of the disease by making girls immune, and also by reducing transmission. However, it may encourage girls to be complacent about protecting themselves against other sexually transmitted infections. Overall, the number of lives saved will probably outweigh the number of girls who are encouraged to try underage or unsafe sex.
1b The research is unethical because the men will not be told whether or not they are infected. The men should be told so that they can discuss with partners the risk of infection. It is not ethical to allow partners to become infected through ignorance. The discovery of infective mechanisms will not justify women becoming infected.

2 The combined magnification of the two lenses in a light microscope is great enough to produce a maximum magnification of only × 1500. More powerful lenses would not be able to resolve objects smaller than the wavelength of light.

3 Magnification is the number of times larger the object appears; resolution is the ability to distinguish between two objects.

4 0.0025 nm (half of 0.005 nm). In practice, the limit due to the wavelength of the electron beam is slightly greater than 0.0025 nm because of technical factors. It is hard, for example, to focus the electron beam perfectly.

5a Nucleus, cytoplasm
5b Mitochondia, ribosomes, endoplasmic reticulum, golgi body, cell membrane

6a The TEM has a higher resolving power than the SEM.
6b The SEM has the ability to image a comparatively large area of the specimen, and to image bulk materials (not just thin films or foils). SEM images are usually much easier to interpret than TEM images.

7a To reduce the activity of enzymes to a minimum
7b So that the organelles do not change size because of osmosis

8a Nucleus, mitochondria, ribosomes
8b The nucleus in the bottom fraction, then mitochondria above, then ribosomes.

9a Measured thickness is about 1.5 mm, which is equal to 1 500 000 nm. Magnification is 190 000. Therefore, the actual thickness is 1 500 000 ÷ 190 000, which equals about 8 nm.
9b By making several measurements of the width, then calculating a mean

10 The length of a phospholipid molecule is half of the 8 nm calculated above, which is 4 nm.

11 Fatty acids and glycerol

12 In the phospholipid, a fatty acid molecule is replaced by phosphoric acid.

13 An unsaturated fatty acid has at least one C=C double bond; a saturated fatty acid has all C–C single bonds.

14 The nuclear pores allow large numbers of mRNA molecules that have been produced within the nucleus to pass out to the endoplasmic reticulum. Here they attach to the ribosomes to take part in protein synthesis.

15 The folded inner membrane gives a large surface area, allowing more enzymes involved in the final stages of respiration to be attached. This increases the potential rate of respiration and therefore energy release.

16 The electron micrograph is a two-dimensional section taken at one point in the specimen. It cannot reveal all the features of the three-dimensional structure from which the section was taken.

17 Cells that manufacture large amounts of protein need large amounts of rough endoplasmic reticulum because this is where the proteins are translated from the mRNA that is transcribed in the nucleus. Often, the polypeptides produced by the ribosomes need to be processed in some way – by the addition of sugar groups, for example – before the protein becomes fully functional. Such processing takes place in the Golgi body.

A1a Image size of diameter of egg = 65.6 mm. Actual size = 0.1 mm. Magnification therefore equals 65.6 ÷ 0.1, which gives × 656.
A1b Using the formula volume = $^4/_3\pi r^3$, volume of egg = 0.000 5 mm^3 volume of sperm = 0.000 000 05 mm^3.
A1c 10 000.

A2 Mean length of single bacteria is

about 10 mm. Actual mean length is therefore 10 ÷ 9240, which gives 0.001 08 mm. This is 1.08 μm.

A3 Bottom left grain is 16 mm diameter. Actual width equals 16 000 ÷ 420, which is 38 μm.

A4 Image length is 24 mm. So length of head is 24 ÷ 25, which is about 0.96 mm.

A5 Length = 81 ÷ 80 000 = 0.00 101 25 mm. This is 1012.5 nm. Width = 28 ÷ 80 000 = 0.000 35 mm. This is 350 nm.

Chapter 3

1a Sodium
1b Sodium
1c Potassium

2 Volume of fluid taken, type of exercise, age, sex, fitness, body build

3a The greater the concentration of NaCl in the external solution, the smaller the volume of the red cell.
3b Swollen red cells might not be able to pass through narrow blood capillaries.

4 The dark colour indicates a region of high concentration of solutes from the tea bag; lighter colours indicate regions of low concentrations of these solutes; there is a concentration gradient of these solutes from the dark colour to the lighter colours.

5 Surface area of spherical cell = 88 μm^2
Surface area of cylindrical cell = 121 μm^2

6a 200 μm per second
6b 2 μm per second

7a Approximately 0.4 milliseconds
7b 2–3 milliseconds

8a Red blood cell: 12.1 milliseconds
8b Spherical cell: 8.8 milliseconds

9 The rate of diffusion across the membrane would increase due to an increase in surface area, but the time taken for oxygen to reach the haemoglobin molecules in the centre would increase by a much larger factor.

10 Osmosis is a special case of diffusion. It involves the movement of water through a partially permeable membrane. This membrane allows water molecules to pass through but not solute molecules. The water molecules move along their concentration gradient, but the membrane prevents the solute molecules moving along their concentration gradient.

11 Those on the right-hand side since they are less free to move

12a

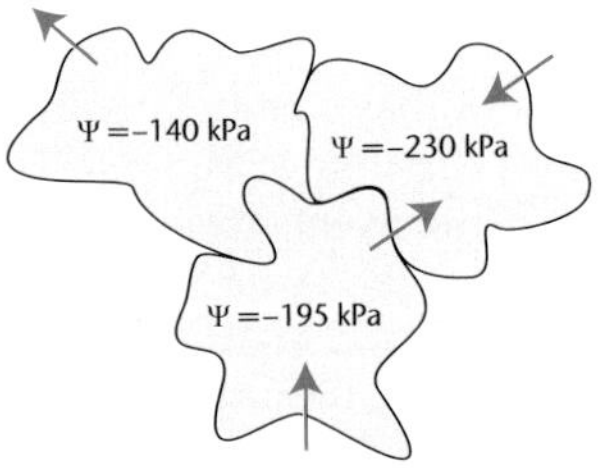

Ψ = –140 kPa
Ψ = –230 kPa
Ψ = –195 kPa

12b Most negative –230 kPa
Least negative –140 kPa

13 Sweating involves the loss of both water and salts from the plasma. If these are replaced only by water then the φ of the plasma will increase. The φ will now be higher than the φ of the brain cells, so water will diffuse into these cells causing them to swell.

14 Yes. Tell the manager that clubbers need to replace salts lost in sweat just like marathon runners, otherwise there is a risk of illness.

15 The folds provide a larger surface area, which increases the rate of diffusion.

16 Molecules move in straight lines but in random directions. Relatively few molecules will be moving along paths that take them between the phospholipid molecules.

17 There is an almost infinite number of gaps between the phospholipid molecules where water molecules can pass through, so the number of spaces does not become a limiting factor to the rate of diffusion. On the other hand, there are relatively few carrier protein molecules for glucose in the membrane. Until all of these are transporting glucose molecules, the concentration of glucose molecules is the limiting factor; when all the carrier molecules are transporting glucose molecules, the number of these carrier molecules becomes the limiting factor for the rate of diffusion.

18 The more mitochondria, the greater the rate of respiration and therefore the more energy available for active transport.

19a Because it is isotonic with blood plasma.
19b Sugars are oxidised to release energy, which is needed for muscle contraction.

20 Since blood takes glucose away from the cell, there will always be a concentration gradient down which glucose will diffuse. The sodium ion concentration in blood is greater than that in the cell, so active transport is needed to move the ions against a concentration gradient.

21 Salt both stimulates the uptake of glucose and contributes to an isotonic solution.

22a Water, glucose, ions
22b No, because it is necessary to maintain a concentration gradient for urea between blood and the dialysis fluid.
22c The dialysis fluid has a more negative water potential than the blood.

23 There is a permanent tube called a catheter which passes into the abdomen. Fluid is introduced through this tube. After a time the fluid is drained out through the same tube.

24a In CPD there is always about 2 litres of dialysis fluid in the abdomen; this is replaced regularly . In IDP, dialysis fluid is introduced, left for one hour, then drained.
24b CPD can be done at home, at convenient times. CPD users can be reasonably flexible with diet and fluid intake. IPD is not done as frequently as CPD, but requires a trip to hospital. The person must be very careful with diet and fluid intake since the IPD fluid is not changed as frequently.

25 A prokaryotic cell does not have a proper nucleus, mitochondria, chloroplasts, endoplasmic reticulum or Golgi bodies.

26a Independent variables: concentrations of glucose and ions
Dependent variable: effect on diarrhoea
26b Initial degree of diarrhoea
26c Age, mass, sex, general health

27 The cells will now have a more negative water potential than the intestinal contents.

28 Voluntary nature, consent, patients never put at risk

A1 Epidemiology

A2 The association of the outbreak with the drinking of water from the pump in Broad Street; the occurrence with people who had drunk coffee made from water from the Broad Street pump; the very few cases in the Workhouse, which had its own pump; the absence of cholera in men from the brewery; the number of cases in the percussion cap factory

Chapter 4

1 Yes, because their combined kinetic energy is higher than the minimum activation energy needed for the reaction.

2 They react because their combined kinetic energy is still as high as the activation energy required.

3a A B D E F
3b C
3c C–D

4a Carbon, hydrogen, oxygen, nitrogen
4b Nitrogen
4c NH_2
4d COOH
4e COO^-, and H^+

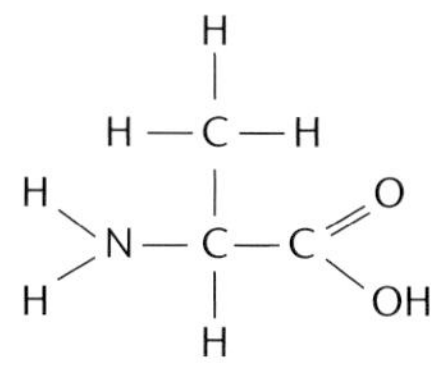

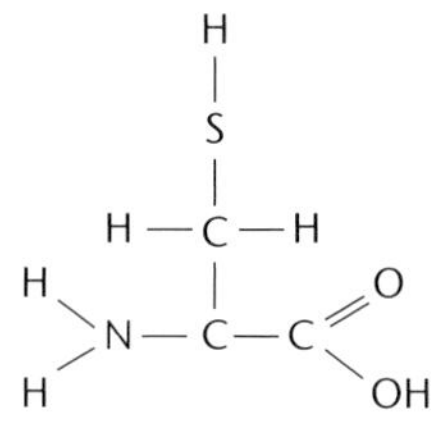

5

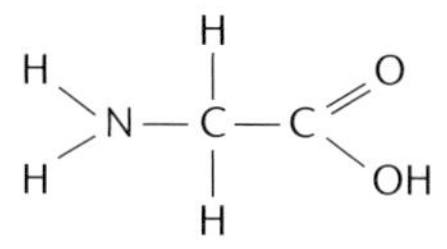

6 $NH_2CHCH_3COOH + NH_2CH_2COOH$
Alanine + glycine,
$\rightarrow NH_2CHCH_3CONHCH_2COOH$
dipeptide

7

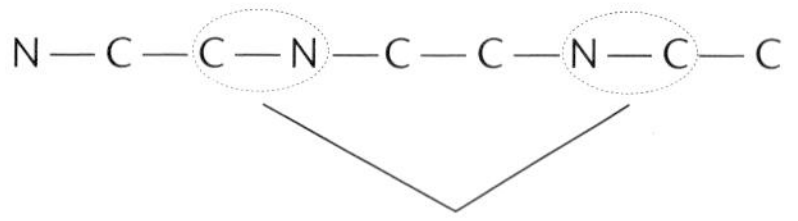

8 The tangle of collagen rods makes it less likely that the cartilage will tear.

9 The cornea has to be transparent. The stacking of the collagen rods prevents light being scattered in all directions.

10 The proteins must be the right size to fit into the membrane; fibrous proteins would be too long. Carrier proteins must be the right shape for the molecules they transport.

11 Each is specific to the particular molecule or ion that it carries.

12 The active site of an enzyme molecule has a particular shape that attracts and accepts only molecules of the correct shape to fit into it. Enzymes are therefore specific.

13 The 'induced fit' hypothesis takes into account the change in shape of the active site as it accepts the substrate molecule(s).

14 Molecules have low kinetic energy, so they move slowly and are less likely to collide and react.

15 2, 4 and 8 units

16 The rate of reaction approximately doubles.

17 Freezing would not denature the enzyme, so the rate of reaction would be unaffected.

18a 55–56 °C
18b Clothes can be washed in warmer water without the enzyme being denatured. There is no need to presoak clothes in the washing powder.

19 Proteins have large molecules and are not soluble in water. The proteases in the washing powder break down the protein into amino acids, which are soluble and are therefore washed away.

20 Your graph should have an optimum around pH 2, and falling to zero at or just before pH 6.

21 The protease is effective over a much wider range of pH, and will work well in alkaline solutions above pH 7.

22 As the concentration of enzyme increases, substrate molecules collide more frequently with active sites and the rate of reaction increases. The maximum rate is reached when all active sites are in use all the time. If the substrate concentration increases so much that there is excess substrate, the rate of reaction cannot increase further. This would normally only occur when enzyme concentration is low, as enzymes work so fast that only small amounts are needed.

23 X = Excess of substrate
Y = Enough substrate to saturate enzyme active sites

24 Increasing the temperature

25 Adding extra enzyme

26a The structure of the molecules is very similar. The malonate is attracted to the active site but no reaction occurs. This reduces the number of sites available for the succinate to react.
26b They do not fit into the active site. Instead they react with a different part of the enzyme molecule.

27 The enzyme being inhibited catalyses a reaction that was unique to the pest; the inhibitor has no toxic effect on humans or other organisms; the inhibitor is stable, and does not break down into harmful substances; it can easily be administered to the pests in the right dose; it does not persist or build up to harmful levels in the environment.

28 Carbohydrates and lipids both contain the elements carbon, hydrogen and oxygen. The proportions of these elements varies widely between different lipid molecules, but the proportions in a carbohydrate molecule are always 1:2:1.

29 Maltose is a reducing sugar, but sucrose is non-reducing.

30 Boiling sucrose breaks it down to glucose and fructose. These are reducing sugars, so Benedict's test now gives a positive result.

31 Prepare glucose solutions of known concentration, e.g. 0.1%, 0.01%. This can be done by diluting a stronger solution. Carry out Benedict's test on each, using the same volumes of both the glucose and Benedict's solutions each time.

32 Diagram should be the reverse of Fig. 16 on page 68.

33 The hydrochloric acid produced by the stomach denatures the enzyme.

34 The molecule of lactose has a different shape to that of sucrose. Therefore it does not fit the active site in sucrase.

35 It is particularly common in people of oriental descent, suggesting that the condition is passed from parent to child.

36 The concentration of lactose in the colon may be high. Therefore the difference in water potential between the contents of the colon and the blood is reduced. Less water is absorbed into the blood by osmosis, so more is contained in the faeces, leading to diarrhoea.

37a B and D have similar concentrations of amylase, at a higher concentration than the standard. C has no amylase.
37b Maltose (or glucose if biological samples also contained maltase).

38 Place samples of each strain of fungus on starch agar plates. Leave under controlled conditions for given time. Test with iodine solution, and measure sizes of clear areas around each fungus.

A1 COOH

A2 The graphs should be a similar shape to that in Fig. 10 in this chapter, but with an optimum temperature of at or just below 60 °C for glucose dehydrogenase and at or just below 100 °C for hydrogenase.

A3 60 °C, since glucose dehydrogenase is denatured above this temperature, so the whole process would cease.

A4 By using the enzyme cellulose which is produced by some species of bacteria.

A5a 200%
A5b 28 000 000 000
A5c Fees for waste disposal – getting rid of an industrial waste product
A5d Less % profit since start-up costs would be the same
A5e Energy costs, cost of enzymes

Chapter 5

1 The moisture will slow down the rate of oxygen diffusion into the blood, since diffusion is much slower in liquids than in gases.

2 As soon as oxygen enters the blood it is moved away from the exchange surface, maintaining the steep concentration gradient.

3 A fairer unit would be litre of oxygen per minute per kilogram of body mass.

4 (50 ÷ 130) × 100 = 38.5%.

5 Line graph since both variables are continuous

6a The winning time for men is always faster than that for women. The times for men and women both show an overall reduction in time, but the decrease between each successive games has not been constant. The performance of women improved more than that for men.
6b One possible reason might be that male athletes have a higher ratio of muscle mass to body mass than female athletes.

7 Pyrogens released by macrophages in response to infection stimulate the temperature-regulating mechanism of the body.

8 Inflammation of lung tissue reduces the rate of gas exchange.

9a Immigration – London is the main entry point for immigrants entering from countries where TB is endemic.

9b Dispersal of some immigrants from London

10 Fibrosis results in a loss of elasticity in the lungs. Since the lungs cannot expand as much, vital capacity is reduced, as is maximum ventilation capacity which is a function of vital capacity and time. The rate of oxygen transfer into the blood is reduced because less oxygen is delivered to the exchange surface.

11 The drugs affect the muscle fibres in the bronchiole walls, causing them to relax, opening up the bronchioles.

12 The map shows a link, since asthma is most prevalent in industrialised countries, where there is likely to be more pollution.

13 There are too many variables to take into account. Ethically, it would be wrong to take a group of children and expose them to pathogens.

14 The differences reported by the Californian team were only slight. It may be that their results were not repeatable in Great Britain and were therefore unreliable.

15 There is an association, because not everyone living near a road develops asthma.

16 The research should look at a wider range of patients – not just the ones with a particular genetic makeup.

17a Chemicals in the cigarette smoke paralyse and destroy cilia in the bronchial tubes. The cilia normally clear the tubes of mucus, which carries air-borne particles such as dust, bacteria and fungal spores. Inflammation of the tubes may occur. Smaller airways become narrower due to the growth of fibrous tissue, and alveoli may become blocked.

17b Trapped microorganisms may cause infections. Inflammation and mucus may block airways, reducing gaseous exchange. Persistent coughing damages the alveoli and reduces lung surface area. All contribute to reduced gaseous exchange and difficulty in breathing.

18 Tobacco smoke irritates the cells lining the bronchi and bronchioles. Increased mucus production and reduced numbers of cilia lead to persistent coughing. The result of coughing is damage to the alveoli.

19 Damage to the alveoli by coughing reduces the surface area of the lung, and loss of elastin reduces the ability of the lung to expel air during expiration. Reduced gaseous exchange fails to supply sufficient oxygen and remove waste carbon dioxide from the blood. Breathing rate increases causing breathlessness, even at rest.

A1 The partial pressure of oxygen in air decreases with altitude.

A2 Diffusion of gases is much slower in liquids than in air.

A3 Allowing body processes to adapt to changed conditions

A4 Shortness of breath, mental confusion and inability to walk

A5 The higher pressure it provides simulates low-altitude conditions.

Chapter 6

1 $60 \div 0.8 = 75$ beats per minute

2 To prevent the valve flaps turning 'inside-out'

3a Blood by-passes the lungs, which are non-functional in the fetus.

3b Much of the blood does not flow through the lungs and therefore is not oxygenated.

4a A fall in systolic blood pressure, since some blood is forced into the atrium rather than the aorta

4b A fall in diastolic blood pressure, since some blood will pass back from the aorta into the left ventricle.

5 The left ventricle pumps blood to the whole body (except the lungs) whereas the right ventricle pumps blood only to the lungs.

6a Multiply your heart rate by 60, then by 24, then by 75 cm^3.

6b Increasing the heart rate, and the volume of blood pumped per beat, increases the blood flow to the muscles, so oxygen and glucose are supplied at a faster rate during activity. This provides the resources for faster respiration in the muscles and hence more energy for contraction. The increased blood flow also carries away carbon dioxide and heat from muscles more rapidly.

7a The contraction of the ventricles, forcing blood against them

7b The contraction of the ventricles, forcing blood against them

7c Ventricular diastole, blood from arteries forcing against them

7d The contraction of the atria, forcing blood against them

8 Sex, smoking, age, blood cholesterol, body mass, blood pressure

9 15–20%

10a At all ages, men have a higher risk of heart disease than women.

10b Smoking increases the risk of CHD at all ages in both sexes; the effect on men is greater than the effect on women; smoking has a greater effect than the other three factors.

10c Age significantly increases the effect of the other three factors; its effect is greater in men than in women.

10d Increased ratio increases the risk at all ages in men, but does not increase the risk for women under 44 years of age.

11 Issues include: were the patients informed of the nature of the investigation? Did they give their consent? The patients who received stents fared better in the long term than those on angioplasty. Is this fair to the angioplasty patients?

A1a A trial in which the participants are placed randomly into an experimental group and a control group. Neither doctors nor patients know which patients receive the treatment and which the placebo.

A1b A pill that looks the same but contains no pravastatin

A2 Pravastatin or placebo

A3 Changes in the thickness of the carotid intima-media, cholesterol and triglyceride levels

A4 Measurements which in themselves could be used to predict the thickness of the carotid intima-media.

A5 Safe, since no serious adverse events were reported; none of the children discontinued treatment because of an adverse event.

A6 Statin treatment effectively reduced reduced total cholesterol (by 22.5%) and LDL-cholesterol (by 29.2%) but triglycerides reduced by only 1.9%. HDL-cholesterol increased by only 3.1%.

A7 Reasonably reliable, since there were 186 children in the trial, but a larger-scale survey would be needed to produce valid data.

A8 The *Daily Telegraph* and the *Daily Mirror* headlines were reasonable summaries of the results, but the *Daily Express* might have led to a conclusion that statins should given to all 8-year-old children.

Chapter 7

1 Enzymes that split up a compound by inserting the components of a water molecule.

2 In the cellular response, T lymphocytes attack pathogens and infected cells. In the humoral response, B lymphocytes produce antibodies to destroy the pathogen.

3 A compound made up of two parts: one part protein, one part carbohydrate.

4 Viral infection requires viruses to replicate their genetic material inside the cells of an organism. This is the first step in viral reproduction and spread within the body. Preventing replication therefore prevents the infection taking hold in the body.

5 Pig insulin is very similar to human insulin. It works in human cells, and does not provoke an immune response. Pig interferon is very different from human insulin and does not work in human cells.

6 Kaposi's sarcoma is an opportunistic disease linked to the breakdown of the immune system in HIV/AIDS.

7 Antibodies are produced by B lymphocytes.

8 The presence of two active sites on each antibody molecule enables the formation of an interlocking antibody–antigen complex that traps the pathogen and prevents it from infecting cells. Trapped pathogens can be engulfed by macrophages.

9 The primary response

10 Following a first infection with a pathogen, lymphocytes known as memory cells are produced and remain in the body. These cells are ready to stimulate a secondary response as soon as the pathogen re-invades the body.

11 The viruses that cause the common cold and influenza exist in a variety of forms, and new forms constantly arise by mutation. Immunity resulting from infection by one type is no use when a different form infects the body.

12 Passive immunity involves the introduction of antibodies into the bloodstream, and these are effective for a few weeks or months. Passive immunity does not result in the production of memory cells, so there is no long-term immunity.

13 As percentage vaccination increased between 1980 and 1982, the number of cases of diphtheria reduced considerably. Percentage vaccination has stayed fairly constant at about 85% since 1990, but the number of cases has fluctuated in that period.

14 The fall to 80% vaccination in the previous 3 years was not high enough to prevent outbreaks.

15 They are produced by competent B lymphocytes, which have been activated by a specific antigen to produce the corresponding antibody. These cells then replicate by mitosis to produce clones – genetically identical cells, all of which produce the same specific antibody.

16 Cytotoxic means 'poisonous to cells'. Cytotoxic substances often kill cells.

17 The monoclonal antibody targets the cell requiring the cytotoxic drug. The cytotoxic drug is therefore delivered straight to the cell and kills it.

18 Radioactive substances emit radiation which can damage the genetic material of cells. A short half-life means that the substance rapidly loses its radioactivity.

A1 Dr Wakefield's study was based on a very small sample – only 12 children. The report could have been biased because it was commissioned by parents of children with autism. Other studies have failed to provide a link, except for a report that the measles virus was found in the gut of some children with autism. The most powerful evidence against the link is the Japanese study, which showed that the number of autism cases has continued to rise even after the MMR vaccine was withdrawn, including children who were not born until after the vaccine was withdrawn. Overall, the evidence points away from a link between the MMR vaccine and autism.

Chapter 8

1a (c) because it shows four distinct phenotypes, corresponding to A, B, AB and O blood groups.
1b (b) because it shows two distinct phenotypes, i.e. wet and dry.
1c (d) because it shows continuous variation. Notice that the distribution curve is not quite normal; it is skewed to the right, meaning that there are more people with an extremely high mass than might be expected. This is because of the environmental factor of 'over-eating' is common in Western societies.
1d (a) because it shows two distinct phenotypes, each showing a range of continuous variation. The continuous variation results from environmental factors that affect growth.

2 Factors such as light, water and nutrient availability; temperature; wind; grazing by animals; disease

3 Siamese cats have an allele of the gene that produces an enzyme needed for synthesis of black pigment. This enzyme is denatured by temperatures above 37 °C. Only the parts of the body where the skin temperature is below 37 °C produce black pigment – the tail, ears and lower legs.

4 A male can mate with several females and can successfully father many young. It may be a more economical use of available food resources for a population to contain a high proportion of females.

5 It may become more difficult to maintain nests below 33 °C, so more males may be hatched. This may increase pressure on food resources and in the longer term reduce the population. Also, fewer females means that fewer young will be produced.

A1a 0% of Native Americans have blood group B.
A1b B and AB: genotypes I^AI^A, I^AI^o and I^oI^o.

A2a No. For example, Nigerian and Japanese people both have 23% with Group B, but their skin colour differs. On the other hand, both Nigerians and native Australians have black skin, but the percentage with group B is quite different.
A2b The black population derives from people of West African ancestry, such as Nigerians, whereas the white population is largely European in origin, where the proportion of group B is around 9%.

A3a They probably migrated to the area where they live from somewhere with a low proportion of group B in the population, such as from the population in Siberia from which some people moved east to North America.
A3b Information from DNA analysis, e.g. from comparing mutations in genes for substances such as haemoglobin

Chapter 9

1a 1
1b Sperm – 1; fertilised egg – 2; young 16-cell embryo – 32.
1c 450

2

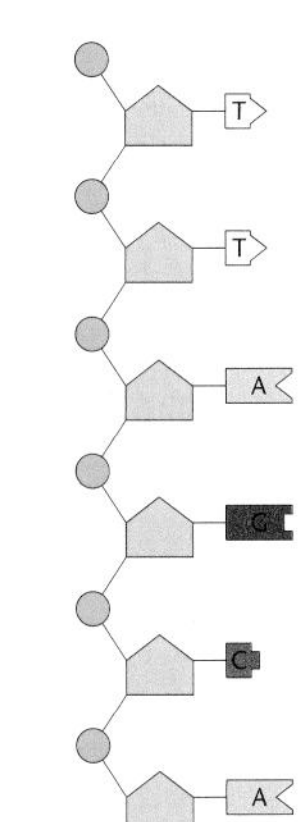

3

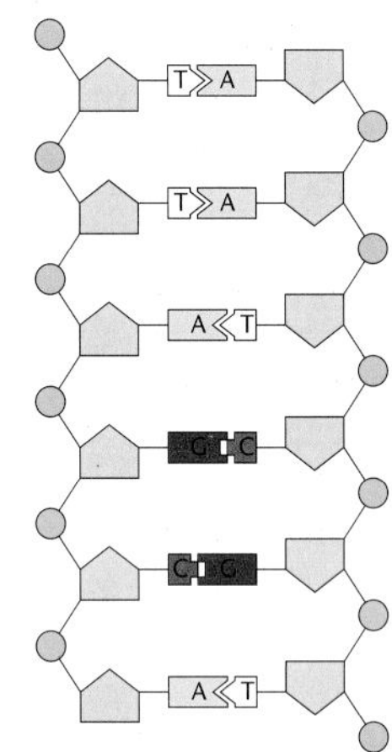

4 In each organism the percentages of adenine and thymine are the same, and the percentages of cytosine and guanine are the same. This is because they fit together as complementary pairs in DNA, so there must be the same number of adenine and thymine bases, and of cytosine and guanine.

5 24% (26% = adenine, so 26% = thymine. The remaining 48% are split equally between cytosine and guanine).

6

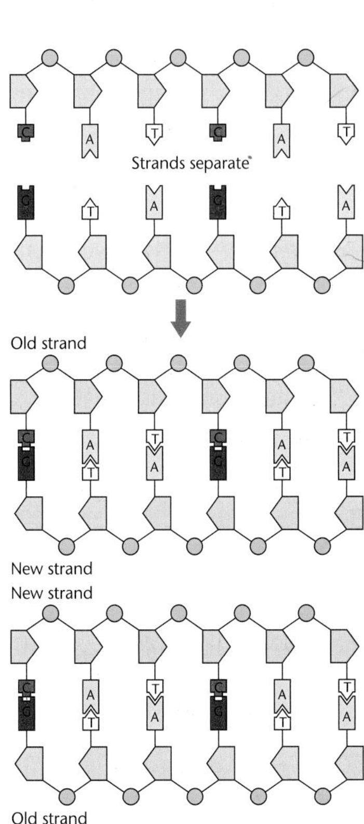

7

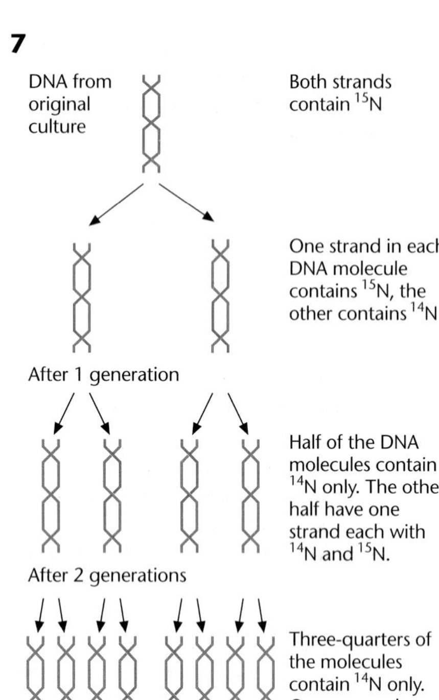
DNA from original culture
Both strands contain ^{15}N
One strand in each DNA molecule contains ^{15}N, the other contains ^{14}N
After 1 generation
Half of the DNA molecules contain ^{14}N only. The other half have one strand each with ^{14}N and ^{15}N.
After 2 generations
Three-quarters of the molecules contain ^{14}N only. One quarter have one strand each with ^{14}N and ^{15}N.
After 3 generations

8 Two equal-sized bands, one at the top level (containing ^{14}N only) and one at the lowest level (containing ^{15}N only).

9 Phenylalanine, valine, asparagine, glutamine

10 Cysteine, asparagine, histidine, valine, histidine

A1a Without the complete DNA some enzymes/proteins would not be coded for, so development would not proceed.
A1b DNA cannot be obtained from protein.
A1c Mammalian red blood cells do not contain nuclei.

A2 DNA polymerase

A3a Because the nucleus contains crocodile DNA.
A3b Nutrients

A4 Obtaining a complete sample of DNA; providing the correct nutrients and conditions to mimic the whole dinosaur egg

Chapter 10

1a There will be twice as much at the end of interphase.
1b 92
1c So that the DNA molecules can easily be pulled to each new nucleus in the later stages of mitosis, without getting tangled.
1d So that they attach to the spindle together. The members of each pair are then pulled to opposite poles. This makes sure that each cell has one copy of each chromosome.

2

Metaphase

Anaphase

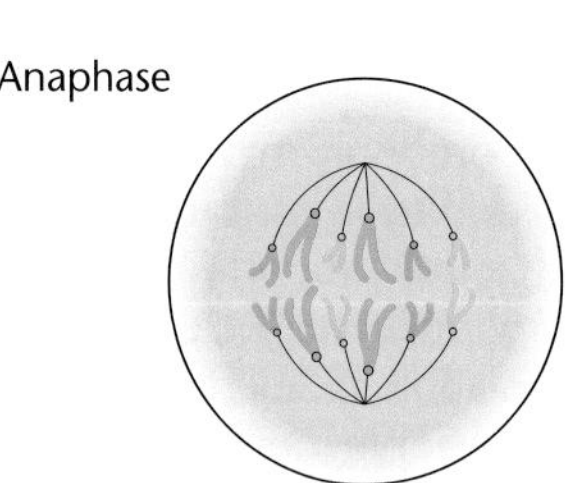

Telophase

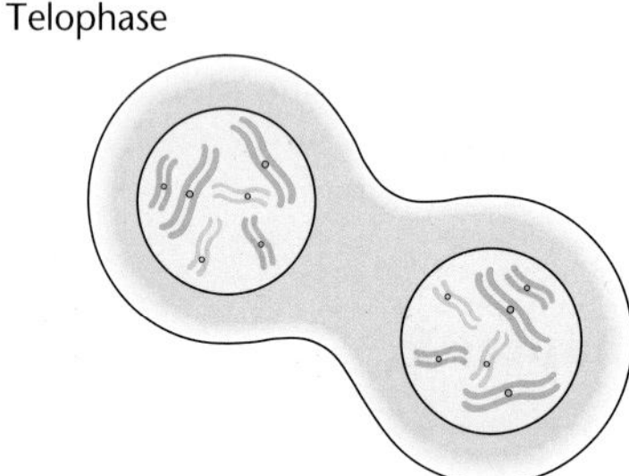

3a 2^4, i.e. 16 combinations
3b 2^{23}, i.e. 8 388 608 combinations

4 Any three from:
Chromosomes replicate in interphase or before the division starts.
Chromatids are joined by centromeres.
Nuclear membrane disappears at end of prophase.
Centromeres attach to equator or spindle in metaphase.
Chromatids are pulled to opposite poles in anaphase.

5

A B C | A b C
a B c | a b c

A1 Cyclin stimulates production of another protein that in turn stimulates the replication of the DNA and initiates mitosis.

A2 The second protein breaks down cyclin.

A3 It is over-produced.

A4 It promotes rapid growth of mammary cells.

A5 Finding drugs to block the action of cyclin D1, which could arrest the growth of human breast cancer without harming normal cells.

A6 As with all experiments involving animals, the pain or distress caused by treating the animals must be balanced against the increased life expectancy of humans receiving the treatment tested in this way. Animal rights activists would argue that the experiments should be done using tissues or models, without the use of live animals.

Chapter 11

1 The plant cell has a cellulose wall, chloroplasts and a vacuole.

2 Glycogen is more compact than starch. Reducing bulk is more important to animals than plants because animals move about whilst plants are stationary.

3 Impregnation with lignin in xylem cells and with suberin in cork cells.

4 Similarities: size; shape; origin; both carry out metabolic processes. Differences: chloroplasts only present in plant cells; chloroplasts contain chlorophyll and are green, mitochondria don't and are not; mitochondria carry out catabolic reactions; chloroplasts carry out anabolic reactions.

A1 The results supported the hypotheses as far as the production of organic chemicals, but not as far as the formation of primitive life forms.

A2 No, they only showed how organic molecules might have been produced. There is no evidence to show that organic materials were actually produced in this way.

A3 The experiments show only that organic materials might be produced in this way on other planets. Both authors hypothesised the formation of 'membranes' that would assimilate organic materials to produce primitive life forms, but the experiments did not support this 'assimilation'.

A4a 4 billion years 3.5 billion years
Primitive prokayotes → Eubacteria
3 billion years 1.5 billion years
→ Cyanobacteria → Eukaryotes

A4b This theory states that the ancestors of modern eukaryotic cells were 'symbiotic consortiums' of prokaryotic cells. For example, aerobic bacteria might have invaded larger amoeba-like anaerobic bacteria. Both organisms would benefit – the anaerobic bacteria would ingest organic material, and aerobic bacteria would oxidise this to provide ATP for both organisms. The embedded aerobic bacteria assumed the role of what we now call mitochondria. These cells were the ancestors of eukaryotic animal cells. Some amoeba-like bacteria also formed symbiotic consortia with cyanobacteria. These cells were the ancestors of eukaryotic plant cells.

Chapter 12

1 Flapping its ears will move a current of air over them, increasing the rate of heat loss via conduction and convection.

2 The small ears present a reduced surface area, which minimises heat loss. Large fat stores under the skin, and fur act as insulations.

3 The shrew is likely to be the more successful since it has the larger surface area in relation to its volume.

4 Shrews have a large surface area to volume ratio so they lose energy as heat comparatively rapidly. They therefore need to eat a large amount of food to replenish their energy stores.

5a Manatee: $4 \times 0.75 = 9.4\ m^{-2}$
Walrus: $3.5 \times 1 = 11\ .0\ m^{-2}$
Elephant: $3 \times 2 = 18.8\ m^{-2}$

5b Manatee: $1000\ kg \div 1000\ kg\ m^{-3} = 1.0\ m^3$
Walrus: $1\ 700\ kg \div 1000\ kg\ m^{-3} = 1.7\ m^3$
Elephant: $5000\ kg \div 1000\ kg\ m^{-3} = 5.0\ m^3$

5c Manatee: $9.4 \div 1.0 = 9.4$
Walrus: $11.0 \div 1.7 = 6.5$
Elephant: $18.8 \div 5.0 = 3.8$

6 In water: water molecules are much closer together than the molecules in air, so conduction of heat is more rapid.

7 The walrus has the smaller surface area to volume ratio. This as an advantage because the walrus living in arctic water will lose less heat. The larger surface area to volume ratio of the manatee is an advantage in warmer climates as it allows the manatee to lose heat more easily.

8a The tapeworm exchanges oxygen, carbon dioxide and soluble food molecules with its host.

8b It does this by diffusion.

9 The tapeworm is long and flat. This ensures that substances do not have to diffuse more than 1 mm to enter any of its cells.

10 Oxygen diffuses in air along the tracheae to the tracheoles, some of which supply the mitochondria direct. All the diffusion is in air, not liquids, so is much faster. In humans, most of the diffusion occurs via liquids. The distance from the outside of a human body cell to mitochondria is comparatively large. Diffusion through cytoplasm is slow because the oxygen is in solution, whereas insect mitochondria are supplied with gaseous oxygen from the tracheoles.

11 When the spiracles are closed, the carbon dioxide concentration in the tracheae increases. When the spiracles open, the concentration of carbon dioxide in the tracheae is greater than that in the burrow, so carbon dioxide will diffuse out of the insect's body.

12a There is a correlation between spiracle opening and water loss in both DGC and cyclic gas exchange. In continuous gas exchange, the rate of water loss is almost constant.

12b Overall, the data appear to show a correlation between spiracle opening and water loss, supporting the hypothesis that spiracle closing is a water-conserving mechanism.

13 Differences: Humans obtain oxygen from air, fish from water; humans have alveoli, fish have gill filaments; ventilation is tidal in humans, unidirectional in fish.
Similarities: Both have a large respiratory surface; both have a good blood supply to the respiratory surface; both ventilate the respiratory surface.

14a Carbon dioxide into the leaf; oxygen out of the leaf; water vapour out of the leaf

14b Oxygen into the leaf; carbon dioxide out of the leaf; water out of the leaf

15 Oxygen

16 Glucose, amino acids, carbon dioxide, water

17 Oxygen, glucose and amino acids pass into the tissues; carbon dioxide passes into the air in the alveoli.

18 In the skin, reduced blood flow through skin capillaries makes the skin look paler and the body conserves heat. In the villi, reduced blood flow through capillaries reduces the rate of absorption of soluble food molecules into the blood.

19 The total cross-sectional area of capillaries is greater than that of arterioles.

20 The total cross-sectional area of veins is less than that of venules.

21 In the lungs and kidneys, because there is a high level of exchange of materials in these organs.

22 The slow flow rate allows more time for diffusion; they are thinner, reducing the length of the diffusion path.

23a $\pi \times (2\ \text{mm})^2 = 12.6\ \text{mm}^2$
23b $\pi \times (4\ \mu\text{m})^2 = 50.3\ \mu\text{m}^2$

24 $12.6\ \text{mm}^2 \div 50.3\ \mu\text{m}^2 = 250$

25a 95%
25b 100%
25c 100%

26 11 kPa

27 100 – 15 = 85%

28 The haemoglobin becomes fully saturated with oxygen even at the low concentrations of oxygen in the surroundings.

29 An S-shaped curve to the left of that for human haemoglobin; the partial pressure of oxygen will be low at high altitude and llama haemoglobin will become saturated with oxygen at these partial pressures.

30

	Red blood cells	White blood cells	Fibrinogen	Other plasma proteins	Glucose, amino acids, ions etc.
Blood	✓	✓	✓	✓	✓
Plasma	✗	✗	✓	✓	✓
Tissue fluid	✗	✓	✗	✗	✓

31

	Capillaries to tissue	Tissue to capillaries
Alveoli	carbon dioxide	oxygen
Intestinal villi	oxygen	carbon dioxide, glucose, amino acids
Brain	oxygen, glucose, amino acids	carbon dioxide
Leg muscles	oxygen, glucose (amino acids)	carbon dioxide
Liver	oxygen, glucose, amino acids	carbon dioxide, urea (glucose when shortage in blood)
Kidney	oxygen, glucose, urea, amino acids	carbon dioxide (glucose and amino acids reabsorbed)

32 Lack of proteins in the blood plasma makes the water potential less negative. Less water returns to the capillaries at the venous end from the capillaries. Therefore, fluid accumulates in the tissues.

33 It reduces the amount of donated blood a patient receives thus minimising the risk of infection; it will stay fresh for 6 months or more; the artificial compounds can be transfused into people of any blood group without fear of provoking a serious allergic reaction.

34 It is cheaper, and may be more acceptable to patients.

35 Haemoglobin works only when intact and when assisted by a cofactor found in red blood cells. Haemoglobin is quickly broken down by enzymes, and the fragments can poison the kidneys.

36 **a and b.** These are ethical issues which have no definitive answer. You could argue that paying donors would increase the supply of blood, but against that you could argue that body parts such as blood should not be for sale.

37 An increase in temperature increases the energy of water molecules, resulting in them moving away from the leaf faster, therefore making the water potential of air next to the leaf more negative. A decrease in humidity decreases the number of water molecules in the air, making the water potential of air next to the leaf more negative. An increase in wind speed moves water molecules away from the leaf faster, therefore making the water potential of air next to the leaf more negative.

38a In very dry soil the stomata do not open as wide during the early morning, and close completely before noon.
38b On a cloudy day the stomata do not open as wide as on a sunny day, and begin to close earlier in the afternoon than on a sunny day.

39 The advantages of opening at night are that carbon dioxide levels in the plant can be replenished without losing much water by transpiration. The disadvantage is that there is no light available for photosynthesis. The advantage of closing the stomata around midday is that less water is lost by transpiration. The disadvantage is that carbon dioxide uptake is reduced at the time when light intensity for photosynthesis is greatest.

40 A plant that lives in very wet conditions, such as a bog plant.

41 Cacti have a reduced surface area to minimise water loss but this also reduces the area for photosynthesis, which produces the carbohydrates needed for growth.

42 The stomata are at the bottom of folds in the leaf, where there is a high humidity because of slower removal of water vapour molecules by air currents. The hairs reduce the flow of air, increasing the humidity at the leaf surface. Both of these make the water potential of the air less negative, reducing the rate of diffusion of water molecules.

43 Screwing up reduces the surface area from which transpiration could occur. Many stomata will be on the inside of the curves, resulting in a higher internal humidity and protection from air currents.

44 Smaller leaves give a smaller surface area and fewer stomata, both of which will reduce the rate of transpiration, but some photosynthesis can still take place.

45 Some roots are nearer the surface of the soil, so any rainfall will reach them quickly. Some roots go very deep, in case there is any water deep in the soil. Other roots grow towards moisture trapped under stones.

46 2500 kPa ÷ 200 kPa per 10 metres, which equals 12.5 × 10 m = 125 m

47 Heat energy from the Sun

48 Tree C: its diameter decreases the least, showing that there is the least tension in the xylem vessels resulting from the transpiration pull.

49 The cell surface membrane of the living cells adjacent to the dead xylem vessel cell

A1a S = Siemens
A1b It measures all ions present. A different chemical test would be needed for each ion.
A1c Given both ions and water

A2 The temperature there is higher, resulting in a greater transpiration rate.

A3a 620 + 30 = 650 cm
A3b The warmer the temperature, the more growth, but fresh water at Besor increased the rate of growth compared with the brackish areas.
A3c The difference was probably due to the different ratio of Na^+ to Ca^{2+} in the water.

A4a The leaves at Neot grew rapidly, reaching maximum size by May. Leaves at Besor grew more slowly, but reached the same size as those from Neot by June.
A4b The difference was probably due to the warmer spring at Neot.

Chapter 13

1 The parents have different numbers of chromosomes, so the sex cells will have different numbers too: 33 in a horse's egg and 31 in a donkey's sperm. The hybrid mule has different numbers of maternal and paternal chromosomes, so two chromosomes will have no partner with which to pair up during prophase of meiosis. The others may also be of different sizes and shapes and be unable to form bivalents. Mules are therefore unable to produce gametes by meiosis.

2 They lived in the sea where conditions may have stayed constant. Successful species would be able to survive largely unchanged while conditions stayed the same.

3 Hydrogen bonds

4 Complimentary base pairing

5 No, they both developed from a common ancestor, shown by the fact that they are the ends of different branches on the tree.

6 So that it can be recognised when hybridised

7 It is unlikely that any sample would show 50% hybridisation, so the results are plotted and the 50% figure determined from the graph.

8 They are at the end of one of the earliest branches in the evolutionary sequence.

9 The more closely related, the fewer the differences in the beta chain. For example, there is only one difference between the human beta chain and that of the gorilla.

10 The same amino acid is found in this position in all the species.

11 Between drosophila and the chicken based on the position of the amino acids Val, Ala and Gln.

12 They do not have a complimentary shape.

13 The sequence of amino acids in the chain

14 Stickleback models can be made with many different colour patterns and shapes. These should be placed near a male fish in a standardised way and the responses of the male fish recorded. The experiment needs to be repeated many times with many different male sticklebacks to obtain reliable results.

A1 Within the hominid population there would be a range of hairiness. Those with the thinnest hair would cool quickest, and would perhaps be able to pursue prey for longer, as well as requiring less water. As a result, hominids with thin hair would be more likely to survive and pass on the alleles favouring thin hair to their offspring. This process would continue for many generations until the bodies of hominids were largely hairless.

A2 The active processes that maintain the brain need a continuous supply of energy from respiration, and this inevitably generates waste heat. The excess heat must be released in order to maintain the blood and body temperature within safe limits.

A3a The fat could would provide buoyancy. It might also help insulation, although in warm waters this might not be necessary.

A3b Aquatic vegetation is much softer. Also shellfish and other animal life found in water is less tough than meat from terrestrial animals.

A3c Species with babies that drowned in water would clearly be at a significant disadvantage in an aquatic environment.

A3d Wasting large amounts of water in urine would be disadvantageous on hot, dry grasslands, but not a problem in aquatic conditions.

A3e They would be able to put their heads under water to search for food.

A4 For example, other aquatic mammals living in shallow water are neither bipedal nor hairless (such as otters).

Chapter 14

1a Antibiotics affect the metabolic activities of bacteria, which are prokaryotic cells. These cells are very different from eukaryotic cells of humans, so antibiotics do not affect human cells.

1b Sulphonamides act by disrupting folic acid synthesis in microbes. Human cells do not produce folic acid.

2 Fungal cells are more similar to human cells than are bacterial cells, so fungal treatments are more likely to damage human cells.

3 Unlike bacterial chromosomes, plasmids can cross species boundaries. This is a major factor in the spread of bacterial resistance.

4 Pathogens mutate spontaneously, producing resistant strains. Antibiotics kill individual pathogens of the non-resistant strain. Individual resistant pathogens survive and reproduce. The population of the resistant strain rises.

5 MRSA stands for methicillin-resistant *Staphylococcus aureus*. Infections caused by MRSA are difficult to treat.

6 That is where most antibiotics are used.

7 Attention to personal hygiene so that bacteria are not passed from an infected person by contact.

8 The rate has quadrupled, with the steepest rises in most recent years.

9 It has increased almost 40-fold with the greatest increase in recent years.

10 No – diagnosis and reporting were not as accurate in the earlier years.

11 There is a greater proportion of deaths in old people because the immune system is not as effective.

12 Eliminating resistant plasmids from bacteria; preventing the expression of resistant genes in the plasmids

A1 Over-use of antibiotics leads to an increase in the number of antibiotic-resistant bacteria, leading to difficulty in treating illnesses caused by the resistant bacteria.

A2 Study A, largest number of participants

A3 A, C and D, since no placebo was used.

A4 There was very little difference between treating with antibiotics, symptomatic treatment and placebo.

A5 Since a majority of parents ask for antibiotics, their efficiency should be investigated. However, there is no information as to whether parents gave their informed consent to their children taking part in a trial. Since colds are rarely serious, giving some children the placebo is not likely to disadvantage them.

Chapter 15

1 Evaporation is reduced close to the hedge. Evaporation reduces because moisture is not dispersed by the wind, so humidity close to the soil increases, thus decreasing the diffusion gradient. Therefore less moisture is lost from the soil. The temperature is increased near the hedge because the cooling effect of evaporation is reduced.

2 At about 14 m from the hedge.

3 The hedge may shade the crop, reducing photosynthesis. There may also be competition for mineral ions and water from the roots of the hedge.

4 The area below the curve for yield showing an increase is greater than the area above the curve showing a decrease.

5 The wind may not blow consistently from the same direction.

6 For timber; space for agriculture; space for housing and industry

7 Many species are found only in rainforests; e.g. 75% of arthropods are only found there.

8 It is likely that the sum total of the different alleles in the isolated population will be less than that for all the animals of that species. Over time, as inbreeding occurs, more and more alleles are likely to be lost.

9 The results show that, for all sizes of trees, mortality was higher on the edge of the forest than in the centre. This effect is greatest for the largest trees. There are several possible reasons for this. Trees on forest edges are more exposed to high winds than in the forest interior, and this is likely to cause more damage to very tall trees than to smaller ones. They get more sunlight, and this may mean that they are more likely to have climbers such as lianas (woody climbing plants) growing up them, which can reduce their life-span. And they are more exposed to drying air than are trees inside the very humid deep forest.

10 Hunting was done mainly by men from the northern hemisphere. The southern group were much less accessible.

11 Being genetically very close, their proteins will be very similar and less likely to be regarded as 'foreign' by macrophages.

12a Because all the present plants are offspring of the one original plant.
12b There is not the range of alleles upon which natural selection can operate in changing environmental conditions.

13 Research needs to be done to find out what conditions this species needs for survival – for example, soil type, water; it is also useful to know other information about it, such as how the flowers are pollinated, how the seeds are dispersed, and what conditions the seeds require for germination. Once this is known, then St Helena can be searched to find suitable sites where the boxwood would be expected to be able to survive. These sites will need protection, perhaps by fencing, to ensure that grazing or human disturbance do not threaten the newly introduced plants.

14 IR-8 plants are much less likely to be blown over.

15 Advantages: Super Rice has fewer but stronger stems and there are many more seeds on each rice flower. Super Rice has a higher proportion of grain to stalk; It has a vigorous root.
Disadvantage: The prototype couldn't supply enough carbohydrates and nutrients to fill the grains.

16 Overall there has been a 40% reduction in the area of rice harvested, but a 20% increase the amount of rice produced.

17 Whilst area under cultivation has decreased, rice production has increased. This could be because of the use of more productive rice strains and the increased use of fertilisers and pesticides.

18 Fast growth, high proportion of meat to bone

19 Students' own discussions

20a Type of chicken – egg-layer or broiler
20b Weight of chicken
20c Same age at start of experiment, same food, same amount of food

21 By using a larger sample size or repeating the experiment

22 Broiler chicken: (2400 g – 50 g) ÷ 45 days = 52.2 g day^{-1}
Layer chicken: (600 g – 40 g) ÷ 45 days = 12.4 g day^{-1}

23 Meat chickens increase mass approximately five times more quickly than layer chickens.

24a

Species	*n*	*n*–1	*n*(*n*–1)
Grass	185	184	34 040
Thistles	28	27	756
Stinging	35	34	1190
Moss	2	1	2
Total (*N*) =	250		35 988

24b $D = \dfrac{250 \times 249}{35\,988}$
$= 1.73$

24c The species diversity in the field is much less than that in the lawn. We can only guess the reasons for this. Some possibilities would be:

- the sheep graze selectively in the field – perhaps they do not eat thistles or nettles, but have grazed so heavily on other plants that they cannot grow;
- perhaps the field has been sprayed with a selective weedkiller, whereas no spray has been used on the lawn;
- perhaps the soil in the field and lawn differ in their mineral content or soil type

24d Reliable since the same method has been used in both places, and the same total number of point quadrat samples.

A1 Wading birds

A2 The measures taken from 1987 onwards have halted the decline in the number of territories occupied by birds, and increased the number for several of the birds, thus helping to preserve biodiversity.

A3 The pesticides get leached in the water, killing insect species, in turn affecting food webs and reducing biodiversity.

A4 Cutting for hay only once a year will allow species that live in grassland to complete their annual life cycles; cutting for silage several times a year will prevent this.

Glossary

α-glycosidic bonds
Bonds formed between two α-glucose molecules.

β-glycosidic bonds
Bonds formed between two β-glucose molecules.

Abdominal muscles
The muscles that make up the walls of the abdomen.

Abscisic acid (ABA)
A plant hormone that is involved in leaf fall.

Acid group (COOH)
A group found in all fatty acids and amino acids, which dissociates producing hydrogen ions.

Aetiology
A list of the characteristic causes of a disease.

Amino group (NH_2)
The group found at one end of every amino acid.

Amylase
An enzyme that hydrolyses starch into maltose.

Antibiotic
A substance produced by a microorganism that kills another species of microorganism.

Antibodies
Protein molecules produced by β lymphocytes to combat infection and provide immunity.

Antibody-antigen complex
The lattice-like structure that forms when a number of antibody molecules combine with a number of antigens.

Antigen
A substance that initiates an immune response.

Antimetabolites
Antibiotics that inhibit key microbial metabolic reactions.

Arterioles
Narrower blood vessels formed by branching arteries.

Artificial fertilizer
A fertilizer made from inorganic components, such as ammonium phosphate.

Artificial selection
Intentional breeding for selected characteristics.

Association
A coming together of two organisms or structures.

Atherosclerosis
Chronic inflammation of the walls of arteries, usually caused by the build-up of fatty deposits in the inner lining.

Atrial diastole
Relaxation of the atria of the heart.

Atrial systole
Contraction of the atria of the heart.

Atrioventricular node
An area of specialised tissue, which conducts electrical impulses from the atria to the ventricles.

B cell
Lymphocyte (a type of white blood cell) that produces antibodies.

Bacteria
Microorganisms composed of prokaryotic cells.

Benedict's solution
A blue solution that turns red when heated with reducing sugars.

Binomial system
A system in which all organisms are identified by a two part name – a generic name and a specific name.

Biodiversity
The variation of life forms in a habitat or ecosystem.

Biuret solution
A colourless solution that turns violet / mauve when warmed with proteins.

Bivalents
The pairing of two homologous chromosomes in an early stage of meiosis.

Bundle of His
Specialised conducting tissue found in the wall between the two ventricles of the heart.

Bursa
A space in a tissue often filled with fluid.

Capsule
Tissue forming a wall around a structure. In bacteria – a slimy protective layer outside the cell wall.

Carbohydrate
A chemical containing carbon, hydrogen and oxygen; the hydrogen and oxygen being in the ratio of 2:1.

Carcinogen
A chemical that causes cancer.

Cardiac output
The volume of blood pumped by the left ventricle of the heart per minute.

Cardiac reserve
The difference between the resting output and the maximum output the heart can achieve.

Cardiovascular system
The heart, arteries, veins and capillaries.

Cell wall
A structure that surrounds the cell in plants and some types of microorganism.

Cell-mediated response
The process by which T cells attack pathogens directly.

Cerebral
Associated with the brain.

Cervical
Associated with the cervix (neck) of the uterus (womb).

Chemotherapy
Medical treatment involving drugs.

Chiasmata
Breaking and crosswise rejoining of homologous chromatids during meiosis.

Class
A taxonomic classification coming between phylum and order.

Colostrum
The first milk produced by the mother after the birth of a baby.

Communicable disease
A disease that is transmissible by infection.

Competent
Term applied to lymphocytes that have been activated and can produce an immune response.

Complement
A plasma protein that can bind with an antibody-antigen complex to destroy a pathogen.

Condensation reaction
A reaction in which two compounds are joined together by removing the elements of a water molecule.

Conjugation
The coming together of two bacteria to exchange genetic material.

Continuous variation
The outcome of a number of factors influencing the expression of a characteristic.

Correlation
The degree to which one phenomenon is associated with another.

Cortex
An outer layer of an organism.

Co-transport
The simultaneous transport of two substances across a cell membrane.

Crossing over
The transfer of a block of genes from one chromatid to a homologous chromatid.

Cuticle
A waxy outer covering.

Cytochrome c
A protein that carries electrons during aerobic respiration.

Cytotoxic
Chemicals that poison cells.

Diarrhoea
Production of very watery faeces.

Diastole
Relaxation of the heart.

Dicotyledonous
A plant that has two 'seed-leaves' in its seed.

Disaccharides
A carbohydrate consisting of two single sugars joined together.

Discontinuous variation
Where only one or a few genes cause variation in a characteristic, resulting in two or a few classes.

Diversity Index
A formula for calculating species diversity.

DNA duplexes
A piece of single-stranded DNA and its complementary DNA sequence.

dsDNA
Double stranded DNA.

Egg albumen
The main protein found in egg white.

Embolus
A sudden blocking of an artery. An embolus is the clot. Embolism is the condition it causes.

Endodermal cells
Cells forming a layer around the vascular tissue in a plant root.

Endotoxins
Toxins released from a bacterium when it dies.

Epidemics
A sudden large increase in the number of people suffering from a disease.

Epidemiologists
People who study the causes of the spread of epidemic diseases.

Epidemiology
The study of causes of the spread of epidemic diseases.

Epidermal cells
Cells forming a layer on the outside of an organism.

Epidermis
A layer of cells forming the outside of an organism.

Exotoxins
Poisons released by bacteria as they grow.

Extracellular
Outside of the cell.

Family
A taxonomic group coming between genus and species.

Fibre
A structure with a long, thin shape.

Founder effect
The establishment of a population from a few original organisms.

Fragmentation
Breaking up into pieces.

Fungi
A group of organisms that absorb food in solution through their cell walls and reproduce by producing spores.

Galactose
A sugar that joins with glucose to form the disaccharide lactose.

Gamma-globulin
Antibody found in human blood plasma.

Gene locus
The position of a gene on a chromosome.

Genetic engineering
The artificial transfer of genes from one organism to another.

Genus
A taxonomic group coming between family and species.

Gills
Structures used for gaseous exchange between an organism and water.

Glycerol
A chemical that combines with three fatty acid molecules to form a lipid.

Guard cells
The two cells that surround each stoma in a plant epidermis.

Habitat
A place where organisms live.

Heart rate
The number of heartbeats per minute.

Herbicides
A chemical that kills plants.

Herpes zoster
A virus that causes a painful, blistery, red rash.

Hierarchy
A classification system with the largest groups at the top and individual species at the bottom.

Homologous pairs
A pair of chromosomes carrying the same genes.

Homozygous
Where both alleles are the same.

Horizontal gene transmission
Transfer of genes between bacteria during conjugation.

Hormones
A substance secreted by cells that affects the metabolism or behaviour of other cells.

Humoral response
An immune response mediated by an antibody.

Hybridoma technique
Using cells that have been genetically engineered to produce large amounts of antibodies.

Hydrogen bonds
A weak chemical bond involving a hydrogen atom.

Hydrogen ions
A hydrogen atom that has lost an electron and is therefore negatively charged.

Hydroxyl groups (OH)
A negatively charged group in which oxygen is bonded to hydrogen.

Immune system
The system that defends the body against infection.

Immunoglobulins
Protein molecules produced by β lymphocytes.

Independent assortment
The random arrangement of chromosomes at metaphase in meiosis.

Infection
What occurs when viable microbes are found in the body.

Interferons
Substances produced by cells that destroy viruses.

Interstitial fluid
The fluid found between cells.

Introns
A non-coding section of DNA.

Iodine
Straw coloured iodine/potassium iodide solution turns blue/black in the presence of starch.

Kaposi's sarcoma
A cancer-type disease that causes abnormal growth of tissue under the skin.

Kingdom
The highest taxonomic group, above phylum.

Lactose
Disaccharide composed of glucose and galactose.

Lactose intolerance
The inability to digest lactose, resulting in abdominal swelling and diarrhoea.

Lamellae
Plates of cells.

Latent period
The time between two events.

Leucocyte
White blood cells that engulf bacteria.

Ligands
A molecule that binds to another molecule.

Lignin
A waterproofing substances found in the walls of xylem cells.

Lymphatic system
The system that helps to drain tissue fluid back into the blood system; also produces some types of lymphocyte.

Lymphocytes
Cells that help to defend the body against infection.

Lysozyme
An enzyme found in tears and saliva that destroys bacterial cell walls.

Macrophages
Lymphocytes that engulf bacteria.

Maltose
A disaccharide composed of two glucose molecules.

Maximum ventilation capacity
The maximum volume of air that can be taken in during one breath.

Mean
Calculated by taking the sum of all the values then dividing by the number of values.

Meiosis
A type of cell division in which the number of chromosomes is halved.

Memory cells
Lymphocytes that respond rapidly to reinfection.

Mesophyll
Tissue that makes up most of the inside of a leaf.

Metabolic rate
The amount of energy released via respiration in a given period.

Microfibrils
The basic unit of plant cell walls, composed of cellulose molecules.

Monoclonal antibodies
Antibodies produced by a single clone of white blood cells.

Monoculture
The growing of large areas of a single crop.

Multiple repeats
Many repeats of a DNA sequence.

Myosin
One of the proteins that forms muscle fibres.

Narrow spectrum antibiotics
Antibiotics effective against only a few types of bacteria.

Natural selection
Evolution of the organisms best suited to the environment.

Naturally induced immunity
Immunity to a disease as a result of exposure to it.

Negative correlation
As the value of one variable increases, the value of the dependent variable decreases.

Neutrophils
A white blood cell capable of engulfing pathogens.

Normal distribution
A bell shaped curve that is defined by the mean and the standard distribution of the data set.

Nucleoli
Structures in the nucleus involved in protein synthesis.

Oral rehydration therapy (ORT)
A solution containing ions and salts to combat the effects of diarrhoea.

Oncogenes
A gene that can release a cell from its normal restraints of growth, possibly leading to tumour formation.

Opsonisation
The coating of a pathogen with a compliment making it easier for phagocytes to ingest the pathogen.

Order
A taxonomic group coming between class and family.

Organic manure
Manure consisting mainly of animal faeces.

Organ
A structure with a specific function made up of different types of tissue.

Osmotic lysis
When a cell bursts due to osmosis.

Pancreas
An organ in the loop of the duodenum that produces hormones and digestive enzymes.

Pandemics
An epidemic that affects a wide geographical area.

Partial pressure
The pressure exerted by a single component in a mixture of gases.

Passive immunity
Immunity produced by giving an individual an injection containing the appropriate antibodies.

Pathogens
Organisms that cause disease.

Percentage cover
The percentage of an area within a quadrat that is covered by a particular species.

Pesticide
A chemical that kills pests such as insects.

Phago-lysosome
The structure formed when a lysosome incorporates an ingested particle.

Phagosome
An ingested particle.

Phenotypes
The total physical appearance of an organism.

Phospholipid bilayer
The membrane that surrounds all living cells; composed of two layers of phospholipid molecules.

Phylogenetic
Based on the natural relationships between organisms.

Phylum
A taxonomic group coming between kingdom and class.

Plasma cell
A B cell that secretes antibodies.

Population bottlenecks
An event that occurs when a large proportion of a population is killed, resulting in a loss of genetic diversity.

Positive correlation
When one variable increases, another increases as well.

Potassium iodide solution
Used to dissolve iodine for the test for starch.

Predators
Organisms that catch and eat other organisms.

Primary response
The antibodies made on the first exposure to an antigen.

Pseudopodia
Temporary cytoplasmic extrusions used in phagocytosis and locomotion.

Pulmonary ventilation
Exchange of air between the lungs and the atmosphere.

Purkinje fibres
Specialised muscle cells that carry the impulses controlling heartbeat.

R plasmid
A plasmid that confers immunity to one or more antibiotics in a bacterium.

Random fertilisation
Where any male gamete can fuse with any female gamete.

Random sampling
Collecting information from a number of randomly selected sites within an area.

Respiratory surface
A surface where oxygen and carbon dioxide are exchanged between an organism and the environment.

Risk factors
An occurrence associated with an increased rate of a subsequent disease.

RNA
A nucleic acid involved in protein synthesis by transferring information from DNA to the sites of protein synthesis.

Saline solution
A solution usually containing 0.9% sodium chloride solution, which is isotonic with blood plasma.

Salivary glands
Glands in the neck that produce saliva containing the enzyme amylase.

Scanning electron microscopes
A microscope that produces high resolution images of the surface of an object.

Secondary response
The response of the immune system to a second infection or injection of vaccine; antibodies are produced more quickly and in greater amounts.

Secondary wall
The part of the cell wall laid down after the cell has ceased to increase in size.

Selective breeding
Artificially selecting the organisms to produce the next generation.

Sino-atrial node
A structure in the right atrium of the heart that initiates heartbeat.

Small intestine
The part of the intestine where digestion is completed and absorption of soluble food occurs.

Species
The taxonomic category comprising individuals with common characteristics that distinguish them from other individuals at the same taxonomic level.

Spiracle
An external opening leading to the air tubes (tracheae) inside an insect.

Spleen
An organ that produces lymphocytes, stores blood cells and destroys old blood cells.

ssDNA
Single-stranded DNA.

Standard deviation
A measure of the variability in a population.

Stem cells
Undifferentiated cells that retain the ability to become specialised.

Stroke volume
The amount of blood pumped out by a single contraction of a ventricle.

System
Multiple organs that work together.

Systole
Contraction of a heart chamber.

T cells
Lymphocytes that attack pathogens directly.

$T_{50}H$
The temperature at which 50% of the DNA duplexes are denatured.

Taxon
A group in a classification scheme.

Taxonomy
A system of classifying organisms.

Theory
An explanation for a phenomenon that is supported by experimental testing.

Thrombus
A blood clot.

Thymus
An organ in which t lymphocytes mature and multiply.

Tidal volume
The volume of air inspired and expired in one breathing cycle.

Tissue
A group of similar cells performing the same specific function.

Toxins
Poisonous chemicals.

Tracheoles
The finest air tubes in the breathing system of an insect.

Transmission electron microscope (TEM)
The type of electron microscope where a beam of electrons passes through the specimen to produce a high resolution image.

Triglycerides
Molecules made up of glycerol molecules each bonded to three fatty acid molecules.

Tumour
An abnormal mass of tissue produced by uncontrolled cell division.

Vaccination
The introduction of a vaccine into the body to induce immunity.

Vaccine
Dead or weakened pathogens introduced into the body to induce immunity.

Valves
Part of the circulatory system that ensure unidirectional flow of blood.

Variation
Deviation in the characteristics of an organism.

Ventilation rate
The number of breathing cycles per minute.

Ventricle
A lower chamber in the heart; pumps blood out of the heart.

Ventricular systole
Contraction of a ventricle.

Vertical gene transmission
Transmission of genes to offspring when a bacterium divides.

Viruses
Pathogens consisting of nucleic acid - surrounded by a protein coat.

Water potential (ψ) gradient
A range of values for water potential.

Xylem cells
Cells in plants concerned with water and ion transport and support.

Index

Notes

Notes

Notes